Canadian Special Publication of Fisheries and Aquatic Sciences No. 123

Water, Science, and the Public: The Miramichi Ecosystem

Edited by

E. Michael P. Chadwick

Department of Fisheries and Oceans
Gulf Region, Science Branch
P.O. Box 5030
Moncton, New Brunswick, Canada
E1C 9B6

NATIONAL RESEARCH COUNCIL OF CANADA
Ottawa 1995

ISBN 0-660-15903-1
ISSN 0706-6481
NRCC No. 38816

Canadian Cataloguing in Publication Data

Chadwick, E.M.P. (Edward Michael Pakenham)
Water, science, and the public: the Miramichi ecosystem

(Canadian Special Publication of Fisheries and Aquatic Sciences No. 123)
Includes abstracts in French.
Includes bibliographical references and index.
ISBN 0-660-15903-1

1. Miramichi River Region (N.B.) — Environmental conditions — Congresses. 2. Water quality — New Brunswick —Miramichi River — Congresses. 3. Human ecology — New Brunswick — Miramichi River — Congresses. 4. Fishery resources — New Brunswick — Miramichi River — Congresses. I. National Research Council of Canada. II. Title. III. Series

QH77.C3C32 1995   333.91'62'0971521   C95-900131-X

This publication is available from:

Subscription Office, Research Journals, National Research Council of Canada, Building M-55
1200 Montreal Road, Ottawa, Ontario, Canada K1A 0R6.

Enquiries: Tel.: 613-993-9084 FAX: 613-952-7656 E-mail: research.journals@nrc.ca

Remittances should be made payable to the Receiver General for Canada, credit NRC.

NRC Monograph Publishing Program
Editor-in-Chief Robert H. Haynes

Publishing Department, NRC
Director Aldyth Holmes
Head, Monograph Publishing Program Gerald J. Neville
Composition Lise Gagnon

Printed on recycled paper.

[*The views expressed are those of the authors and not necessarily those of the National Research Council of Canada.*]

Correct citation for this publication:

Chadwick, E.M.P. [*editor*]. 1995. Water, science, and the public: the Miramichi ecosystem. Can. Spec. Publ. Fish. Aquat. Sci. 123: 300 p.

Contents

Contents (*Concluded.*)

Evaluating Human Impacts

Summary and Selected Abstracts

Abstract

This volume is a summary of all the current research on Miramichi's aquatic ecosystem. Much of this information was exchanged between scientists and the public at a workshop in Newcastle, April 13-14, 1994. The papers are organized in four sections: history, physical environment, biological environment, and human impacts. From an historical view, many of the world's first studies on the harmful effects of mining and pesticides were done by scientists working on the Miramichi. In the physical environment, fresh water has a dominant role. At both short-term and seasonal time scales, the flow of fresh water is most important for determining the position of the salt wedge. The estuary is stratified because there is always sufficient river discharge to overcome the effects of mixing by wind and tide. Over 60% of discharge comes from precipitation. In the biological environment, over 90% of the 20 000 tonnes of annual fish production comes from smelt (*Osmerus mordax*) and gaspereau (*Alosa pseudoharengus* and *A. aestivalis*). Miramichi River is the northern limit of the known spawning distribution of striped bass (*Morone saxatilis*). The estuary is the site of greatest biological productivity, home for over 47 species of fish, and an important overwintering area. Abundance of some of the less-migratory fish species associated with the inner estuary appears to be declining. In terms of human impacts, the estuary is most affected. Dredging, contaminants and release of suspended organic matter have an impact on life in the estuary. Discharge from the pulp mills has probably the greatest impact because a large volume of material is released; it is chronic, occurring year round; and it takes place near the salt wedge, blanketing the entire upper estuary with a layer of turbid water. Heavy metal pollution from mining will continue to be important, particularly because the acidity of airborne precipitation seems to be increasing. Dredging has unknown impacts and will need to be monitored closely. Untreated sewage affects the suitability of aquatic animals as food for humans, particularly shellfish in the outer estuary. A monitoring station that records year-round movements of water layers in the estuary would improve our understanding of human impacts and any biological-sampling programs.

Résumé

Le présent ouvrage résume les recherches en cours sur l'écosystème aquatique de la Miramichi. Une bonne partie de cette information a fait l'objet d'un échange entre les scientifiques et le public lors d'un colloque qui a eu lieu à Newcastle les 13 et 14 avril 1994. Les articles sont rassemblés en quatre sections : histoire, milieu physique, milieu biologique et impacts de l'activité humaine. Du point de vue historique, bon nombre des premières études menées dans le monde sur les effets néfastes de l'extraction minière et des pesticides ont été réalisées par des scientifiques qui étudiaient la Miramichi. Dans le milieu physique, l'eau douce joue un rôle dominant. À court terme comme à l'échelle des saisons, l'apport d'eau douce est extrêmement important car il détermine la position du coin salé. L'estuaire est stratifié parce que le débit du fleuve est toujours suffisant pour l'emporter sur les effets de brassage du vent et des marées. Plus de 60 % de cet apport d'eau douce provient des précipitations. Dans le milieu biologique, plus de 90 % des 20 000 tonnes de poisson produites chaque année sont constituées d'éperlan (*Osmerus mordax*) et de gaspareau (*Alosa pseudoharengus* et *A. aestivalis*). La Miramichi marque la limite septentrionale de l'aire connue de reproduction du bar rayé (*Morone saxatilis*). L'estuaire, lieu d'une très grande productivité biologique, abrite 47 espèces de poissons, et constitue une zone d'hivernage importante. L'abondance de certaines espèces moins migratrices associées au fond de l'estuaire semble en déclin. Pour ce qui est des impacts de l'activité humaine, l'estuaire est extrêmement touché. Le dragage, les contaminants et la libération de matières organiques en suspension ont un impact sur la vie estuarienne. Ce sont probablement les effluents des usines de pâtes et papiers qui ont l'incidence la plus forte à cause du volume des matériaux rejetés; c'est un phénomène chronique, qui dure toute l'année; de plus, le rejet se fait près du coin salé, et toute la partie supérieure de l'estuaire se trouve recouverte d'une couche d'eau trouble. La pollution par les métaux lourds provenant de l'extraction minière va rester importante, d'autant que l'acidité des précipitations atmosphériques semble en hausse. Le dragage a des effets mal connus et devra être surveillé de près. Les eaux usées non traitées nuisent à la comestibilité des animaux aquatiques, particulièrement les coquillages dans la partie extérieure de l'estuaire. L'installation d'une station de surveillance qui consignerait tout au long de l'année les mouvements des couches d'eau dans l'estuaire améliorerait notre connaissance sur les impacts de l'activité humaine et enrichirait les programmes d'échantillonnage biologique.

Foreword and Acknowledgements

When the water supply in Newcastle[1] was contaminated with organic toxins, the public outcry resulted in the formation, January 1989, of the Miramichi River Environmental Assessment Committee (MREAC) to assess the health of Miramichi River. The public was concerned about drinking water, swimming, declining fish stocks, and the suitability of seafoods for eating. The public wanted to know the effect of pulp and paper mills, industrial development, mining pollution, spraying of pesticides, untreated sewage, and dredging of the river channel.

In November, 1990, Graham Daborn organized a workshop on the possible long-term impacts of maintenance dredging. One of the outcomes of this workshop was that experts were in no position to evaluate the long-term effects of dredging. They were uncertain if one of the largest dredging projects in the world had had any deleterious effects on the Miramichi ecosystem. Many questions were raised about why Miramichi was a productive watershed. The salt wedge was important but what caused it to move up and down the estuary: wind, tide, or freshwater discharge? What organisms should be looked at? Where should the samples be taken? How many samples were needed? How long should the sampling continue? There were no good answers to any of these questions.

At the same time, André St-Hilaire was interested in studying Miramichi Estuary for a masters degree and David Booth, an expert in coastal oceanography, wanted to work in an estuary with environmental problems. My role was to bring these two people together. In our search for support, Public Works Canada agreed to finance a study that would calibrate a numerical model of tidal currents in Miramichi Bay. The model would be used to predict the direction of sediments disturbed by dredging. The contract from Public Works was sufficient to pay for André's field work and afterwards to support Caroline Lafleur in her careful analysis of the various current meters.

The oceanographic work attracted other projects. Simon Courtenay solicited Green Plan money to study the uptake of low levels of contaminants in tomcod. Andrea Locke received a postdoctoral fellowship to study larval fish and secondary production in Miramichi Bay. Mark Hanson started an intensive sampling program of estuarine fisheries to better understand the recruitment of groundfish in the southern Gulf. Together with the work by Rick Cunjak, Daniel Caissie, Gérald Chaput and Kim Robichaud, a credible research program had sprouted in Miramichi.

Lack of publicity, however, can have drawbacks. MREAC organized a large public meeting in 1992 at Newcastle to announce the results of the report written by Dr. Michael Burt. Government departments were criticized for poor coordination, poor communication, and lack of commitment to the Miramichi watershed. Part of the blame was directed towards Science Branch, Department of Fisheries and Oceans. It was clear that the public was not involved with some of the research and a workshop that brought scientists and MREAC together would be appropriate. Coincidently, MREAC was considering a workshop to update the public about what had happened on the Miramichi since 1992. MREAC was delighted to expand their idea and include ours.

There were several challenges. How could we mix the layman with the expert? We often hear that experts cannot communicate with each other, so it seemed unlikely that they could communicate with the public. We wanted to bring the experts together in non-technical language that everyone could understand and show that science has a useful role. Many members of the public are sceptical of research. They are tired of being told that more information is needed before a decision can be made. They feel that issues

[1]Readers should note that the towns within the Newcastle, Chatham, and Douglastown area were amalgamated January 1, 1995, under the name Miramichi City.

are often investigated to death. Hopefully, this document will show that experts and the public can communicate and that science provides a useful framework for improving our society.

Many people have contributed to this volume. I thank Vern Goodfellow who promoted the workshop and who keenly supported the research even to the extent of spending part of one summer cleaning the algae from current meters at the murky bottom of Miramichi River. I also thank the other members of MREAC for their support. I thank Harry Collins who quietly and competently put the whole workshop into action. He coordinated the speakers and abstracts and enlisted many helpers from the New Brunswick Community College in Chatham who I also thank. Harry's help and guidance were invaluable. Michael Burt did a masterful job at chairing the workshop. Ross Alexander helped moderate the session on environmental impacts. Frank Gallant methodically video-taped the entire workshop. Louise Robichaud helped to organize the papers and assisted by sending manuscripts back and forth to referees and authors and keeping track of the whole process. Gerry Neville provided his professional expertise and patience to ensure that the publication was high quality. John Loch and Department of Fisheries and Oceans gave their complete support to the project from its conception. Julia Chadwick has encouraged me to pursue the idea of blending together water, science and the public. I thank you all.

Peer review is an important part of science. Each paper in this volume has been reviewed by at least two referees and I would like to acknowledge the following people for their assistance with this task:

Addison, R.
Bugden, G.
Caron, F.
Castonguay, M.
Chiasson, A.
Couillard, C.
Daborn, G.
Dadswell, M.
de Lafontaine, Y.
Dempson, B.
Drinkwater, K.
Ducharme, A
Gibson, J.
Gobeil, G.
Grant, J.
Jessop, B.
Kite, G.
Koutitonsky, V.
Lambert, T.
Mallet, V.
Meerburg, D.
Melvin, G.
Parker, R.
Peterson, R.
Pollock, T.
Randall, R.
Scarratt, D.
Scrivener, J.
Silverberg, N.
Therrien, F.
Wildish, D.
Zitko, V.

E. Michael P. Chadwick
January 11, 1995

Preface

MREAC is a widely-based citizen's committee representing all sectors of the Miramichi. This committee was formed in the fall of 1988; its mission is to clean-up and preserve this beautiful river system. The name "Miramichi River Environmental Assessment Committee" is currently a bit of a misnomer. For the first 3 years of the committee's existence, its work was strictly to study (assess) the health of the river. However, after that initial study phase, the committee switched to a more active watchdog and action-oriented role.

As stated in our official Modus Operandi, "...MREAC decisions and environmental programs will need to be based upon the best available science. Where scientific information is not available, studies will be recommended." With this scientific focus as a crucial part of our work, the Committee organizes a yearly meeting to coordinate the scientific studies that take place on the river.

We were thus delighted when we were approached by Dr. Michael Chadwick with a proposal to expand our annual meeting of scientists to a full-blown conference on the current state of the science pertaining to the health of the Miramichi. Not only did Dr. Chadwick come with a proposal but he also came with funding to help defray the costs of organizing this conference. We are deeply appreciative of the help that he and DFO have provided.

Dr. Michael Burt, who originally authored our "MREAC Final Report", has remained an active member of MREAC and has also played a crucial role in planning this conference.

Throughout the 5 years' experience of MREAC, we have greatly benefited by the support that government agencies have given to us. Neil Brodie (NBDOE) and George Lindsay (DOE) provided their continued excellent advice and support throughout the planning process. Our appreciation also goes to officials from the federal program "The State of Environmental Reporting" who provided funding in support of this conference.

We continue to be indebted to our local municipalities for their willingness to share both facilities and equipment. The Town of Newcastle was very helpful in providing us with the Town Hall as a venue and we thank Mayor Peter Murphy for opening the conference for us.

The work of the scientists provides the understanding necessary to formulate actions and directions for the efforts of all involved with MREAC.

Vernon Goodfellow, Chair
Miramichi River Environmental Assessment Committee

INTRODUCTION

"The health of our river is an issue near and dear to all Miramichiers. Unfortunately with so many sick waterways on this continent and around the world, it's not something we can afford to take lightly, or put off until tomorrow. This is what makes this technical workshop so important, because it's clear we need the most up-to-date scientific information available today to guide our planning and decision-making for the future."

Premier Frank McKenna
April 13, 1994, Miramichi Environmental Workshop

Dr. Paul Elson was one of the research pioneers on Atlantic salmon, much of his research was done on the Miramichi watershed.

CHAPTER 1

Why Study the Miramichi Ecosystem?

E. Michael P. Chadwick

Department of Fisheries and Oceans, Science Branch, Gulf Region, P.O. Box 5030, Moncton, New Brunswick, Canada E1C 9B6

Introduction

Miramichi, a Montagnais word for country of the Micmac (Kranck 1988), has seen many changes since the First Nations settled beside the river. The first known environmental issue was probably the Great Miramichi Fire in 1825 where 200-500 people died and 25% of New Brunswick's forests were destroyed (Kranck 1988). Today, the issues are more insidious than forest fires. Logging, shipping, trucking, forest spraying, mining, fishing and development all interact, however subtly. The question is: How can these industries co-exist without destroying outdoor living, hunting, fishing, fresh air, clean water, quality of life; those features which attract people to live in the area?

An important reality is that in order for the non-specialist or the average Miramichier to deal with their environmental problems, they must understand what the issues are. Science is important for understanding problems, but it does little to resolve them. The challenge of scientists, therefore, is not only to conduct research, but to make sure that this research is accessible to those who make the decisions. One type of exchange is achieved at face-to-face meetings between the public and scientists. Such a meeting took place on the Miramichi, April 13-15, 1994. A primary objective of this volume is to present the information that was exchanged between scientists and the public during this meeting.

Another reality is that money for research is limited. No longer is it the role of academics or specialists to decide what is important. Today it is the public who sets the direction of research; after all, it is the public who pays the bill. The public must sort through information prepared by specialists and decide: What are the environmental issues? What type of knowledge is necessary to understand these issues? How much money do we need? Should we be spending any money at all? A second objective of this volume, therefore, is to present information to residents of Miramichi to help answer these questions.

Environmental questions are never simple; they often don't happen over night. Environmental problems are often chronic, occurring at low levels over long time periods. Consequently, we sometimes don't know what questions to ask until its too late. I hope that the approach we have taken on the Miramichi has broad application because similar questions are being asked in watersheds and coastal areas around the world.

Solving the Puzzle

Many environmental problems are like a jigsaw puzzle and the analysis begins with all the pieces upside down on the table. Specialists or scientists are engaged by the public to help solve the puzzle by turning over pieces. Unfortunately, specialists can show only the piece that relates to their own discipline. On rare occasions, different specialists are able to turn over their pieces together, in front of the public. In this monograph we have such an opportunity: knowledge of many specialists is collected together to describe the Miramichi watershed, giving as complete a picture as possible. The papers, or puzzle pieces, are organized in four sections: history, physical environment, biological environment, human impacts.

History

The first section looks at what has happened historically in the Miramichi. Alyre Chiasson, a professor in biology at Université de Moncton who is actively involved with watershed management and the influence of suspended sediments on the behaviour of fish brings his insight on the main features of the watershed. We learn about the geology, the First Nations, settlement by Europeans, and the development of scientific research. We learn that features like the Barrier Islands, which greatly influence the nature of the estuary, are always changing (Chiasson 1995).

A point of interest is that many of the world's first studies on mining pollution were done by scientists working on the Miramichi. Most of these scientists were located at the St. Andrews Biological Station, N.B. We are fortunate to have a key scientist, Vlado Zitko, to describe his perception of the highlights of his career at the station. He worked closely with scientists like Elson, Sprague, and Carson and can explain the significance of their discoveries on the combined influences of organic matter, acidity, and temperature on the toxicity of heavy metals to salmon parr. We also learn that pioneering research on the affect of organic pesticides like DDT was done on the Miramichi (Zitko 1995).

A key lesson from this section is that specialists are capable of solving problems. Spraying of DDT was discontinued in the 1960's partly because of the research at St. Andrews Biological Station. Strict guidelines were imposed on heavy metal pollution because of the research undertaken on the Miramichi.

Although we do not discuss expense, I mention it now because cost of research is very important today. The research programs on DDT and heavy metals were expensive. But there are few people today who would argue that this research was not worth it. The programs were effective probably because stakeholders agreed that the questions were important to answer, the objectives were clear, and the research was well-focussed. Maybe it is these pieces from the past that we must bring forward to our puzzle of the future.

Physical Environment

In second section we turn over pieces concerning the recent physical environment, freshwater and marine. Generally, the physical environment is the framework within which all the organisms live, including humans. The framework can be considered as the pieces with at least one straight edge, those that make the outside of the puzzle. We know, however, that Miramichi's aquatic environment is not fixed, the framework moves. For example, if we were to sample organisms known to live in brackish water, where would we look and when?

The all-important salt wedge moves upstream and downstream according to freshwater discharge, tides and wind, but until recently, we didn't know which of these factors was most important, neither did we know how quickly materials were flushed from the system. Caroline Lafleur working with David Booth at University of Quebec at Rimouski established the role of these factors in determining the structure of the estuary. It appears that at both short-term and seasonal time scales, the flow of fresh water is most important for the position of the salt wedge (Lafleur et al. 1995). The salt wedge migrates. During the spring it is downstream of Chatham. During low river discharge, it is upstream of Newcastle. It also migrates with the tide and appears to be pushed upstream every second week during spring tides.

Biological productivity results when nutrients and sunlight are brought together. Mixing by wind from the top down, or by the tides from the bottom up, are usually important forces in shallow estuaries like the Miramichi. André St-Hilaire, a graduate student at Université de Moncton who studied the estuary during the summer of 1991 found that it was almost always stratified. There was always sufficient river discharge to overcome the effects of mixing by wind and tide (St-Hilaire et al. 1995). Nevertheless, both wind and tide influence the temperatures and salinities of the estuary.

Appreciating the dominant role of fresh water, we turn to Daniel Cassie, a DFO hydrologist who has examined discharge at Catamaran Brook for the past 5 years. He shows that 63% of discharge in the Miramichi comes from precipitation, which averages about 100 mm per month and over 1 m yr-1 (Caissie and El-Jabi 1995). We also learn that precipitation influences conductivity, stream temperature, and the distribution of suspended sediments. Sediments are an important limiting factor to many organisms, both directly by smothering their respiratory surfaces or by removing oxygen from the water column, and indirectly by blocking sunlight and hiding prey from predators, and vice versa (M. Brylinski, Acadia University, pers. comm.).

Biological Environment

The third section looks at living organisms, the pieces inside the puzzle without any straight edges. Miramichi Estuary is an important nursery area for over 20 species of fish, which Andrea Locke, a scientist at Gulf Fisheries Centre, has divided into two communities: one with an affinity for fresh, turbid and warm water; the other preferring salty, clear and cold water. But she also shows that these communities are not static (Locke and Courtenay 1995). As they mature, larval fish move downstream. Thus, environmental impacts for any species will vary throughout the life history. Her work also indicates that larval smelt is a keystone species, being important in the food chain for many others.

Because Miramichi is so famous for Atlantic salmon, we may be surprised to hear from Gérald Chaput who is responsible for stock assessments in the Miramichi that salmon is not an abundant species. Over 90% of the 20 000 t yr-1 of fish production comes from smelt and gaspereau (Chaput 1995). Chaput identifies other paradoxes: eel and Atlantic salmon make the longest migrations at sea and also make the most extensive migrations into fresh water and populations of these species are healthy; by contrast, species that live in and around the salt wedge don't migrate far and they appear to have declining populations. He makes the conclusion that the environment associated with the estuary has become less favourable than that in fresh water.

Mark Hanson and Simon Courtenay, who are scientists at Gulf Fisheries Centre, have assembled information on variations in the structure of fish communities throughout the year at various sites along the Miramichi Estuary. We learn that at least 47 fish species live in the estuary; the Outer Bay is an important overwintering site; and, Sheldrake Island located below the limit of freshwater influence is the most abundant site during all four seasons (Hanson and Courtenay 1995). In another paper by Gérald Chaput and Kim Robichaud, a graduate student at UNB, we learn that the Miramichi River is the northern limit of known spawning distribution of striped bass (Chaput and Robichaud 1995).

Human Impacts

Discarded wastes, whether intentional or not, are the most obvious pieces of the puzzle. Beginning in the headwaters, metals like zinc, lead, and cadmium enter the river occasionally during heavy rainfalls when leachate is washed from slag heaps and old mines or when a coffer dam breaks as happened just prior to the workshop (N. Brodie, N.B. Department of Environment, Fredericton, N.B., pers. comm.). Such influxes of metals are not usually a chronic problem: they are toxic but short term, the slug gets flushed downstream to eventually settle on the bottom. Between Newcastle and Chatham the sediments are low in oxygen and contain lots of organic material, perfect conditions to bind metals and keep them out of the food chain (Buckley 1995).

Air pollution plays an important role in chronic metal contamination. Precipitation that falls on the Miramichi drainage basin is about 1000 times more acid than the groundwater. About

20% of this acid is airborne sulphur dioxide and nitric acid produced at New Brunswick's power plants (D. Albin, NB Department of Environment, Fredericton, N.B., pers. comm.). Generally, the watershed is well buffered by sedimentary bedrock, but the leaching of heavy metals during snow melt, when the acid conditions are worst, merits more study (M. Burt, University of New Brunswick, Fredericton, N.B., pers. comm.).

The most abundant form of waste is suspended organic material. More than 99% of organic particulate matter originates from the two pulp and paper mills with the larger mill at Newcastle contributing about 90% (N. Brodie, pers. comm.). On the positive side, this organic effluent prevents heavy metals from entering the food chain. But because the effluent enters the river near the head of tide at the salt wedge and in a steady dose of about 10-20 t day^{-1} year round, it likely has many negative impacts: it blocks light and prevents photosynthesis (M. Brylinski, Acadian University, pers. comm.); it smothers benthic plants and animals (M. Brylinski, pers. comm.); it removes oxygen at the rate of about 12 t day^{-1} (N. Brodie, pers. comm.); and because predators and prey cannot be seen, it probably disrupts the feeding activities and early life history of about 20 species of fish that live in the salt wedge (Locke and Courtenay 1995).

Other impacts result from toxic compounds like dioxins and furans that are part of the effluent. Courtenay has studied how these toxins affect the cell biology of some fishes, mainly tomcod and smooth flounder. He can show that the nucleic acids of tomcod are altered after only brief exposures to mill effluent (Courtenay et al. 1995). The normal functioning of cells is restored when fish are removed from the effluent. This type of research may lead to precise monitoring of dangerous organic pollutants that occur at very low levels, although the human health hazard from these chemicals is not believed to be a problem (C. Dalpé, Health and Welfare Canada, Ottawa, pers. comm.).

River sediments, which contain heavy metals, toxic organic compounds and other dangerous materials, are also a concern when they are disturbed. There are natural movements of these sediments. Dale Buckley, a marine geochemist, shows that about 100 000 t yr^{-1} of sediments are moved naturally by the river into the estuary (Buckley 1995). Most of this transport takes place during the spring floods. As an aside, 5-10% of these sediments are the organic wastes from pulp and paper mills described above. His analysis shows that Miramichi sediments are relatively unaffected by heavy metal contamination, probably because the flushing capacity of the river is so great.

There is also an unnatural movement of sediments. Dredging has taken place for a century but the intensity has increased. In just 2 years, 1981-1982, 6.6 million m^{-3} of sediments were moved, three times more material than was dredged in the previous 75 years (Daborn 1990). This amount of material is about 30 times the natural rate of sediment transport. Earlier work suggested that most dredge spoils were unstable and became resuspended (Kranck and Milligan 1989). But recent work shows that the dump sites are stable with the exception of organic sediments, which hold water, become fluid, and are mobilized even by the passage of a ship (M. Brylinski, Acadia University, pers. comm.).

The other impacts on the Miramichi arise from forestry and fishing. For the past 40 years, almost the entire province of New Brunswick has been sprayed with insecticides every 2-3 years, earning it the reputation of the longest and largest-scale spray program in the world. The legacy of DDT spraying during the 1950's and 1960's continues to live with us in the tissue of fish, where levels of DDT are as high today as they were 30 years ago (Zitko 1995). Fenitrothion is also gaining a tainted reputation. Recent work shows that this compound, once widely acclaimed as safe, has many detrimental side effects (W. Ernst, Environment Canada, Dartmouth, N.S., pers. comm.): it persists for >1 yr in bogs; it kills pollinators and birds; it causes high mortality of insects and mites that are natural predators of spruce budworm; it accumulates in fish and affects reproduction; and it drifts 5 times more than biological insecticides.

The most complete approach to understanding the long-term impacts of logging on the environment is through experimental harvesting. Rick

Cunjak, a scientist with DFO, leads a 15-year experiment that will examine the impact of clearcutting on a small, third-order stream, Catamaran Brook (Cunjak 1995). The pre-logging phase of this project has been completed. The major sources of variability, such as the impact of sediments on egg-to-fry survival, the impact of increased nutrients on growth and survival, and the impact of winter ice, will be watched closely in the second and third phases.

Fishing can also have unpredictable impacts. Dave Moore and colleagues show that the size and sea age of Atlantic salmon can be affected by commercial exploitation. Closures of commercial fisheries throughout Atlantic Canada have resulted in an increased return of repeat spawners to the river (Moore et al. 1995). Currently, more than 40% of the large salmon that enter the river had spawned previously. We also know that fishing gear set for one species can have important impacts on other species. Rod Bradford, a scientist on contract with DFO, shows that gaspereau, smelt and eel traps can catch large numbers of juvenile and adult striped bass (Bradford et al. 1995). With this information, it may now be possible to make slight adjustments to these traps and avoid any by-catch.

Where to from here?

Many pieces of the puzzle still lie face down on the table, but enough of the picture is exposed to give us some direction. The freshwater pieces tell us that heavy metal pollution from mining will continue to be important, particularly because the acidity of airborne precipitation seems to be increasing. Insecticide spraying is also a problem that could be diminished with biological insecticides like Bt. The 25% greater cost for Bt would not seem to outweigh its advantages over fenitrothion (W. Ernst, pers. comm.).

The puzzle shows that there is more activity in the estuary than other parts of the watershed. The estuary is the site of greatest biological productivity and most human activity. Dredging, contaminants and release of suspended organic matter have an impact on life in the estuary. Discharge from the pulp mills has probably the greatest impact because a large volume of material is released; it is chronic, occurring year round; and it takes place near the salt wedge, blanketing the entire upper estuary with a layer of turbid water. The toxicity of sediments has been tested against some marine organisms and found not to be a problem (R. Parker. Environment Canada, Halifax, NS, pers. comm.), but it seems to be important that the same tests are repeated on freshwater organisms, which may be more sensitive. Certainly, the impact of suspended matter merits further investigation.

Part of this investigation might include a comparison with the neighbouring Kouchibouguac estuary, which although smaller than Miramichi, probably shows what the biological community would look like in a more pristine state. Emphasis should be placed on the seasonal distribution of organisms, including plants. All sampling should be done in relation to the salt wedge. In this regard, it would be useful to maintain a long-term monitoring station that would record the movements of the water layers throughout the year. These monitoring sites would provide the framework for any sampling of organisms or human impacts.

In the marine environment, dredging and contamination will be the main issues. Dredging has unknown impacts and will need to be monitored closely, again a good reason to initiate a long-term monitoring program. Untreated sewage is an impact that affects the quality of seafoods in the outer estuary, particularly shellfish (Dalpé, pers. comm.). There is considerable public interest in remedying this problem (MREAC 1992).

Assembled in this volume is a summary of all the current research on the Miramichi's aquatic ecosystem. The Premier has stressed in his opening remarks the importance of having up-to-date information. You may ask if there is enough up-to-date information to make any reasonable decisions? Some have argued that the Miramichi has suffered from neglect and that scientific information is wanting (MREAC 1992). Others have argued that Miramichi is one of the best studied rivers in Canada (Buckley 1995) and there is plenty of information to work with. Reader, we must ask you to decide for yourself.

References

Bradford, R.G., Robichaud, K.A., and Courtenay, S.C. 1995. By-catch in commercial fisheries as an indicator and regulator of striped bass (*Morone saxatilis*) abundance in the Miramichi River Estuary, p. 249–259. *In* E.M.P. Chadwick [*editor*]. Water, science, and the public: the Miramichi ecosystem. Can. Spec. Publ. Fish. Aquat. Sci. 123.

Buckley, D.E. 1995. Sediments and environmental quality of the Miramichi Estuary: new perspectives, p. 179–190. *In* E.M.P. Chadwick [*editor*]. Water, science, and the public: the Miramichi ecosystem. Can. Spec. Publ. Fish. Aquat. Sci. 123.

Caissie, D., and El-Jabi, N. 1995. Hydrology of the Miramichi River drainage basin, p. 83–93. *In* E.M.P. Chadwick [*editor*]. Water, science, and the public: the Miramichi ecosystem. Can. Spec. Publ. Fish. Aquat. Sci. 123.

Chaput, G.J. 1995. Temporal distribution, spatial distribution and abundance of diadromous fish in the Miramichi River watershed, p. 121–139. *In* E.M.P. Chadwick [*editor*]. Water, science, and the public: the Miramichi ecosystem. Can. Spec. Publ. Fish. Aquat. Sci. 123.

Chaput, G.J., and Robichaud, K.A. 1995. Size and growth of striped bass, *Morone saxatilis*, from the Miramichi River, Gulf of St. Lawrence, Canada, p. 161–176. *In* E.M.P. Chadwick [*editor*]. Water, science, and the public: the Miramichi ecosystem. Can. Spec. Publ. Fish. Aquat. Sci. 123.

Chiasson, A.G. 1995. Miramichi Bay and Estuary: an overview, p. 11–27. *In* E.M.P. Chadwick [*editor*]. Water, science, and the public: the Miramichi ecosystem. Can. Spec. Publ. Fish. Aquat. Sci. 123.

Courtenay, S.C., Williams, P.J., Vardy, C., and Wirgin, I. 1995. Atlantic tomcod (*Microgadus tomcod*) and smooth flounder (*Pleuronectes putnami*) as indicators of organic pollution in the Miramichi Estuary, p. 211–227. *In* E.M.P. Chadwick [*editor*]. Water, science, and the public: the Miramichi ecosystem. Can. Spec. Publ. Fish. Aquat. Sci. 123.

Cunjak, R.A. 1995. Addressing forestry impacts in the Catamaran Brook basin: an overview of the pre-logging phase, p. 191–210. *In* E.M.P. Chadwick [*editor*]. Water, science, and the public: the Miramichi ecosystem. Can. Spec. Publ. Fish. Aquat. Sci. 123.

Daborn, G. 1990. Proceedings of workshop on the impacts of maintenance dredging of the Miramichi channel on biological resources of the Miramichi system. Acadia Centre for Estuarine Research, Wolfville, N.S. (mimeo.)

Hanson, J.M., and Courtenay, S.C. 1995. Seasonal abundance and distribution of fishes in the Miramichi Estuary, p. 141–160. *In* E.M.P. Chadwick [*editor*]. Water, science, and the public: the Miramichi ecosystem. Can. Spec. Publ. Fish. Aquat. Sci. 123.

Kranck, K. 1988. Miramichi River. *In* J. H. Marsh [*editor*]. The Canadian encyclopedia, Second Edition, Hurtig Publ., Edmonton, Alta. 2736p.

Kranck, K., and Milligan, T.G. 1989. Effects of a major dredging program on the sedimentary environment of Miramichi Bay, New Brunswick. Can. Tech. Rep. Hydrogr. Ocean Sci. No. 112: 61p.

Lafleur, C., Pettigrew, B., St-Hilaire, A., Booth, D.A., and Chadwick, E.M.P. 1995. Seasonal and short-term variations in the estuarine structure of the Miramichi, p. 45–71. *In* E.M.P. Chadwick [*editor*]. Water, science, and the public: the Miramichi ecosystem. Can. Spec. Publ. Fish. Aquat. Sci. 123.

Locke, A., and Courtenay, S.C. 1995. Ichthyoplankton and invertebrate zooplankton of the Miramichi Estuary: 1918-1993, p. 97–120. *In* E.M.P. Chadwick [*editor*]. Water, science, and the public: the Miramichi ecosystem. Can. Spec. Publ. Fish. Aquat. Sci. 123.

Moore, D.S., Chaput, G.J., and Pickard, P.R. 1995. The effect of fisheries on the biological characteristics and survival of mature Atlantic salmon (*Salmo salar*) from the Miramichi River, p. 229–247. *In* E.M.P. Chadwick [*editor*]. Water, science, and the public: the Miramichi ecosystem. Can. Spec. Publ. Fish. Aquat. Sci. 123.

MREAC. 1992. Summary final report, 1989-1992, Miramichi River Environmental Assessment Committee. 157p. (mimeo.)

St-Hilaire, A., Booth, D., Bettignies, C., Chadwick, E.M.P., and Courtenay, S.C. 1995. Is the Miramichi a stratified estuary?, p. 73–82. *In* E.M.P. Chadwick [*editor*]. Water, science, and the public: the Miramichi ecosystem. Can. Spec. Publ. Fish. Aquat. Sci. 123.

Zitko, V. 1995. Fifty years of research on the Miramichi River, p. 29–41. *In* E.M.P. Chadwick [*editor*]. Water, science, and the public: the Miramichi ecosystem. Can. Spec. Publ. Fish. Aquat. Sci. 123.

HISTORICAL OVERVIEW

"In 2014, what will be the attitude towards today's activities?"
Will we appear as foolish?"

Dr. Vlado Zitko
April 13, 1994, Miramichi Environmental Workshop

This instrument is used to measure conductivity (salinity) and temperature at various depths of the water column. Salinity profiles are used to describe the location of features like the salt wedge in Miramichi Estuary.

CHAPTER 2

The Miramichi Bay and Estuary: An Overview

Alyre G. Chiasson

Département de biologie, Univerité de Moncton, Moncton, New Brunswick, Canada E1A 3E9

Chiasson, A.G. 1995. Miramichi Bay and Estuary: An overview, p. 11-27. *In* E.M.P. Chadwick [*editor*]. Water, science, and the public: the Miramichi ecosystem. Can. Spec. Publ. Fish. Aquat. Sci. 123.

Abstract

The Miramichi Estuary is one of the larger estuaries located on the east coast of New Brunswick, Canada. Most of the research over the past four decades has described its physical features, among the most interesting being the presence of a salt wedge and a system of barrier islands located at the mouth of the bay. In the 1980's it was the location of one of the largest dredging operations in the Atlantic Region. The natural resources of the estuary and its associated river basin have for over a century been vital to the economic health of the region. This continued prosperity depends on a vision that extends beyond the boundaries of the estuary to include the whole Miramichi ecosystem.

Résumé

L'estuaire de la Miramichi est l'un des plus grands estuaires de la côte orientale du Nouveau-Brunswick, au Canada. La plus grande partie de la recherche menée depuis quatre décennies a décrit ses caractéristiques physiques, parmi lesquelles les plus intéressantes sont la présence d'un coin salé et un système d'îles-barrières situées à l'entrée de la baie. Dans les années 80, on y a effectué l'une des plus grandes opérations de dragage de la région atlantique. Les ressources naturelles de l'estuaire et du bassin du fleuve jouent depuis plus d'un siècle un rôle vital dans la santé économique de la région. Pour préserver cette prospérité, il faut une vision qui dépasse les limites de l'estuaire et qui couvre l'ensemble de l'écosystème de la Miramichi.

Introduction

Early explorers to the Miramichi region found an area rich in natural resources. The ocean provided bountiful fisheries and the forests timber for a growing population. Despite considerable growth since the turn of the century, natural resources remain the backbone of the local economy. In 1978, 50 000 residents made their living from the forests, fisheries, or mining (Philpott 1978). Equally important, the Miramichi also serves as a shipping connection for export of ore, pulp, and paper products and delivery of fuel oil. Recognition of the natural wealth of the region and the potential as a shipping port probably led to the first scientific surveys of the Miramichi Bay and Estuary.

As early as 1837, Captain W.H. Bayfield of the Royal Navy conducted a bathymetric survey of the Miramichi Outer Bay which was later published in 1841 (Reinson 1977). However, systematic studies of the Miramichi Bay are regarded as beginning in the 1950's (Buckley and Winters 1983) with the work by Fothergill (1953) and Bousfield (1955). Since then scientific research has taken several directions: (1) hydrology investigations of the river and the estuary, (2) the study of fish populations, (3) the effects of channel dredging, and (4) the impacts of pollution.

More recently, the effects of years of industrial and urban development in the Miramichi have come under evaluation. In 1989, concern over the environmental health of the system led to the formation of the Miramichi River Environmental Assessment Committee (MREAC). This organization continues to be a motivating force behind an ongoing assessment of the environmental status of the Miramichi River and Estuary.

The Miramichi Bay and Estuary have been the subject of several extensive and well documented publications, notably: Bousfield (1955), Philpott (1978) and Vilks and Krauel (1981). Although these authors deal primarily with the estuary component, the influence of the river is also covered. In general, there has been little evaluation of the interplay between resources and the users. Studies in this area have been restricted to cause and effect, addressing problems associated with pollution or dredging of the shipping channel.

A fully integrated management plan for the estuary and river has yet to evolve and is beyond the scope of this paper. However, the building blocks are provided through a review of the physical and biological features of the Miramichi Estuary. Finally, this paper gives a perspective on the effect of urban and industrial development in the region.

Physical Features

Excellent descriptions of the major features of the Miramichi Bay and Estuary can be found in the publications of Howells and McKay (1977), Philpott (1978) and Vilks and Krauel (1981). Except where otherwise noted the following information was obtained from these publications. However, the particular authors are cited where key or contrasting features are of particular importance. The descriptions are supplemented with reference material from other sources where appropriate.

The Miramichi Bay is one of the larger bays located along the northeastern coast of New Brunswick, Canada (Fig. 1 and 2). It lies in the Maritime Plain which is an area of low-relief extending from Chaleur Bay southward to Cape George. Roughly triangular in shape, the bay is approximately 45 km along its north and south coasts and 32 km wide at its seaward end . The surface area is greater than 300 km^2.

The river basin itself is approximately 14 000 km^2 in area and measures 190 km east to west and 95 km north to south. It encompasses part of the New Brunswick Highlands, the Chaleur Uplands and the Gulf of St. Lawrence Maritime Plain. The principal subwatersheds are: Northwest Miramichi, Sevogle, Little Southwest Miramichi, Northwest Millstream, Renous, Dungarvon, Barnaby, Bartholomew, Southwest Miramichi and Cains (Fig. 3). Large lakes are scarce but a number of small pools exist in the upper reaches. River flow varies from less than 25 $m^3 \cdot s^{-1}$ to more than 6000 $m^3 \cdot s^{-1}$ with an average of 306 $m^3 \cdot s^{-1}$.

Monthly flow values may peak at 6 times the annual mean. During the summer and the winter, flow rates may be as little as 1/10 the annual mean. Total suspended sediment load to the bay is approximately 100 000 $t \cdot yr^{-1}$.

Two major rivers flow into the Inner Bay, the Northwest Miramichi and the Southwest Miramichi with drainage areas of 3900 km^2 and 7700 km^2, respectively. In New Brunswick, the combined drainage area of both tributaries is only exceeded by that of the Saint John River (Table 1).

Table 1. Drainage basins for a number of major rivers in New Brunswick (Environment Canada 1974).

River	Area(km^2)
St. John	55 164
Miramichi	11 680
St. Croix	4 247
Petitcodiac	1 812
Richibucto	1 054

Fresh water also flows into the bay from

seven minor rivers (Table 2).

Table 2. Drainage areas of Miramichi tributaries (Vilks and Krauel 1981).

River	Area(km^2)
Bartibog	550
Burnt Church	120
Nappan	110
Black	260
Bay du Vin	280
Eel	70
Portage	26

The estuary is generally very shallow with a minimum average depth of 4 m and a maximum of 5 m (Bousfield 1955). A navigation channel with an average depth of 6 m runs through the center of the Inner Bay. Several areas exceed 10 m in depth: the meander bends in the river channel in the upper estuary; the channels that exist between the coastal barrier islands; and, an isolated channel adjacent to the southwest side of Bay du Vin Island (Fig. 2) (Schafer and Smith 1983).

According to terminology introduced by Pritchard (1967), the Miramichi Estuary is a bar-built estuary, comprising an outer embayment

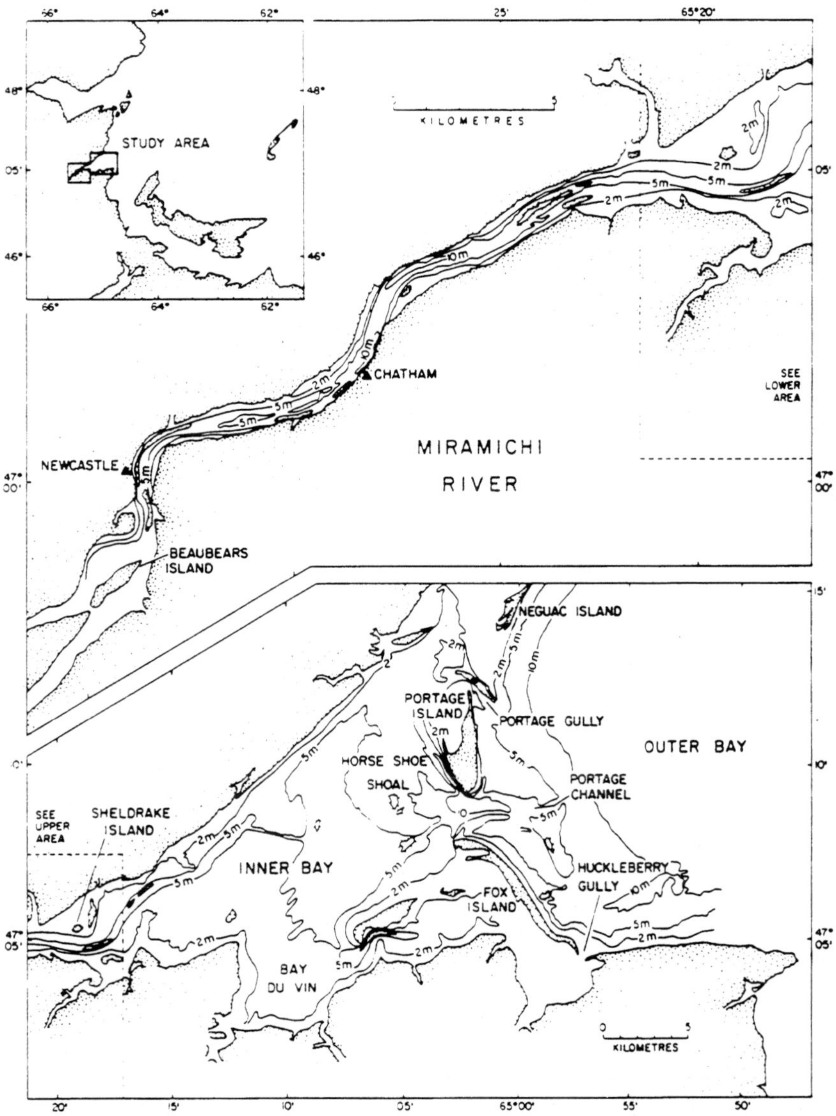

Fig. 1. Study area showing bathymetry of the Miramichi River Estuary and Inner Bay. From Vilks and Krauel (1981).

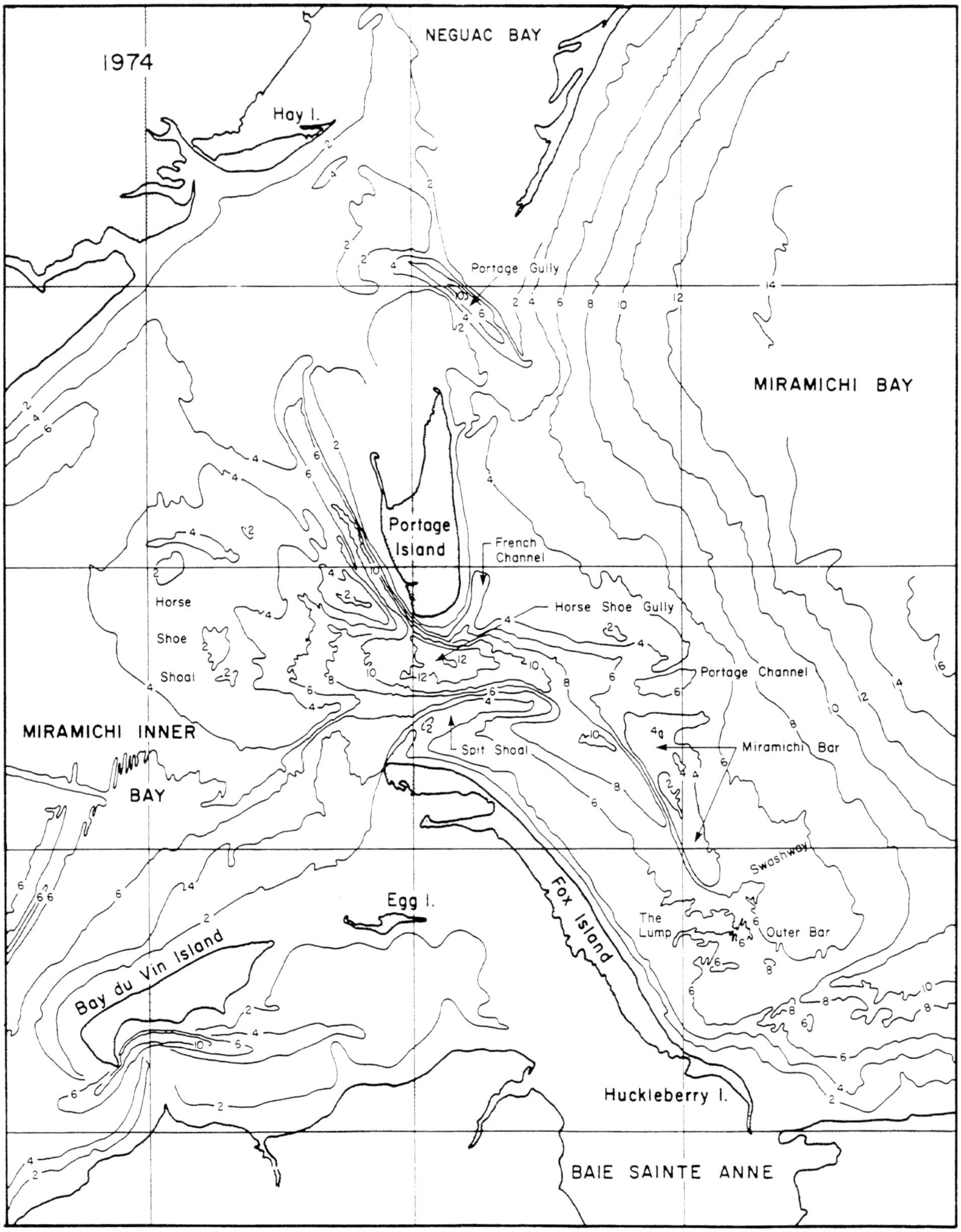

Fig. 2. Detailed map of Miramichi Inner and Outer Bay. From Philpott (1978). Used with permission of Public Works and Government Services Canada.

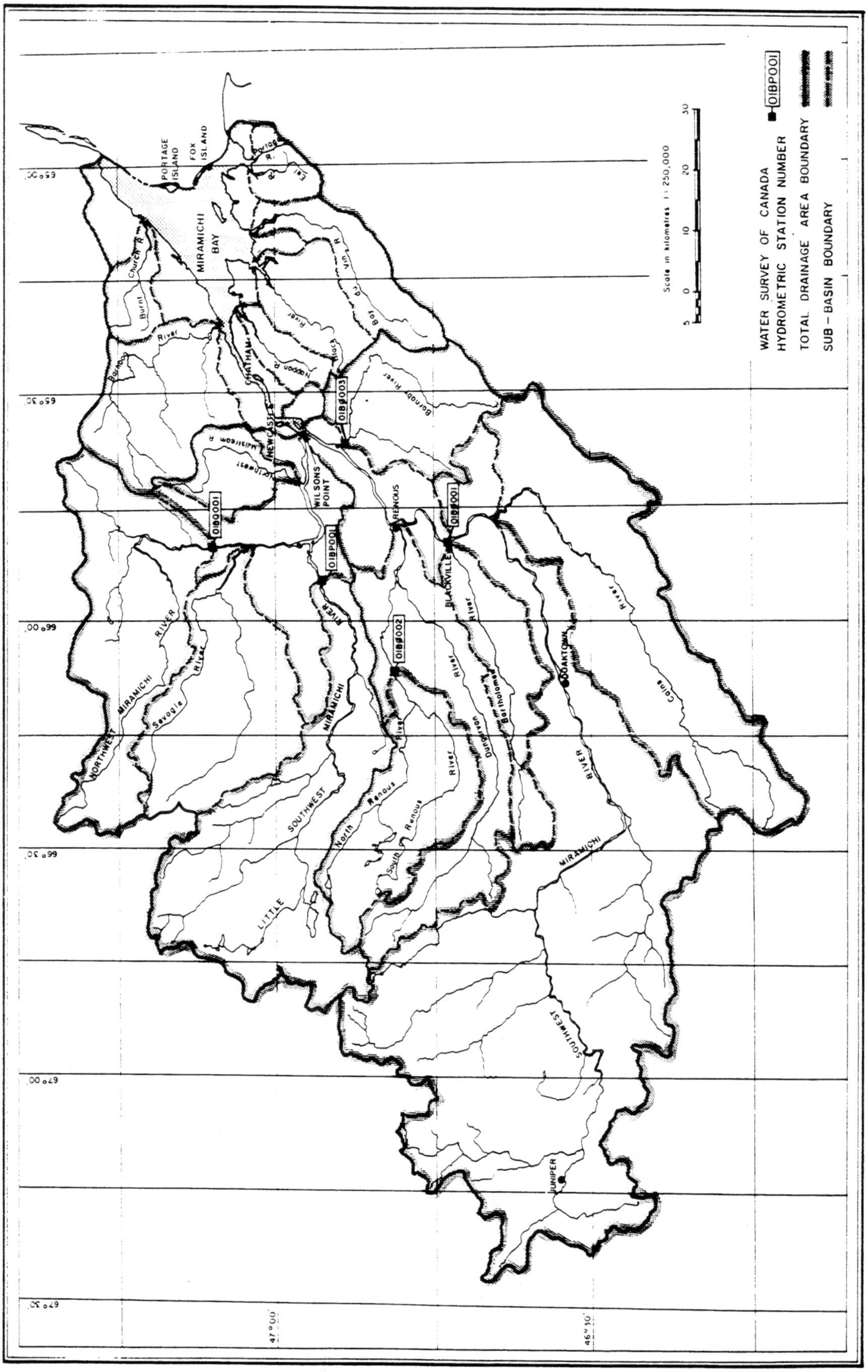

Fig. 3. Map of the Miramichi River watershed. From Ambler (1976).

enclosed by a coastal barrier with an inner, drowned-river valley. The coastal barrier is a series of islands that stretches across the mouth of the bay. The gullies between the barrier islands contain fan-like deposits of sand and fine gravel which form tidal deltas. Both barrier islands and the tidal deltas will be dealt with in greater detail in a later section. Despite Pritchard's classification, Vilks and Krauel (1981) point out that the river estuary is actually a coastal plain estuary. This is in agreement with information presented earlier in this paper, that the Miramichi Bay lies on the Maritime Plain. The Miramichi Estuary is also classified as a microtidal estuary with large inlets and well-developed deltas (Reinson 1977). By definition, microtidal estuaries have a tidal range between 0 and 2 m.

Geology

The highland region of the Miramichi basin is underlain by Ordovician, Silurian, and Devonian age rocks. In the upper reaches of the Highlands, bedrock consists primarily of granite, quartz, monzite, granodiorite and similar rocks of the Devonian period or earlier. The portion of the basin that lies in the Maritime Plain contains sandstones, conglomerates and siltstones of the Pennsylvanian period or earlier.

As with the two major tributaries, the Little Southwest Miramichi is underlain by granitic rocks, whereas the headwaters of the Northwest Miramichi have less resistant rocks with some deposits of zinc, copper, lead, silver, and gold. Flat-lying rocks of Palaeozoic sandstones form a low cliff shoreline along most of the lower river channel except on the inner sides of meander bends where "Pleistocene?" to Holocene alluvial deposits occur (Schafer and Smith 1983). The Inner Bay is bordered by low cliffs of flat lying Palaeozoic sandstone.

The Miramichi Bay is underlain and bounded by "flat-lying or gently-dipping, undifferentiated, red and grey arkosoic sandstones, conglomerates, mudstones and minor shales of the Pictou Group of the Pennsylvanian Series," (Philpott 1978). Seismic profiling of the bay by Howells and McKay (1977) revealed two reflectors. These authors interpreted the lower reflector as probably the Pennsylvanian bedrock surface. The upper reflector appeared to be a 3000 to 4000 yr old marine terrace overlain with glacial tills and proglacial sediments. Superimposed on the upper reflector were soft muds with sand deposits near shore. A "gas reflector" was also detected in the soft muds of the Inner Bay and was thought to represent methane released from decaying organic matter. The soft muds can be further divided into two layers, a loose-textured, highly-mobile layer a few centimetres thick, and a more permanent bottom jelly layer that shows considerable resistance to shear stress.

The glacial tills and proglacial sediments were deposited when the ice retreated about 12 000 yr ago (Schafer 1977). The barrier islands system separating the inner and outer bay probably developed 4400 to 3600 yr ago (Wagner and Schafer 1980). The previous authors state that the sea level in 1966 was probably much the same as it was 4000 yr ago.

Climate

The climate of the Miramichi area is considered continental rather than maritime. The reduced influence of the ocean is attributable to the general movement of air masses from west to east. The air reaching the area carries the imprint of its passage over the continent. Average annual air temperature is 4.3°C. Precipitation occurs during an average of 160 days in the year, totalling more than 995 mm. Maximum water temperatures of 18 to 22°C are reached in July and August, cooling rapidly in October and November to reach freezing point in December.

The Miramichi Estuary is usually ice covered in winter. Average time of freeze up is the middle of December with an average ice thickness of 0.5 m but up to 1.0 m thick. Spring breakup occurs from the middle of March to the end of April with an average date of April 10. Minimum rainfall is during the months of April and July. Maximum rainfall is in November. River flow rates are maximum in the spring with snow melt in April and May and again with a smaller peak in November (Fig. 4).

Annual mean river flows are in the order of 305 $m^3 \cdot s^{-1}$. Peak monthly flows may be six times the annual mean; by contrast low summer and winter flows are about 1/10 of the annual mean.

Subdivisions of the Estuary

The estuary can be divided into several different compartments depending on physical or functional characteristics. Buckley and Winters (1983) divide the estuary into two physiographic parts: (1) the Inner Bay, characteristic of a broad shallow bar-built estuary separated from the Gulf by a series of Holocene-age barrier islands; and (2) a connecting drowned-river channel which is typical of a coastal plain estuary.

Based on sediment distribution patterns, the estuary can be divided into three environmental subdivisions: (1) the river section, (2) the river mouth and (3) the Inner Bay (Philpott 1978). Reinson (1976) identifies five major sedimentary environments: (1) the river channel, (2) the inner bay, (3) the tidal-delta complex, (4) the coastal barrier and (5) the outer bay (Fig. 2).

Definitions of particular components can be found in a number of publications. The Inner Bay is described by Schafer and Smith (1983) as an interface between the river-dominated estuarine and tidally-dominated marine environments. Rogers (1940) defines the upper estuary as that part of the river just above the region of greatest mixing of salt and fresh water. Clearly, various points of view and definitions have evolved to meet the needs of the researchers.

Salinity

Salinity in the estuary has received considerable attention. Due to differences in densities, the presence of salt and fresh water, unless mixed by an outside force such as wind, lead to stratification with the saline water forming the bottom layer. In an estuary, the presence of a lower layer of saline water subject to upstream and downstream movements with the tides is referred to as the "salt wedge". Over the course of a year the wedge migrates up and down the channel between Newcastle and Oak Point located 26 km farther downstream. However, high and low tides account for less than a 2-3 ‰ difference in salinity for a given location. In general, salinity is less than 3 ‰ above the confluence of the Northwest and Southwest Miramichi, less than 10 ‰ above Sheldrake Island and less than 23 ‰ above Portage Island (Fig. 1 and 2). The difference in salinity is greatest, exceeding 10 ‰, at the mouth of the river which lies in a 10 km section of the river extending downstream from Bartibog (Fig. 5). The vertical movement of the halocline, the transitional zone between fresh and salt water, depends on location, tidal range, freshwater inflow, and wind. The depth of the halocline can be as large as 8 m for high inflows and less than 1 m for low inflows. High inflow rates also result in the salt wedge being displaced downstream and greater stratification (Fig. 5c,d). The salinity of the bottom water

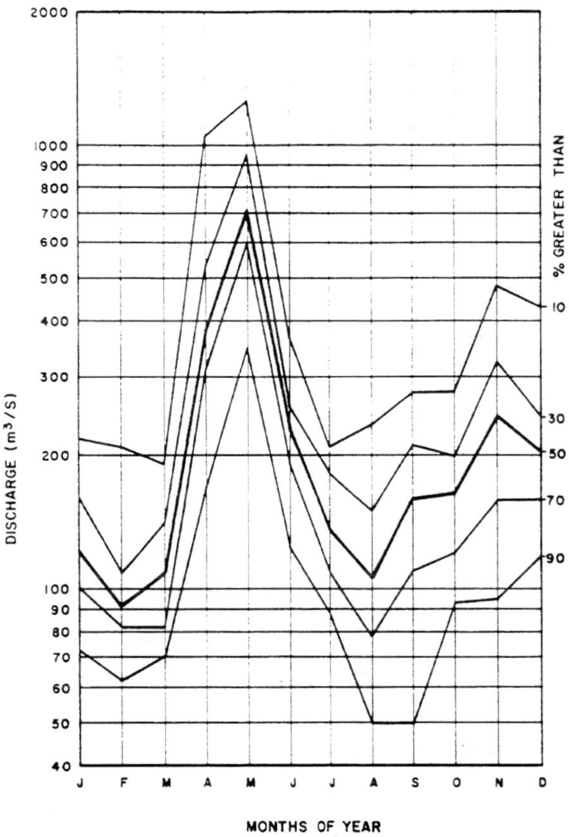

Fig. 4. Freshwater inflow at Wilson Point. Estimated from a partly synthetic streamflow model. Values for the Miramichi Estuary can be obtained by adding approximately 25%. From Ambler (1976).

decreases less than the surface water resulting in an increase in stratification. Increases in freshwater inflow to the upper Inner Bay results in a dilution of the entire water column with no increase in

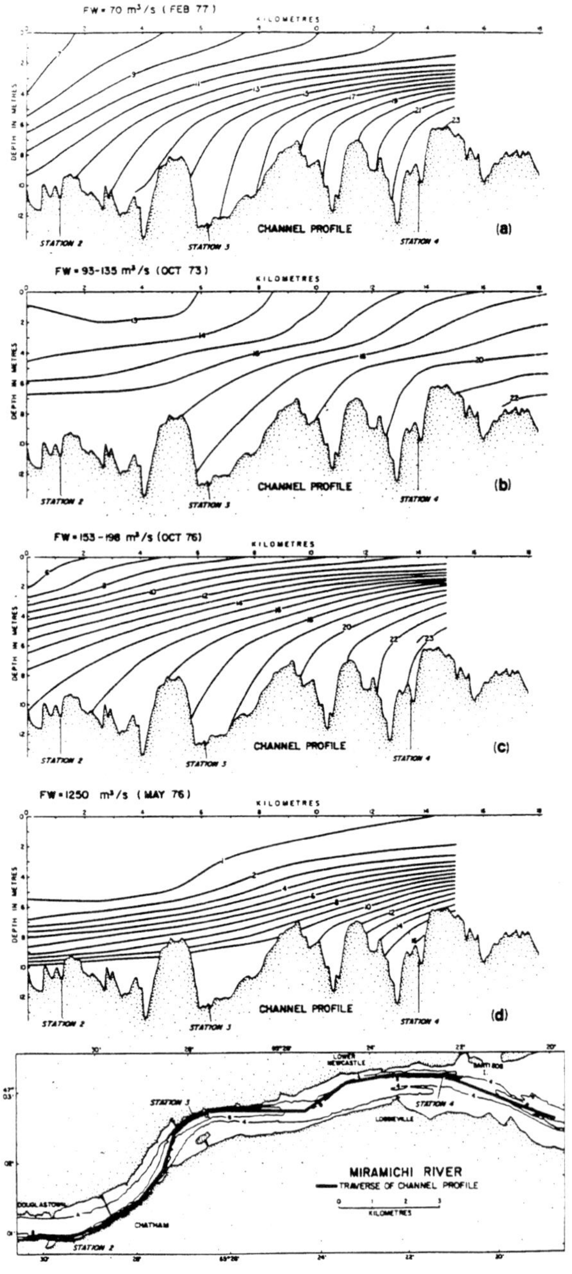

Fig. 5. Mean longitudinal salinity distribution in the river estuary for various freshwater inflow rates. (a) 70 $m^3 \cdot s^{-1}$, ice cover; (b) 93-135 $m^3 \cdot s^{-1}$, no ice cover, (c) 153-198 $m^3 \cdot s^{-1}$, no ice cover and (d) 1250 $m^3 \cdot s^{-1}$, no ice cover. From Vilks and Krauel (1981).

stratification or salinity ranges (Fig. 5b-d). In the middle Inner Bay, the channels and Outer Bay, high inflows result in only a slight reduction in surface salinities compared to bottom salinities. A small increase in stratification in these areas is therefore observed.

No halocline is present in the river part of the estuary at extremely high inflows and salinity intrusion is less than 10 km. At low flows, tidal mixing becomes more predominant in the river estuary and the halocline is nearer the surface. Under such conditions saline water extends through the Inner Bay and upriver almost to the head of tide.

Low flow rates are associated with winter conditions. In the presences of ice cover the upper estuary is well stratified and salt water extends upriver west of Newcastle (Fig. 5a)

Tides

As previously indicated, the Miramichi Estuary is a microtidal estuary. At Portage Island, a reference point in the *Canadian Tide and Current Tables*, mean tidal range is 1.0 m. Tides in the Miramichi area and the Gulf of St. Lawrence are of the "mixed type" though mainly semi-diurnal. At neap tides, "little tides", when the declination of the moon is low the two semi-diurnal tides are nearly of equal height. When the declination of the moon is high, "large tides", there is only one predominantly low and high water period with a secondary oscillation that can be regarded as the remains of the previous neap tide. However, the higher low water of the secondary oscillation (neap tide) becomes nearly equal to that of the lower high water of the dominant tide. Therefore as the neap cycle reaches its highest point, the second tidal cycle commences, giving the appearance of a continuous rise in water level for a consecutive period of 18 hours with a slight levelling off at the midway point. This type of tide is called declinational or mixed semi-diurnal. Tidal reach extends as far as Mitchell's Rapids, which are located about 1 km upstream from the mouth of the Little Southwest Miramichi River (Fig. 3), the head being located in an area referred to as the "Oxbow"

(Discovery Consultants 1989).

Flows in the Inner Bay are dominated by ebb-tide flows (Schafer and Smith 1983). Mean ebb and flood tidal current profiles in this area tend to be uniform with depth and contain no layering. In the river estuary ebb currents are maximum at the surface and flood currents are maximum at mid-depths. However, the presence of ice cover and low inflow conditions results in current maxima being located at midwater depths. Greater freshwater inflow results in stronger ebb currents but flood currents remain unchanged.

It is interesting to compare the volume of water entering or leaving the estuary in one tidal cycle with the typical input of fresh water from the river. The total base volume of the estuary at low water is 1500×10^6 m^3. The volume of water entering or leaving the estuary with one run of the tide can be as little as 31×10^6 m^3 or as much as 470×10^6 m^3. Over a diurnal cycle(1 day) these values are $250 - 550 \times 10^6$ m^3. In comparison, the average freshwater input from the river is 26×10^6 m^3. Freshwater inflow is therefore only 5 – 10% of the volume of tide water.

Tidal currents in the central part of the Inner Bay are rotary in nature and usually follow a counter-clockwise circulation pattern. Discharge from the river estuary is deflected south across the Inner Bay to Portage Channel on the ebb, with the flood favoring Portage Gully and along the north shore to the river estuary (Vilks and Krauel 1981). Schafer and Smith (1983) are in general agreement with this pattern except that flood waters enter through Portage Channel rather than Portage Gully. The authors point out that Portage Channel accounts for up to three quarters of the tidal flow entering the Inner Bay.

The deflection to the south is caused by the Coriolis force arising from the rotation of the earth which causes objects or currents in the northern part of the hemisphere to be defected towards the right in a clockwise pattern. Under high-inflow conditions, the discharge from the river becomes more important than the Coriolis force and water flows along the north side of the Inner Bay, and saline bottom water moves westward along the north shore of Bay du Vin Island on the south side of the bay (Schafer and Smith 1982). Speed of currents in the Inner Bay rarely exceeds 30 $cm \cdot s^{-1}$. Between the barrier islands much higher flows are observed, ranging from 50 to 80 $cm \cdot s^{-1}$. The total flushing time for the Inner Bay has been calculated as 43 tide cycles for an average inflow rate of 150 $cm \cdot s^{-1}$ to 17 tide cycles for a high inflow of 1250 $cm \cdot s^{-1}$.

Secondary currents arise from forces other than the tides. They provide a mechanism for lateral distribution and mixing of water and suspended or dissolved solids. In the Inner Bay the maximum variation in secondary currents is most often observed: (1) during moderate wind speeds (10-15 $m \cdot s^{-1}$); (2) in the absence of surge conditions; and (3) during stable density stratification and slack water or minor tide cycles.

Tidal movement of water through the gullies between the barrier islands separating the Inner and the Outer Miramichi Bays has resulted in the formation of tidal deltas (Fig. 6). These are shallow, sand-dominated areas located on both sides of the coastal barrier. Tidal exchange between the Gulf of St. Lawrence and the Miramichi occurs primarily through two major inlets, Portage Channel and Portage Gully.

Tidal deltas are of two types, flood and ebb (Fig. 6). Flood-tidal deltas are located on the landward side of the inlet and are the result of flood currents. Ebb-tidal deltas are located on the seaward side and are the result of ebb currents. In the Miramichi, flood-tidal deltas are bigger than ebb-tidal deltas indicating the predominant or net movement of marine sand and gravel into the estuary (Schafer and Smith 1983).

Horse Shoe Shoal, a large flood-tidal delta is located to the west of Portage Island and its adjacent channel. It is a large shallow area covering about 1/4 of the Inner Bay and is completely subtidal (less than 3 m deep). The flood-tidal delta associated with Portage Gully is much shallower than Horse Shoe Shoal.

Sedimentation

According to Reinson (1976) there are three sources of sediments in Miramichi Estuary: (1)

marine, (2) fluvial, and (3) shoreline erosion of sandstone bedrock. Philpott (1978) lists six sources of sediments, some of which are subdivisions of the previous three: (1) littoral sand which has contributed to the formation of the barrier islands and delta systems; (2) fine-grained sands from the river basin, most of which are deposited in the Inner Bay; (3) shoreline erosion, mainly sand which remains in the estuary; (4) offshore fine-grained material from the Gulf of St. Lawrence; (5) biogenic materials which are the skeletal remains of living animals; and (6) solid industrial wastes.

Suspended sediments can be categorized by their nature, existing either in solution, as particulate matter in suspension or as solid particles rolling along the bottom referred to as "bed-load". One of these categories, particulate matter, is frequently subdivided into three other components: (1) a non-volatile fraction, inorganic and largely the remains of diatoms, (2) a volatile fraction, and (3) a fraction associated with water.

Buckley and Winters (1983) observed that

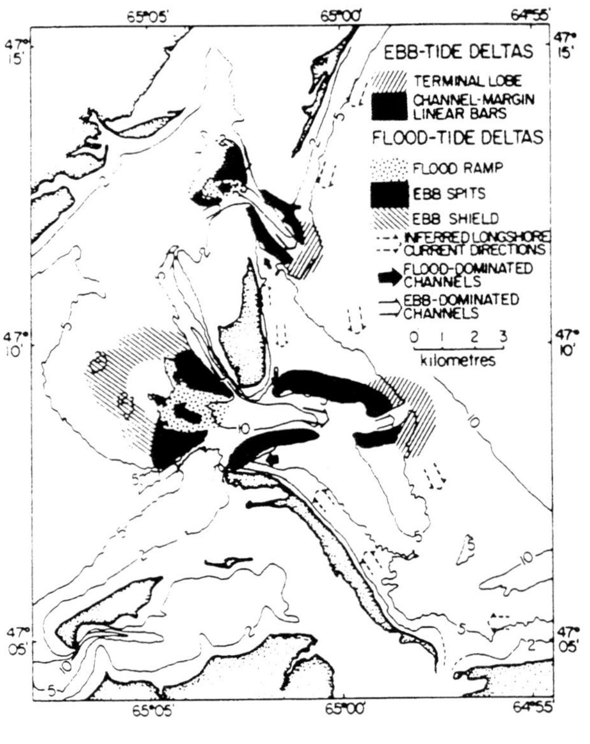

Fig. 6. Location of tidal deltas in Miramichi Estuary. From Reinson (1977).

85% of the annual supply of suspended particulate matter enters the estuary from the Miramichi River during the spring freshet. Values in the estuary ranged from 12 945 to 24 247 $\mu g \cdot L^{-1}$ in May to 320 to 1550 $\mu g \cdot L^{-1}$ in February. Deposited sediments remain rather mobile during the summer months and are redistributed during the fall and winter. In the Inner Bay, an algal mat entraps the sediments which are easily resuspended by wave action and subsequently transported upstream by the salt wedge or swept out of the bay. Wind also plays an important role in lateral transport across the bay.

Overall average concentration for total dissolved solids in the Miramichi River Basin is 22 $mg \cdot L^{-1}$. When combined with average concentration of suspended solids, the mean total solids concentration carried by the Miramichi is approximately 33 $mg \cdot L^{-1}$. Philpott (1978) reported suspended sediment concentrations of 5 to 15 $mL \cdot m^{-3}$ in the river section above Sheldrake Island and 1 to 5 $mL \cdot m^{-3}$ in the Inner Bay. These values would be 38% less if given by weight in ppm. Most of the particulate matter passes through the river estuary during the year when river inputs are low (Buckley and Winters 1983). These authors also determined that during spring flood the amount of suspended particulate matter almost doubles and about half is deposited in the river estuary.

The distribution of sediment types in the estuary are determined by environmental conditions such as water depth and tidal currents. In a study of the surficial sediment in the Miramichi Estuary, Reinson (1976) describes three locations: (1) the upper reaches of the river channel, (2) the Inner Bay, and (3) the area between the flood-tidal delta and the Inner Bay. The first area is dominated by "moderately- to poorly-sorted, medium to coarse-grained sands of fluvial origin." Muddy deposits only occur along the thalweg of the channel or deep adjacent side channels. The Inner Bay is characterized by muddy sediments with very little sand and is the main depositional sink for material coming downstream. The area between the flood-tidal delta and the inner bay basin contains a mixture of fine-grained river deposits and marine sand. In the following year, 1977, Howells and McKay undertook a study similar to Reinson's but with greater focus on the Inner Bay. Their findings although more detailed showed similar results with the Inner Bay consisting mainly of sandy muds and muds (Fig. 7).

The presence of trace metals in the sediments in the Miramichi has been of concerns for two main reasons, the potential for the salt wedge to increase level of contamination and the release of previously deposited materials through dredging. The mechanism behind the salt wedge phenomena has been explained by MacKnight (1994). When river water intermixes with incoming sea water there are changes in pH and in ionic strength. These promote flocculation (formation of aggregates) of organic compounds, detritus, inorganic clays and silts. The movement of the salt wedge upstream presents the opportunity for accumulation of these materials at the head of tide. However, Philpott (1978), based on dredging records prior to 1974, found no evidence for such deposits. Willy and Fitzgerald (1980) did find a significantly higher concentration of trace metals in the sediments of the river portion of the estuary compared to the Inner Bay. Evidently, more research is required to clarify the role of the salt wedge or other factors in the accumulation of trace metals in the upper estuary.

Some trace metals have received more attention than others. Studies by Willy and Fitzgerald (1980) as well as MacKnight (1994) report concentrations of zinc in sediments from the Miramichi to be significantly higher compared to sediments from other areas of eastern Canada. The sources of zinc (Zn) include mining, pulp and paper effluent, and municipal sewage.

The concentrations of iron (Fe), manganese (Mn), Zn and copper (Cu) in the Miramichi sediments were examined in a study by Buckley and Winters (1983). Concentrations of Fe, Mn, Zn, and Cu in the river estuary exceeded the concentrations normally predicted from mixing of fresh and salt water in this area. The excess of Fe was attributed to high turbidity in the river estuary whereas Mn, Zn, and Cu were associated with the area downstream from the turbidity maximum. However, more Fe, Mn, Zn, and Cu are exported from the Miramichi Estuary than enter from the river. During the spring the river supplies 30-35% of the annual amount of labile metals (chemically active Fe, Mn, Zn, and Cu) to the estuary. Buckley and Winters (1983) concluded that much of the influx of suspended particulate matter and metal is temporarily stored in the river estuary but is eventually flushed into the Inner Bay, and eventually flushed out of the estuary over the remainder of the year. Ray and MacKnight (1983) report average concentrations in the estuary of Cu, Zn, cadmium (Cd), and lead (Pb) as 1165, 2600, 26, and 604 ($\mu g \cdot g^{-1}$ dry weight). However, the range of values was so wide that any changes in concentrations due to dredging were masked.

Dredging

The Miramichi has historically served as a port facility for the northern and eastern sections of New Brunswick. However, the entrance channel is not very deep. Prior to 1978, a total of 1 484 956 m^3 of sediment was dredged from the Inner Bay and the river (Philpott 1978). In the early 70's, it was decided that the Miramichi would serve as a centrally-located port for eastern New Brunswick. In addition to upgrading docking facilities, dredging of the navigation channel was proposed. Volumes removed were 4.9 million m^3 in 1981, 1.7 million m^3 in 1982 and 58 000 m^3 in 1983 (MacKnight 1994). Release of previously-accumulated contaminants from the sediments was of primary concern. The results of the various monitoring programs are reviewed by MacKnight (1994) and summarized below.

The first 15 cm of sediments in the river section was described as contaminated, containing concentrations above background values for Pb, Zn, Cu, and Cd (background values, 40, 90, 20, and 0.4 $mg \cdot kg^{-1}$ respectively). Contaminated sediments consisted of clayey silts with a high proportion of organic matter, mostly wood residues. Sediment cores taken in the Loggieville section of the estuary by MacLaren Marex (1978) contained metal concentrations in sediments as deep as 30 cm that were well above standards for ocean disposal. Mercury contamination in this area was very heterogenous and probably reflected contamina-

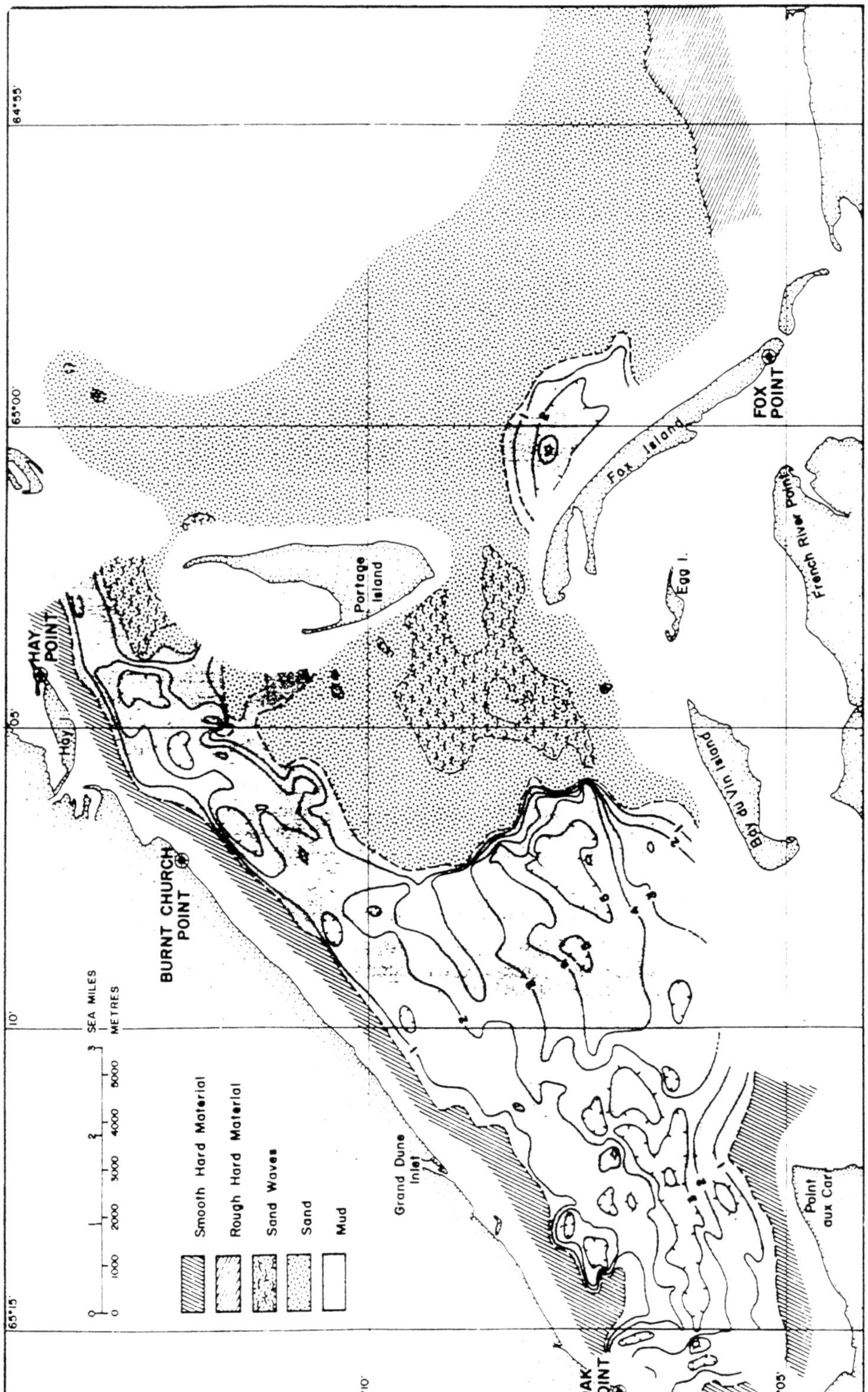

Fig. 7. Distribution of surface sediments in Miramichi Inner and Outer Bay. From Howells and McKay (1977).

tion prior to 1972.

The concentrations of PCB, pentachlorophenol and DDT were analyzed in the estuary section in 1978. An increase in contamination throughout the estuary was noted, with the first 0–5 cm containing the highest concentrations followed by a sharp drop to background concentrations of 5 $ng \cdot g^{-1}$ below this layer. Concentrations of trace organics in the Inner and Outer Bay were well within ocean disposal limits.

Unfortunately, as pointed out by MacKnight (1994) no monitoring of contaminant loadings in suspended particulate matter have been made either during dredging or disposal since 1983 when a maintenance schedule was begun. In a review of 1982 dredging operations by MacLaren Plansearch (1983), several observations were made. The contaminants at both dumping sites for the dredging operations contained substantially reduced concentrations from predredging levels. Levels of cadmium in oysters, clams and crabs were approximately ten times the levels observed in 1981, though some short term recovery in the water column and benthic region was noted. No siltation of shellfish beds or fouling of gear was noted. Finally, no major long-term changes in fish and epibenthos diversity were found based on trawl data.

Changes in the Estuary Over Time

Sedimentation and geomorphological evidence suggests that circulation patterns in the estuary have changed significantly between the years 1967 and 1977 (Scott et al. 1977). Analysis of foraminiferal assemblages by Schafer and Scott (1976) suggest a decrease in flushing rate between 1964 and 1975. They suggest that changes to circulation and sediment patterns appear to be related to shifting and infilling of inlets between the barrier islands.

Prior to 1837, there were a total of five gullies at the mouth of the bay. They are from north to south: Neguac, Portage, Horse Shoe, Fox and Huckleberry (Fig. 2). At present only three gullies remain. Somewhere between 1837 and 1974, Portage Gully supplanted Neguac Gully and Neguac Beach was extended by 4 km. By 1974, Portage Gully was the only northern entrance to the Inner Bay and had developed a very pronounced ebb-tidal delta, as well as adding material to the inner flood-tidal delta. Between 1945 and 1975, Portage Gully had also moved a further 700 m south. By 1974, Portage Island itself had moved 3 km south through accretion at its southern extremity. However, loss from the northern extremity resulted in an overall decrease in area. In 1837, Portage Island comprised 578 hectares and in 1974, 404 hectares. Fox Gully which was immediately north of Huckleberry Gully closed early in 1972. Huckleberry Gully has maintained position but has shallowed considerably.

Changes to the large flood-tidal delta (Horse Shoe Shoal) are described by Reinson (1977). It has maintained its present location over the past 140 years but has grown on the seaward side, contributing to a shallowing of the Inner Bay. The flood ramp (edge of the shoal) and the main ebb channel have changed from a northwest-southeast orientation to an east-west orientation. There has also been the growth of a large spit on the seaward side, parallel to the rotated channel axis.

The barrier islands are therefore dynamic in nature. Littoral drift is an important factor in this process. Net southward drift due to wave action is 140 000 $m^3 \cdot yr^{-1}$ (Philpott 1978). The sea provides almost three times the amount of sediment to the estuary than the river.

Biota

Many commercially important species of fish, shellfish and crustaceans can be found in the Miramichi Bay (Table 3). Species of recognized commercial importance are smelt, cod, blueback herring, alewife, eel, flounder, herring, mackerel, shad, and striped bass (Chaput 1995; Hanson and Courtenay 1995). Atlantic salmon and brook trout are important recreational species. Important spawning sites for herring occur in the Escuminac area located in the southwestern portion of the bay. There are locally important lobster and crab fisheries. Both juveniles and adult lobsters are found in the inner bay (Philpott 1978) with productive grounds in the Outer Bay off Fox Island and the north shore of Escuminac. Crab are restricted to deeper water in the Outer Bay. Oysters and quahaugs are found but because of bacterial contamination, closure areas exist in both the north and southeast sections of the bay. The bay is home to other forms of wildlife such as waterfowl, nesting terns, gulls, ospreys, herons, and the endangered piping plover.

Table 3. Primary species from Miramichi Estuary (from Philpott 1978).

Species	Scientific name
Fish	
Smelt	*Osmerus mordax*
Cod	*Gadus morhua*
Alewife	*Alosa pseudoharengus*
Blueback herring	*Alosa aestivalis*
Tomcod	*Microgadus tomcod*
Shad	*Alosa sapidissima*
Striped bass	*Morone saxatilis*
American eel	*Anguilla rostrata*
Herring	*Clupea harengus*
Mackerel	*Scomber scombrus*
Smooth flounder	*Liopsetta putnami*
Yellowtail flounder	*Pseudopleuronectes americanus*
Sand flounder	*Limanda ferruginea*
Atlantic salmon	*Salmo salar*
Brook trout	*Salvelinus fontinalis*
Crustaceans	
American lobster	*Homarus americanus*
Bivalves	
Oyster	*Crassostrea virginea*
Softshell clams	*Mya arenaria*
Quahaug clam	*Venus mercenaria*
Bar clam	*Spisula solidissima*

Human Activities

Based on the last census taken in 1991, the population in the Miramichi region has increased by 2948 since 1961 (Table 4). A growing population will continue to put additional stress on the natural resources which are vital to the economy of the area. However, there is no evidence of increased shipping over the past years (Table 5).

Table 4. Population growth in the Miramichi region. Reference site is Northumberland County, New Brunswick.

Year	Population
1961	50 035
1971	51 561
1981	54 134
1986	52 981
1991	52 983

Sources: (1961-1981) Philpott (1978); (1986,1991) Statistics Canada.

Table 5. Total cargo through-put in metric tons for the Port of Miramichi.

Year	Thousands of tons
1971	630
1972	656
1973	658
1974	723
1975	560
1976-1987	not available
1988	224
1989	290
1990	531
1992	93
1993	127

Sources: 1971-1975 (Philpott 1978); 1988-1993 Miramichi Port Manager.

Table 6. Top five major industrial groups in the Miramichi.

Sector	Number of firms
1992	
Wood industries	12
Manufacturing industries	6
Clothing industries	11
Printing and publishing	4
Food industries	3
1993	
Wood industries	21
Manufacturing industries	10
Mineral products industries	6
Printing and publishing	6
Food industries	5
1994	
Wood industries	19
Manufacturing industries	10
Food industries	7
Printing and publishing	6
Metal products	5

Source: New Brunswick, Economic Development and Tourism

Table 7. Miramichi sewage treatment plants.

Location	Year built	Average flow ($m^3 \cdot day^{-1}$) 1994
Newcastle	1972	2545
Douglastown	1979	1364
Chatham	1971	4545
Nordin	under study	
Nelson-Miramichi	1982	1091[a]
Chatham Head	shared with Nelson-Miramichi	1364
CFB[b] Chatham	unavailable	unavailable

[a] Flow specification, average flow not available.
[b] CFB, Canadian Forces Base.

Source: UMA et al. (1994).

Forestry-associated industries continue to lead the way in term of number of businesses, followed by manufacturing (Table 6). Miramichi Pulp and Paper Incorporated owns a kraft mill in Newcastle and a groundwood mill at Nelson-Miramichi. In the late 80's two light weight paper mills were added at Newcastle. The Kraft mill was first established in 1949 by the Fraser Company and later purchased by REPAP, the actual owners in the 80's. The groundwood mill was established in 1964 by Acadia Forest Products and purchased in the mid-70's by REPAP. The effluent from these mills, although treated, continues to be of concern and the focus of current effects on fish populations.

In 1956, a mine was developed in the Tomogonops in the Northwest Miramichi River watershed. The mine operated in 1957 and part of 1958. It resumed operation again in 1960, with another shutdown period in the 1980's. Release of mine wastes into the Miramichi appears to have been accompanied by a decline in the return of adult salmon, though the association has not been clearly established (Zitko 1995). Diversion of the Little South Tomogonops by Heath Steele Mines in 1971 resulted in a considerable drop in concentrations of copper and zinc in the Northwest Miramichi. The 1990 report by MREAC Committee lists another 20 potential industrial pollution sites. Their finding suggested that the overall water quality of the Miramichi River was good but further study of point sources of industrial and domestic sewage was required.

In a report by ASA Consulting Ltd., "Assimilative Capacity of the Miramichi for Organic Pollutant Loads," the authors concluded that fecal coliform contamination and not organic contamination was the major threat to the health of the Miramichi system (UMA et al. 1994). At present there are five sewage treatment plants in the Miramichi area with a sixth currently under study (Table 7). A number of these communities are experiencing problems with additional inflows to the system other than wastewater and are seeking solutions to reduce the effects.

Conclusion

Although the Miramichi Estuary can be defined by physical boundaries, it is affected by processes both natural and human that occur throughout the watershed. Past research has identified the major physical features of the estuary, giving evidence of a dynamic and unique environment. For over a century, the natural resources of the Miramichi have contributed to the economy of the region. To date, industrial and population growth have not caused serious long-term damage. The challenge in the future is to maintain and enhance the health of the river. Establishing a management plan for the whole Miramichi River basin that incorporates all the stakeholders may be the first step.

Acknowledgements

My sincere thanks to Graham Daborn who graciously provided much of the reference material for this paper and to Ingrid Tollefsen who assisted with the revisions.

References

Ambler, D.C. 1976. Surface water sediment investigations, Miramichi River navigation channel study. Water Survey of Canada, Atlantic Division, Halifax, Nova Scotia. Unpublished. 71 p.

Bousfield, E.L. 1955. Some physical features of the Miramichi estuary. J. Fish. Res. Board Can. 12: 342-361.

Buckley, D.E., and Winters, G.V. 1983. Geochemical transport through the Miramichi Estuary. Can. J. Fish. Aquat. Sci. 40 (Suppl. 2): 162-182.

Chaput, G.J. 1995. Temporal distribution, spatial distribution, and abundance of diadromous fish in the Miramichi River watershed, p. 121–139. *In* E.M.P. Chadwick [*editor*]. Water, science, and the public: the Miramichi ecosystem. Can. Spec. Publ. Fish. Aquat. Sci. 123.

Environment Canada. 1974. New Brunswick Flood, April-May 1973. Inland Water Directorate Atlantic Region, Halifax, Nova Scotia. Tech. Bull. No. 81: 114p.

Discovery Consultants. 1989. Tidal survey at Miramichi River and analysis of data to determine location of "upstream limit of tidal influence". Discovery Consultants Ltd. P.O. Box 599, Wolfville, N.S., Canada B0P 1X0. 32 p.

Fothergill, N.O. 1953. Tidal circulation in Miramichi Bay. Tidal and Current Survey. Department of Mines and Technical Surveys, Ottawa, Ont. 18p.

Hanson, J.M., and Courtenay, S.C. 1995. Seasonal abundance and distribution of fishes in the Miramichi Estuary, p. 141–160. *In* E.M.P. Chadwick [*editor*]. Water, science, and the public: the Miramichi ecosystem. Can. Spec. Publ. Fish. Aquat. Sci. 123.

Howells, K., and McKay, A.G. 1977. Seismic profiling in Miramichi Bay, New Brunswick. Can. J. Earth Sci. 14: 2909-2927.

MacKnight, S. 1994. Contaminants in the Miramichi Estuary system. Land and Sea Environmental Consultants Ltd. Suite 320, 33 Alderney Dr., Dartmouth, Nova Scotia, Canada, B2Y 2N4. 41p.

MacLaren Marex Inc. 1978. Miramichi Channel Study. Supplementary environmental assessment report on additional work report to Public Works Canada, Atlantic Region, 84 p.

Miramichi River Environmental Assessment Committee. Interim report. 1990. 34p.

Philpott, K.L. 1978. Miramichi Channel Study. Public Works Canada. 18 p.

Pritchard, D.W. 1967. What is an estuary: physical viewpoint, p. 3-5. *In* G.L. Lauff [*editor*]. Estuaries, American Association for the Advancement of Science. No. 83.

Ray, S., and McKnight, S. 1983. Trace metal association with suspended particle matter due to dredging of Miramichi estuary, New Brunswick, Canada, p. 976-979. In Proceedings of the International Conference: Heavy Metals in the Environment. Heidelburg, West Germany. 6-9 September.

Reinson, G.E. 1976. Surficial sediment distribution in the Miramichi estuary, New Brunswick. Report of activities. Part C. Geol. Surv. Can. Pap. 76-1C: 41-44.

Reinson, G.E. 1977. Tidal-current control of submarine morphology at the mouth of the Miramichi estuary, New Brunswick. Can. J. Earth Sci. 14: 2524-2532.

Rogers, H.M. 1940. Occurrence and retention of plankton within the estuary. J. Fish. Res. Board Can. 5: 164-171.

Schafer, C.T. 1977. Distribution and deposition history of sediments in Baie des[sic] Chaleurs, Gulf of St. Lawrence. Can. J. Earth Sci. 14: 593-605.

Schafer, C.T., and Scott, D. 1976. Temporal changes in major foraminiferal assemblages. Can. J. Earth Sci. 14: 1566-1587.

Schafer, C.T., and Smith, J.N. 1983. River discharge, sedimentation, and benthic environmental variations in Miramichi Inner Bay, New Brunswick. Can. J. Earth Sci. 20: 388-398.

Scott, D.B., Medioli, F.S., and Schafer, C.T. 1977. Temporal changes in foraminiferal distributions in the Miramichi River estuary, New Brunswick. Can. J. Earth Sci. 14: 1566-1587.

UMA Engineering Ltd., Codfrey Associates Ltd., and Brunswick Engineering Group Ltd. 1994. Miramichi wastewater collection and treatment plan. 14 p.

Vilks, G., and Krauel, D.P. 1981. Environmental geology of the Miramichi Estuary: Physical oceanography. Geol. Surv. Can. Pap. 81-24: 53p.

Wagner, F.J.E., and Schafer, C.T. 1980. Upper Holocene paleoceanography of the Inner Miramichi Bay. Marit. Sed. 16: 5-10.

Willy, J.D., and Fitzgerald, R.A. 1980. Trace metal geochemistry in sediments from the Miramichi estuary, New Brunswick. Can. J. Earth Sci. 17: 254-265.

Zitko, V. 1995. Fifty years of research on the Miramichi River, p. 29–41. *In* E.M.P. Chadwick [*editor*]. Water, science, and the public: the Miramichi ecosystem. Can. Spec. Publ. Fish. Aquat. Sci. 123.

CHAPTER 3

Fifty Years of Research on the Miramichi River

Vlado Zitko

Department of Fisheries and Oceans, Physical and Chemical Sciences Branch,
Marine Chemistry Division, Biological Station,
St. Andrews, New Brunswick, Canada E0G 2X0

Dedicated to the memory of Dr. Paul F. Elson

Zitko, V. 1995. Fifty years of research on the Miramichi River, p. 29-41. *In* E.M.P. Chadwick [*editor*]. Water, science, and the public: the Miramichi ecosystem. Can. Spec. Publ. Fish. Aquat. Sci. 123.

Abstract

The research by scientists from the St. Andrews Biological Station (SABS) on the Miramichi watershed is reviewed and a complete bibliography is attached as an Appendix. The research includes early descriptive studies of the watershed and its fisheries, the detection and documentation of the effects of DDT on fish, and the impact of mining pollution by copper and zinc on the migration of Atlantic salmon (*Salmo salar*). The levels of DDT and metabolites and of PCB have not changed appreciably over a period of over 20 years. Major sources of pollution in the Miramichi Estuary in the 1970's included a wood preserving plant discharging creosote and pentachlorophenol, and pulp mills. Numerous oil spills occurred in the estuary and high levels of heavy oil residues were detectable in the sediments.

Résumé

Nous faisons le point sur les recherches menées par des scientifiques de la Station de biologie de St. Andrews sur le bassin versant de la Miramichi, et présentons en annexe une bibliographie complète. Parmi les recherches, on note les premières études descriptives du bassin et de ses pêches, la détection et l'étude des effets du DDT sur le poisson ainsi que l'impact de la pollution causée par l'extraction du cuivre et du zinc sur la migration du saumon de l'Atlantique (*Salmo salar*). Les concentrations de DDT et de ses métabolites ainsi que de BPC n'ont pas changé de façon notable sur une période de vingt ans. Les principales sources de pollution de l'estuaire de la Miramichi dans les années 70 étaient une usine de traitement du bois, qui rejetait de la créosote et du pentachlorophénol, et des usines de pâte et papier. Il y a eu de nombreux déversements d'hydrocarbures dans l'estuaire, et on a détecté dans les sédiments des concentrations élevées de résidus de mazout lourd.

Introduction

Atlantic salmon in the Miramichi first attracted the attention of scientists in the 1920's. Somewhat later came an investigation of the smelt fishery, followed by that of the shellfish fisheries. The studies were also supplemented by an investigation of algae and oceanography of the Miramichi Estuary. However, all these studies were dwarfed by the subsequent work on Atlantic salmon.

The bibliography in the Appendix is arranged chronologically and documents the sequence of the investigations. Not all the references are mentioned in the text. Regardless of their age, the publications are well worth reading since they often contain baseline information very valuable for the assessment of changes over the years.

The Miramichi has always been known as one of the best, if not the best Atlantic salmon river in North America. Two pollution events have also made the Miramichi famous worldwide. One was the forest spraying with DDT. The not anticipated deleterious effects of DDT on fish were noticed and documented by scientists from St. Andrews Biological Station (SABS) and were included by Rachel Carson in her book Silent Spring (Carson 1962). The other event was the copper-zinc pollution by mining activity in the Northwest Miramichi watershed.

After these two events, the Atlantic salmon catches continued to decline and the attention of SABS scientists shifted to the estuary, as well as to the possibilities of national and international overfishing. Later, as a result of shifting priorities and a reorganization, the direct involvement of SABS in research on the Miramichi was terminated. In the late 1980's SABS became again involved in the activities of the Miramichi River Environmental Assessment Committee (MREAC).

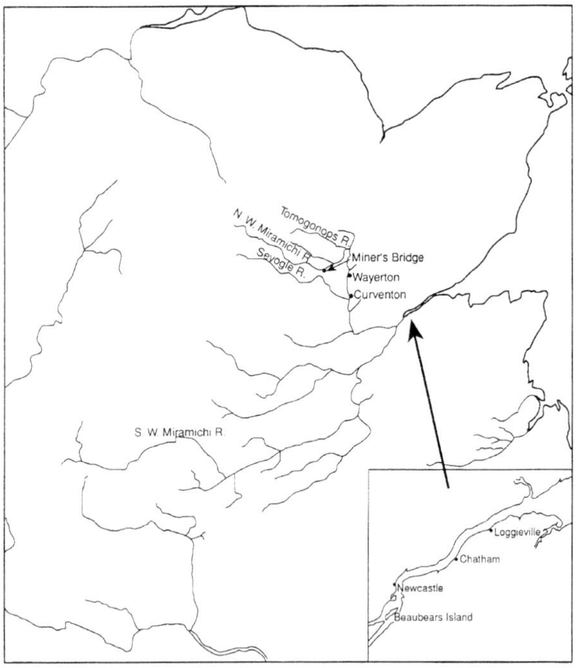

Fig. 1. Map showing the Miramichi area.

The Miramichi Watershed

The drainage area of the Miramichi is about 12 000 km^2 (Fig. 1). The Miramichi has a meander length of some 250 km and is capable of supporting about 200 000 returning adult Atlantic salmon (*Salmo salar*). A few additional characteristics are given in Table 1. Extensive wood harvesting in the

Table 1. Some characteristics of the Miramichi watershed (Elson 1967).

	Upper reaches			Lower reaches			Salmon angling no. fish 1949-59	
	Length (km)	Bed-rocks	Conductance (µmhos)	Length (km)	Bed-rocks	Conductance (µmhos)	Average	SD
Main								
Southwest	104	Igneous	45	80	Igneous	55	22 453	7659
Northwest	24	Igneous	45	80	Sediment	60	3 689	1847

Table 2. Movement of fish through the Curventon counting fence (Kerswill and Edwards 1967).

		Brook trout		White suckers	
Year	Spraying	Upward	Downward	Upward	Downward
1953	None	254	237	608	290
1954	Heavy upriver	406	254	1010	2982
1955	None	7	75	30	160
1956	Both up and downriver	24	115	30	127
1957	Some	2	646	10	470
1958	None	46	185	20	280
1959	None	384	317	124	325

Northwest Miramichi watershed in 1951-57 has resulted in some widening and shallowing of the river and the disappearance of small pools. Water temperature has also increased somewhat in the lower part of the Northwest Miramichi (Elson et al. 1972a).

Miramichi salmon spawn in October/ November. The young emerge from gravel beds in May/June. They grow to 5–7 cm in the first, 7–10 in the second, and to 10–14 cm, in the third year. Parr over 10 cm in the fall, generally migrate to sea in the next spring. The normal population estimates are: underyearlings 29, small parr 24, large parr 14, all per 100 m^2 (Elson 1967).

St. Andrews Biological Station scientists studied Atlantic salmon movements in the Northwest Miramichi River since 1950. A counting fence at Curventon has been used to keep track of the fish.

Curventon counting fence.

Table 3. Abundance of salmon in the Miramichi and of other fish in the NW Miramichi (Keenleyside 1959).

	Number/100 m^2		
	Underyearling	Small parr	Large parr
Before spraying	29	24	14
After spraying			
same year	4	11	5
after 1 yr	61	1	4
after 2 yr	29	53	4
after 3 yr	33	57	36

	Area sprayed			Area not sprayed	
	1954	1956	1957	1953	1955
Trout	2	10	7	7	4
Eel	0.4	0.2	0	2	0.5
Minnows	12	72	18	14	18

Table 4. DDT and PCB residues in Canadian Atlantic fishes and shellfishes.

	Tissue	Size (cm)	DDE	DDD	DDT
Sprague and Duffy (1971) - 1967 data					
Mussel (*Mytilus edulis*)	Whole	7.6	0.05	<0.01	<0.01
Clam, soft-shelled (*Mya arenaria*)	Whole		(0.03-0.11)	<0.01	<0.01
			0.01		
Quahaugs (*Venus mercenaria*)	Whole		(-0.04)		<0.01
Atlantic mackerel (*Scomber scombrus*)	Whole	41.8	0.01	0.08	0.38
	Muscle	59.0	0.09	0.01	0.02
Atlantic salmon (*Salmo salar*)	Viscera		0.03	0.06	0.08
	Whole	16.6	0.16	0.01	<0.01
American smelt (*Osmerus mordax*)	Whole	24.5	0.01	<0.01	<0.01
Tomcod (*Microgadus tomcod*)			<0.01		

	μg/g wet weight			
	PCB	DDE	DDD	DDT
Zitko (1971) - October 1970 data				
Mussel (*Mytilus edulis*)	0.14	0.02		
Elson et al. (1973) - 1971 data				
Salmon parr				
Headwaters	0.35	0.33	0.046	0.032
Lower reaches	0.39	0.36	0.027	0.053
Hatchery-reared (South Esk)	0.10	0.01		
Salmon fry	0.21	0.07		
Grilse	0.30	0.146		
Salmon (2 sea-winter)	0.47	0.052		

Populations of juvenile salmon in the Northwest Miramichi were measured from 1951 to the late 1960's in 10 locations. Both DDT spraying in 1954-57, as well as the mining pollution in 1960-63 severely depressed the juvenile salmon populations.

Runs of shad (*Alosa sapidissima*), gaspereau (*Alosa pseudoharengus*), and rainbow smelt (*Osmerus mordax*) above the Curventon counting

Table 5. Recent DDT and PCB residues in fish from the Miramichi (samples from MREAC and Mr. B. Baldwin, and Dr. G. Hare 1990, Zitko, unpublished).

			µg/g wet weight				
					PCB		
		Tissue	DDE	DDD	153	138	180
Tomcod		muscle	0.0005	0.0002			
		liver	0.117	0.067	0.024	0.032	
Flounder		muscle	0.001	0.002			
		Liver	0.025	0.018	0.015	0.014	
Eel 'TCDD homogenate'			0.053	0.032	0.014	0.025	
NW Mir. 1.6.90		muscle	0.051	0.027	0.008	0.013	
		liver	0.014	0.008			
		muscle	0.058	0.025	0.010	0.012	
Atlantic salmon							
Sex & Treatment	Length (cm)	Weight (kg)					
MT	58	1.54	0.046	0.040[a]	0.028	0.071	0.053
MT	57.6	1.54	0.057	0.030[a]	0.039	0.089	0.065
F?	86	6.49	0.064	0.026	0.033	0.042	0.029
MT	57	1.81	0.038	0.022[a]	0.026	0.059	0.051
MT	57.4	1.41	0.043	0.022[a]	0.031	0.080	0.060
FT	57.6	2.09	0.027	0.013[a]	0.020	0.047	0.042
FN	78	2.04	0.029	0.006[a]	0.012	0.012	0.013
MT	61	1.99	0.021	0.016[a]	0.017	0.042	0.034
FR	61	2.49	0.068	0.050	0.049	0.058	0.050
			0.057	0.036	0.042	0.049	0.037
MR	61	2.72	0.078	0.042	0.048	0.060	0.046
MR	61.5	2.81	0.081	0.035	0.048	0.058	0.043
MR	58.5	2.27	0.078	0.052	0.041	0.055	0.047
FR	103.5	9.52	0.118	0.045	0.047	0.061	0.071
			0.104	0.042	0.053	0.064	0.080
FN	84	3.63	0.044	0.013	0.013	0.020	0.022
			0.062	0.022	0.030	0.037	0.051
G			0.023	0.010	0.008	0.010	0.006
F			0.018	0.008	0.009	0.011	0.007

[a]Additional peaks in the DDD area;
M - male; F - female; R - reconditioned, N - not reconditioned; T - trap; G - grilse.

fence virtually disappeared in 1955 and did not return for the next 13 years. No data are available after 1968 (Elson et al.1972b).

Forest Spraying with DDT

In 1952, DDT was used against the spruce budworm epidemic. Initially, the applications rates were high, up to 0.56 kg/ha. Already in 1954, mortality of juvenile salmon following the spraying was reported. Various means of avoiding the fish kills were tried, such as the lowering of the application rate and the use of other pesticides in buffer zones, but the use of DDT went on for another 14 yr (Elson et al. 1973).

The deleterious effects of the spraying, particularly on juvenile Atlantic salmon, are documented in a series of publications and going into details is beyond the scope of this paper. Only as an example the data of Kerswill and Edwards (1967) on the relationship between spraying and movements of fish are given in Table 2, and between spraying and abundance of fish are presented in Table 3. Total DDT residues in salmon parr decreased according to the equation: total residues [ppm] = 1.91/(years since last spraying). The decrease of DDE was considerably slower, with a half-life of about 4 yr (Sprague et al. 1971).

Forest spraying was not the only source of DDT. Many varied applications have lead to a world-wide dissemination of DDT and its metabolites, DDE and DDD. The levels of DDT, its metabolites, as well the levels of PCB, as they were about 20 years ago, are summarized in Table 4. Sprague and Duffy (1971) have not measured PCB and, consequently, their levels of DDE, DDD, and DDT may be somewhat overestimated. The Elson et al. (1973) parr and fry data were recalculated on a wet weight basis assuming a lipid content of 5%. The grilse and salmon data were converted from body burden to concentration, assuming body weights of 1.5 kg for grilse and 5 kg for salmon, respectively.

As the 1990 data (Table 5) show, the levels of DDE and DDD have not changed appreciably during the last 20 years. PCB levels given in Tables 4 and 5 are not directly comparable because of different techniques, but, in any case, the values are low. The absence of less chlorinated congeners in detectable concentrations indicates no recent input of PCB. On the other hand, the data show that salmon reconditioning results in a slight and toxicologically insignificant, but measurable increase of DDE, DDD and PCB in the salmon. As one would expect, higher environmental levels of these compounds are present in the estuary than in the open ocean. Feed is a possible source of DDE and DDD, but PCB may be coming from elsewhere. Salmon from the 'trap' also contain additional unidentified compounds. It is not known at the moment whether these compounds are in the fish or are contaminants from the trap material or from the sample handling. The subject deserves further attention.

Mining Pollution

A base metal mine was developed in the Tomogonops/Northwest Miramichi River watershed in 1956, operated in 1957 and during a part of 1958. The operation was resumed in June 1960, when large volumes of mineshaft water were pumped into the Tomogonops. Associated with the

Table 6. Population indices (no./100 m^2) and number of migrating smolts (Elson, unpublished).

	1968	1969	1970
Below the Tomogonops			
Underyearlings	10.3	0.8	0.8
Parr	3.1	2.1	0.2
Above the Tomogonops			
Underyearlings	19.5	1.1	23.8
Parr	19.5	25.2	10.3

Table 7. Humic acid correction factors (Zitko et al. 1973).

Humic acid (mg/L) Hardness as $CaCO_3$ (mg/L)	Multiply LC by		
	5	10	20
15	1.82	2.37	6.60
30	1.81	2.57	4.05
90	1.65	1.77	3.20

Table 8. Thallium in the South Tomogonops (Zitko et al. 1975).

	Minimum	Maximum
Tl	0.7	20 µg/L
Cu	1	21 µg/L
Zn	10	2500 µg/L
Hardness	250	890 mg/L
Flow	4.2	108 m^3/min

resulting concentrations of copper and zinc in the Northwest Miramichi, elevated to 0.77 of the Incipient Lethal Level (ILL), the proportion of migrating salmon, returning downstream increased from 1–3 to 22%. Most of the fish moving downstream did not return that season (Sprague et al. 1965).

These field observations were the impetus for research at SABS by John B. Sprague, W. Victor Carson, and W.G. (Bill) Carson, on the toxicity to Atlantic salmon, and on the analytical chemistry of copper and zinc. It was determined that, in soft water (water hardness of 20 mg/L as calcium carbonate), the lethal thresholds or ILLs, below which mortality does not occur, of copper and zinc were 50 and 600 µg/L, respectively, and that the toxicity of copper and zinc was additive. Consequently, the overall toxicity can be expressed in toxic units. These are the sums of the respective fractions of the lethal thresholds. Salmon parr also avoid waters with elevated (to about 0.1 toxic unit), concentrations of copper and zinc ions, in the laboratory. In the field, downstream movement of adult salmon was observed at about 0.4 toxic unit. A level of 0.8 toxic unit seemed to have completely inhibited upstream migration in August and September 1963 (Sprague et al. 1965).

The effects of mining pollution on the population indices persisted for a number of years (Table 6). However, there is not a clear relationship between parr and smolts.

After field observations of unusual downstream movement of salmon and other fish at the Curventon counting fence in 1960 and the assumption that it is linked with heavy metals, a chemical laboratory was established at SABS in 1961 by W. Victor Carson. From this time until 1973, this laboratory monitored the Northwest Miramichi at four locations: Curventon/Wayerton, Miner's Bridge, Sevogle, and the South Tomogonops. The data were presented in 12 reports and include pH, temperature, conductance, hardness,and the concentration of copper and zinc ions. River flow data provided by the Department of Energy, Mines and Resources were also given. The monitoring

Table 9. Methyl mercury in fish from the Miramichi (Zitko et al. (1971).

	Weight or length	Methyl mercury (µg/g wet weight)
American eel (*Anguilla rostrata*)	317 g	0.07(0.01-0.12)
	114 g	0.08
Atlantic salmon (*Salmo salar*)	47 cm	0.09
White sucker (*Catostomus commersoni*)	143 g	0.09

frequency was daily at Curventon from April to October and three times a week at Wayerton during the winter. The other sites were monitored at first on a weekly, and later on a monthly basis.

In 1971 the Little South Tomogonops was diverted by Heath Steele Mines. This resulted in a considerable drop in the concentration of copper and zinc ions in the Northwest Miramichi (Carson 1974). In 1973, the yearly average toxic unit value at Curventon was 0.08, and the monitoring by SABS was terminated. The monitoring by the mine has continued for a number of years. The effects of the pollution, however, have persisted in the most affected locations for a long time. For example, in 1974-75, mayflies were still almost absent from the Tomogonops, whereas the number of caddisfly larvae had increased (Peterson 1978).

As a final remark on assessing the risk of mining pollution to the aquatic environment, one has to consider that the toxicity of heavy metals to fish decreases with increasing water hardness. The toxicity of copper is also decreased by increasing concentration of humic acid (Table 7). In the Miramichi, humic acid concentration may vary from 5 to 15 mg/L. At the same time, other aquatic species may have different sensitivities to heavy metals. Very little is also known about the effects of concentration 'spikes' which may result, for example, from spills or other accidents. There are also no established criteria by which to prove that fish were killed by heavy metals.

Metals other than copper, zinc or lead may cause mining pollution problems. An interesting observation was the presence of thallium in the South Tomogonops (Table 8).

Thallium is a slow acting toxin for juvenile Atlantic salmon, with a possible 'No observed effect concentration' below 30 µg/L. Its action is not likely to be affected by water hardness or by humic acid, since it does not form complexes of this type. The toxicity of thallium and copper or zinc, to juvenile Atlantic salmon is not additive (Zitko et al. 1975). There have been no follow-up studies on thallium in the Miramichi system.

Another metal of environmental interest is mercury. It is widely distributed in the environment, both by natural sources and by human activities. Mining is not a source of mercury in the Miramichi watershed. The prevalent form of mercury in fish is methyl mercury. The relatively limited data on methyl mercury in fish from the Miramichi suggest that mercury is not a problem here (Table 9).

Estuarine Pollution

In connection with sonic tag tracking of salmon movements in the estuary, the distribution of organic matter was followed by UV spectra (Elson et al. 1972b). The absorbance at 250 nm is a measure of the concentration of organic matter, both natural (humic and fulvic acid) and anthropogenic (for example pulpmill effluent), but does not distinguish between them. The concentration of organic matter decreased sharply with the increase of salinity to about 5 parts per thousand, due to precipitation by increasingly saline water. This occurred somewhere below Newcastle. A further decrease with increasing salinity was only slight. In the Newcastle area, the UV absorbance was about 0.020, at a river discharge of 167 m3/s. The absorbance decreased to about 0.010 around Loggieville. Upstream movement of salmon, followed by sonic tags, was relatively slow up the

Sonic tracking of Atlantic salmon in the Miramichi Estuary.

estuary, and no fish migrated through the channel on the northwest side of Beaubears Island.

Attention was also paid to specific sources of pollution in the estuary. After detecting pentachlorophenol in the effluent from the Domtar wood preserving plant in Newcastle, the effluent was monitored during the plant's operation from March 17 to May 13, 1969. During that time, about 600 kg of pentachlorophenol and an estimated 60 t of oil were discharged into the estuary. The amount of oil may have been overestimated because of sampling problems (Zitko and Carson 1969).

Salmon and trout avoid the wood treatment plant effluent in concentrations starting around 5 µg/L. A considerable leaching of organic compounds related to the effluent, from a marsh area into the Miramichi Estuary, was detected by UV spectra. The concentration of chloroform-extractable substances was 24 mg/L, as compared to 1200 mg/L in a brook, receiving the drainage from the plant (Zitko et al. 1969).

The debarker effluent from the (then) Fraser's pulp mill in Newcastle contained 180 mg/L of resin acids (expressed as abietic acid), 173 mg/L of petroleum ether extractable compounds, and an additional 70, 44, and 60 mg/L of ether, carbon-chloroform, and carbon-ethanol extracts, respectively. Aeration of the effluent for 14 d decreased the petroleum ether-extractable compounds to about 30%, the ether-extractable compounds to about 60% of the original level. The concentrations of the carbon-chloroform, and carbon-ethanol extracts have not changed appreciably. Foam collected at various points contained the above fractions in proportions similar to those in the original effluent. The effluent from the (then) Acadia groundwood mill at South Nelson contained 34, 93, 82, 23, and 76 mg/L of the above fractions, respectively (Zitko and Carson 1971). The petroleum ether fraction was acutely toxic to juvenile Atlantic salmon. The toxicity was caused by resin acids as well as by other, unidentified compounds.

In the late sixties and early seventies, the estuary was the scene of frequent oil spills: four spills were documented in 1969-70 (Zitko and Carson 1970). Two of them were spills of a heavy

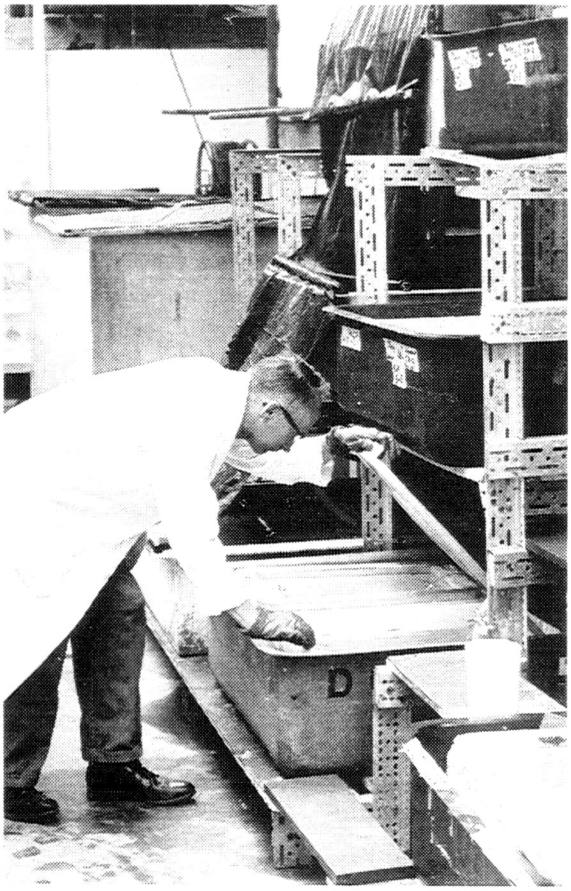

W. (Bill) G. Carson performing toxicity tests with juvenile Atlantic salmon.

fuel oil and two were spills of light fuel oils. Some light fuel oil was probably present in one of the heavy fuel oil spills and a gasoline might have been spilled together with the second heavy fuel oils spill, since a dye, tentatively identified Oil Red Dye IR 1340, as well as heavy fuel oil were isolated from water samples. The dye was then used in gasoline. A similar red dye was isolated from water samples associated with with a light fuel oil spill from a boat. Another light fuel oil spill occurred from the oil companies storage tanks and we were able to identify the source (Zitko and Carson 1970).

In 1970, samples of Miramichi Estuary sediments, taken above and below Chatham, at Loggieville, at the mouth of Miramichi Bay, contained 0.8, 1.9, 6.8, and 2.5 µg/g of heavy fuel oil, as measured by fluorescence. In the bay, the concentration decreased from 2.7 µg/g on the north side, to 0.3 µg/g in the center, and was practically not detectable in the south (Zitko and Carson 1970).

Conclusions

A considerable amount of research effort was devoted to the Miramichi, particularly during the last 45 yr. Studies of Atlantic salmon and other fish in the Miramichi have played an important role in the identification of unintentional consequences of the indiscriminate use of DDT. Similarly, anomalous behaviour of salmon in the Miramichi, detected by observant individuals, was instrumental in the development of aquatic toxicology of copper and zinc and in eventual solution of mining pollution problems. Studies of salmon movement in the estuary may have been a factor leading to the closure of the wood preserving plant. All these problems, highlighted by research, contributed to the creation of environmental awareness of the 1960's and 1970's.

One cannot imagine that forest spraying, mining or wood preservation would be carried out in such a manner today. Environmental sciences have developed and attitudes have changed. This does not mean that 30 yr from now our present actions and attitudes will not appear equally unimaginable. It is more than likely that the discharge of untreated sewage, indiscriminate disposal of solid waste, insufficiently treated effluents, and many other activities will appear irresponsible and primitive to future generations.

At the same time, we must not go to the other extreme and consider all environmental changes and activities as undesirable. I think that the status of the Miramichi watershed corresponds to its population density and industrial activity. On the basis of the available information, there are no problems with organochlorine compounds and heavy metals in the biota. Unfortunately, our information base is not very good. It is certainly inferior to those of Western European countries for watersheds of similar importance. We need more measurements, verified models for better predictions of environmental impacts of our actions, and much more effort both in development and in implementation in the area of waste treatment. With the latter we can start locally, on a personal, household, and community level; the sooner the better.

Acknowledgments

I thank all present and former employees of the Fisheries Research Board of Canada and Fisheries and Oceans Canada, whose work forms the basis of this overview, particularly to Mr. W. Victor Carson, for his pioneering work in the trace analytical chemistry of heavy metals, to members of the former Anadromous Fish, and Pollution, Investigations, particularly to Mr. Emerson J. Schofield and his team at Curventon, to members of MREAC, particularly to Mr. Jack Hunter for sampling and logistic support, to Dr. Gerry Hare and workers of the kelt reconditioning project, and to Mr. Ben Baldwin for samples of fish. Many of the toxicity studies were carried out skillfully and meticulously by the late Mr. William G. Carson. Ms. Brenda Best formatted the manuscript and provided the figure and the photographs.

References

Carson, R. 1962. Silent Spring. Houghton Mifflin Company, The Riverside Press, Cambridge, MA.

Carson, W. V. 1974. Chemical conditions in the Northwest Miramichi River during 1973. Fish. Res. Board Can. MS Rep. 1323.

Elson, P. F. 1967. Effects on wild young salmon of spraying DDT over New Brunswick forests. J. Fish. Res. Board Can. 24: 731-767.

Elson, P. F., Saunders, J. W., and Zitko, V. 1972a. Impact of forest-based industries on freshwater-dependent fish resources in New Brunswick. Fish. Res. Board Can. Tech. Rep. 325: 26 p.

Elson, P. F., Lauzier, L. N., and. Zitko, V. 1972b. A preliminary study of salmon movements in a polluted estuary, p. 325-340. *In* Marine Pollution and Sea Life, Fishing News (Books) Ltd., West Byfleet, Surrey.

Elson, P. F., Meister, A. L., Saunders, J. W., Saunders, R. L., Sprague, J. B., and Zitko,V. 1973. Impact of chemical pollution on Atlantic salmon in North America. Atlantic Salmon Symposium, St. Andrews, N.B., 83-110, The International Atlantic Salmon Foundation.

Keenleyside, M. H. A. 1959. Effects of spruce budworm control on salmon and other fishes in New Burnswick. Can. Fish Cult. 24: 17-22.

Kerswill, C. J., and Edwards, H. E. 1967. Fish losses after forest sprayings with insecticides in New Brunswick, 1952-62, as shown by caged specimens and other observations. J. Fish. Res. Board Can. 24: 709-729.

Peterson, R. H. 1978. Variations in aquatic insect densities associated with copper-zinc concentrations. Fish. Mar. Serv. Manuscr. Rep. 1470: 3p.

Sprague, J. B., Elson, P. F., and Saunders, R. L. 1965. Sublethal copper-zinc pollution in a salmon river – a field and laboratory study. Int. J. Air Water Pollut. 9: 531-543.

Sprague, J. B., and Duffy, J. R. 1971. DDT residues in Canadian Atlantic fishes and shellfishes in 1967. J. Fish. Res. Board Can. 28: 59-64.

Sprague, J. B., Elson, P.F., Duffy, J.R. 1971. Decrease in DDT residues in young salmon after forest spraying in New Brunswick. Environ. Pollut. 1: 191- 203.

Zitko, V. 1971. Polychlorinated biphenyls and organochlorine pesticides in some freshwater and marine fishes. Bull. Environ. Contam. Toxicol. 6: 464-470.

Zitko, V., Carson, W. G., and Carson, W. V. 1969. Wood preserving plant effluent: Chemical composition, toxicity to salmon and trout. Fish. Res. Board Manuscr. Rep. 1042: 26 p, 15 figs.

Zitko, V., and Carson, W. V. 1969. Analysis of the effluent from the Domtar wood preserving plant at Newcastle, N.B. Fish. Res. Board Manuscr. Rep. 1024: 14 p, 5 figs.

Zitko, V., and Carson, W. V. 1970. The characterization of petroleum oils and their determination in the aquatic environment. Fish. Res. Board Can. Tech. Rep. 217: 29 p.

Zitko, V., and Carson, W. V. 1971. Resin acids and other organic compounds in groundwood and sulfate mill effluents and foams. Fish. Res. Board Manuscr. Rep. 1134: 28 p.

Zitko, V., Carson, W. V., and Carson, W. G. 1973. Prediction of incipient lethal levels of copper to juvenile Atlantic salmon in the presence of humic acid by cupric electrode. Bull. Environ. Contam. Toxicol. 10: 265-271.

Zitko, V., Carson, W. V., and Carson, W. G. 1975. Thallium: occurrence in the environment and toxicity to fish. Bull. Environ. Contam. Toxicol. 13: 23-30.

Zitko, V., Finlayson, B. J., Wildish, D. J., Anderson, J. M., and Kohler, A. C. 1971. Methylmercury in freshwater and marine fishes in New Brunswick, in the Bay of Fundy, and on the Nova Scotia Banks. J. Fish. Res. Board Can. 28: 1285-1291.

Appendix

Miramichi Bibliography in Chronological Order

M'Gonigle, R.H. 1929. Report of investigations into the cause of mortality of salmon fry at the Miramichi hatchery. Fish. Res. Board Manuscr. Rep. Ser B 86: 13 p.

Huntsman, A.G. 1931. Big catch of Miramichi salmon every three years. Prog. Rep. FRB Atl. Coast Stn., Issue No. 2: 18-19.

Kerr, R.B. 1933. Investigations on the Atlantic salmon (*Salmo salar*) of the Miramichi River, New Brunswick, Canada. Fish. Res. Board Manuscr. Rep. Ser B 43: 12 p.

Blair, A.A. 1933. Migrations of Atlantic salmon (*Salmo salar*) in Miramichi River. Fish. Res. Board Manuscr. Rep. Ser B 59: 15 p.

Blair, A.A. 1935. Ages at migration of Atlantic salmon in Miramichi River. J. Biol. Board Can. 1: 159-169.

Klugh, A.B. 1945. The ecology of the algae of the Miramichi region of New Brunswick. Fish. Res. Board Manuscr. Rep. Ser B 341(2): 14 p.

McKenzie, R.A. 1945. Smelt tagging, 1943-1944, Miramichi area. Fish. Res. Board Manuscr. Rep. Ser B 369(1-4): 29 p.

Medcof, J.C., and Morrison, E.I. 1943. Report on 1943 shellfish investigations. Fish. Res. Board Manuscr. Rep. Ser B 370: 65 p.

Bousfield, E.L. 1954. Some physical features of the Miramichi estuary. Fish. Res. Board Manuscr. Rep. Ser B 574: 19 p.

Bousfield, E.L. 1955. Some physical features of the Miramichi estuary. J. Fish. Res. Board Can. 12: 342-361.

Kerswill, C.J., and Elson, P.F. 1955. Progress Reports of FRB Atlantic Coast Stations, Issue No. 62: 17-24.

Kerswill, C.J. 1955. Recent developments in Atlantic salmon research. Atl. Salmon J. 26-30.

Kerswill, C.J. 1955. Effects of black salmon angling on Miramichi salmon stocks. Atl. Salmon J. 30-31.

Keenleyside, M.H.A. 1959. Effects of spruce budworm control on salmon and other fishes in New Burnswick. Can. Fish Cult. 24: 17-22.

McKenzie, R.A. 1961. The smelt fishery of the Miramichi 1960-61. Can. Dep. Fish. Trade News 14: 13-16.

Carson, R. 1962. Silent Spring. Houghton Mifflin Company, The Riverside Press, Cambridge, MA.

Keenleyside, M.H.A. 1962. Skin-diving observations of Atlantic salmon and brook trout in the Miramichi River, New Brunswick. J. Fish. Res. Board Can. 19: 625-534.

Henderson, E.B., Saunders, R.L., and Kerswill, C.J. 1965. Daily counts of Atlantic salmon at the Curventon and Camp Adams counting fences on the Northwest Miramichi River, New Brunswick, from 1950 to 1963. Fish. Res. Board Manuscr. Rep. Ser B 805: 78 p.

Hart, J.L., and Ferguson, R.G. 1966. The American smelt. Trade News 18: 22-23.

Sprague, J.B., Elson, P.F., and Saunders, R.L. 1965. Sublethal copper-zinc pollution in a salmon river - a field and laboratory study. Int. J. Air Water Pollut. 9: 531-543.

Gibson, R.J. 1966. Some factors influencing the distributions of brook trout and young Atlantic salmon. J. Fish. Res. Board Can. 23: 1977-1980.

Allen, K.R., and Saunders, R.L. 1967. A preliminary study of the influence of the Greenland salmon fishery on the salmon stocks and fishery of the Miramichi River system, New Brunswick, Canada. Int. Comm. North Atl. Fish. Redb. Part III: 159-180.

Allen, K.R. 1967. A revised estimate of the rate of growth between Greenland and home waters of salmon from the Miramichi River, New Brunswick, Canada. Int. Comm. North Atl. Fish. Redb. Part III: 66-68.

Allen, K.R. 1967. Results of Atlantic salmon tagging in the Maritime Provinces of Canada, 1964-66. Int. Comm. North Atl. Fish. Redb. Part III: 69-73.

Saunders, R.L. 1967. Seasonal pattern of return of Atlantic salmon in the Northwest Miramichi River, New Brunswick. J. Fish. Res. Board Can. 24: 21-32.

Kerswill, C.J. 1967. Studies on the effects of forest spraying with insecticides, 1952-1963, on fish and aquatic invertebrates in New Brunswick streams; Introduction and summary. J. Fish. Res. Board Can. 24: 701-708.

Kerswill, C.J., and Edwards, H.E. 1967. Fish losses after forest sprayings with insecticides in New Brunswick, 1952-62, as shown by caged specimens and other observations. J. Fish. Res. Board Can. 24: 709-729.

Elson, P.F. 1967. Effects on wild young salmon of spraying DDT over New Brunswick forests. J. Fish. Res. Board Can. 24: 731-767.

Ide, F.P. 1967. Effects of forest spraying with DDT on aquatic insects of salmon streams in New Brunswick. J. Fish. Res. Board Can. 24: 769-805.

Keenleyside, M.H.A. 1967. Effects of forest spraying with DDT in New Brunswick on food of young Atlantic salmon. J. Fish. Res. Board Can. 24: 807-822.

Grant, C.G. 1967. Effects on aquatic insects of forest spraying with phosphamidon in New Brunswick. J. Fish. Res. Board Can. 24: 823-832.

Forsythe, M.G. 1967. Analysis of the 1965 smolt run in the Northwest Miramichi River, New Brunswick. Fish Res. Board Can. Tech. Rep. 4: 73 p.

Saunders, R.L, and Sprague, J.B. 1967. Effects of copper-zinc mining pollution on a spawning migration of Atlantic salmon. Water Res. 1: 419-432.

Forsythe, M.G. 1968. Analysis of the 1966 smolt run in the Northwest Miramichi River, New Brunswick. Fish. Res. Board Can. Tech. Rep. 91: 33 p.

Elson, P.F. 1969. Utilization of three stocks of Atlantic salmon tagged and liberated as smolts in the Northwest Miramichi River from 1964 to 1967. Int. Comm. North Atl. Fish. Redb. Part III: 71-77.

Zitko, V., and Carson, W.V. 1969. Analysis of the effluent from the Domtar wood preserving plant at Newcastle, N.B. Fish. Res. Board Manuscr. Rep. 1024: 14 p. + 5 figs.

Zitko, V., Carson, W.G., and Carson, W.V. 1969. Wood preserving plant effluent: Chemical composition, toxicity to salmon and trout. Fish. Res. Board Manuscr. Rep. 1042: 26 p. + 15 figs.

Pippy, J.H.C., and Hare, G.M. 1969. Relationship of river pollution to bacterial infection in salmon (*Salmo salar*) and suckers (*Catostomus commersoni*). Trans. Amer. Fish. Soc. 98: 685-690.

Pippy, J.H.C. 1970. Fish mortalities in the Northwest Miramichi River in 1969. Fish. Res. Board Can. Tech. Rep. 226: 7 p.

Besch, W.K., Ricard, M., and Cantin, R. 1970. Utilisation des diatomees benthiques comme indicateurs de pollution minière dans le bassin de la Miramichi N.W. Fish. Res. Board Can. Tech. Rep. 202: 72 p.

Besch, K.W., and Roberts-Pichette, P. 1970. Effects of mining polution on vascular plants in the Northwest Miramichi River system. Can. J. Bot. 48: 1647-1656.

Zitko, V., and Carson, W.V. 1970. The characterization of petroleum oils and their determination in the aquatic environment. Fish. Res. Board Can. Tech. Rep. 217: 29 p.

Sprague, J.B., Duffy, J.R. 1971. DDT residues in Canadian Atlantic fishes and shellfishes in 1967. J. Fish. Res. Board Can. 28: 59-64.

Sprague, J.B., Elson, P.F., and Duffy, J.R. 1971. Decrease in DDT residues in young salmon after forest spraying in New Brunswick. Environ. Pollut. 1: 191-203.

Zitko, V., Finlayson, B.J., Wildish, D.J., Anderson, J.M., and Kohler, A.C. 1971. Methylmercury in freshwater and marine fishes in New Brunswick, in the Bay of Fundy, and on the Nova Scotia Banks. J. Fish. Res. Board Can. 28: 1285-1291.

Zitko, V. 1971. Polychlorinated biphenyls and organochlorine pesticides in some freshwater and marine fishes. Bull. Environ. Contam. Toxicol. 6: 464-470.

Zitko, V., and Carson, W.V. 1971. Resin acids and other organic compounds in groundwood and sulfate mill effluents and foams. Fish. Res. Board Can. Manuscr. Rep. 1134: 28 p.

Elson, P.F., Lauzier, L.N., and Zitko, V. 1972. A preliminary study of salmon movements in a polluted estuary. Marine Pollution and Sea Life, Fishing News (Books) Ltd., West Byfleet, Surrey, 325-340.

Elson, P.F., Saunders, J.W., and Zitko, V. 1972. Impact of forest-based industries on freshwater-dependent fish resources in New Brunswick. Fish. Res. Board Can. Tech. Rep. 325: 26 p.

Montreal Engineering Company, Limited. 1972. Northeastern New Brunswick Mine Water Quality program, Volume One, Summary of Findings. Fredericton, N.B.

Elson, P.F., Meister, A.L., Saunders, J.W., Saunders, R.L., Sprague, J.B., and Zitko, V. 1973. Impact of chemical pollution on Atlantic salmon in North America. Atlantic Salmon Symposium, St. Andrews, N.B., 83-110, The International Atlantic Salmon Foundation.

Zitko, V., Carson, W.V., and Carson, W.G. 1973. Prediction of incipient lethal levels of copper to juvenile Atlantic salmon in the presence of humic acid by cupric electrode. Bull. Environ. Contam. Toxicol. 10: 265-271.

Carson, W.V. 1974. Chemical conditions in the Northwest Miramichi River during 1973. Fish. Res. Board Can. Manuscr. Rep. 1323.

Zitko, V., Carson, W.V., and Carson, W.G. 1975. Thallium: Occurrence in the environment and toxicity to fish. Bull. Environ. Contam. Toxicol. 13; 23-30.

Peterson, R.H. 1978. Variations in aquatic insect densities associated with copper-zinc concentrations. Fish. Mar. Serv. Manuscr. Rep. 1470: 3 p.

THE PHYSICAL ENVIRONMENT

"The act of observing is more important than the state of knowledge."

Dr. David Booth

April 13, 1994, Miramichi Environmental Workshop

A plankton net is used to collect small organisms in the upper parts of the water column. This type of equipment was used to collect larval fish and zooplankton throughout the Miramichi Estuary.

CHAPTER 4

Seasonal and Short-Term Variations in the Estuarine Structure of the Miramichi

Caroline Lafleur

Centre Océanographique de Rimouski, 310, allée des Ursulines, Rimouski, Québec G5L 3A1

Bernard Pettigrew

*Ministère des Pêches et des Océans, Institut Maurice-Lamontagne, C.P. 1000
850, route de la Mer, Mont-Joli, Québec G5H 3Z4*

André St-Hilaire

*Department of Fisheries and Oceans, Science Branch, Gulf Region, P.O. Box 5030,
Moncton, New Brunswick E1C 9B6*

David A. Booth

Centre Océanographique de Rimouski, 310, allée des Ursulines, Rimouski, Québec G5L 3A1

and E. Michael P. Chadwick

*Department of Fisheries and Oceans, Science Branch, Gulf Region, P.O. Box 5030,
Moncton, New Brunswick E1C 9B6*

Lafleur, C., Pettigrew, B., St-Hilaire, A., Booth, D.A., and Chadwick, E.M.P. 1995. Seasonal and short-term variations in the estuarine structure of the Miramichi, p. 45–71. *In* E.M.P. Chadwick [*editor*]. Water, science and the public: the Miramichi ecosystem. Can. Spec. Publ. Fish. Aquat. Sci. 123.

Abstract

We analyzed three years of temperature and salinity data from the Miramichi to examine the factors that modify the estuarine structure at scales between one day and one year. At the seasonal scale, variations in stratification are determined by river discharge and heat flux. At the monthly scale, vertical mixing by tides is the dominant process. At the diurnal scale, variations in stratification are determined, in the main, by tidal shear and short events in wind and precipitation.

Résumé

Dans le but de déterminer les processus qui causent des modifications dans la structure estuarienne de la Miramichi, nous avons analysé, sur des échelles de temps de une journée à un an, trois années de données de température et de salinité. À l'échelle saisonnière, les changements de stratification sont causés par des variations de débit d'eau douce et d'apport calorifique. À l'échelle mensuelle, le processus dominant est le mélange vertical engendré par le mouvement des marées. Enfin, à l'échelle journalière, les changements de stratification sont surtout causés par le cisaillement vertical du courant de marée et également par des variations sporadiques des vents et des précipitations.

Introduction

The Miramichi, located on the east coast of New Brunswick (Fig. 1) is a major estuary supporting both commercial fishing and industry. Despite numerous studies of various aspects, including the physical and chemical environment, fish populations, and sedimentary processes, several basic questions remain concerning the physical processes of the estuary. Causes of variations in the structure of the salt wedge is one such question.

The Miramichi system separates into three distinct parts: the bay, the river estuary, and the river. The bay is triangular in shape, 20 km long by 5–20 km wide and protected from the Gulf of St. Lawrence by a series of barrier islands. This section is shallow with depths ranging from 3 to 5 m except within the shipping channel, which has been dredged to a depth of 8 m. In general, waters in the bay are vertically well mixed (Vilks and Krauel 1982). The river estuary, however, is the region of mixing between fresh and salt water. It has a funnel-shaped section and meanders up to Beaubears Island where it separates into two branches. The river estuary is shallow but a channel of 10 m depth may occupy up to half its width. This section is stratified but the upper limit of the salinity intrusion varies with time. Upstream of the river estuary is the river itself which is composed of a network of tributaries with a total catchment area of 1.4×10^4 km^2.

This study concerns the physical oceanography of the estuarine section of the Miramichi. Between the summer of 1991 and the autumn of 1993, various oceanographic instruments were moored at different locations along the estuary. These instruments provided winter and summer measurements of salinity and temperature and some measurements of current during the ice-free season. Using these data, we studied in particular, the salinity structure and its dependence on variations of environmental factors such as winds, tides, and river inflows.

The equilibrium distribution of water masses in an estuary depends on the dimension of the estuary, the inflow of fresh water, tidal mixing, and wind mixing (particularly in shallow estuaries) (Bowman 1988). In a salt-wedge estuary (Fig. 2) such as the Miramichi (Vilks and Krauel 1982), the river runoff is sufficient to maintain a vertical salinity gradient, implying that the stratifying effects of freshwater inflow prevail over tidal and wind mixing (Geyer and Farmer 1989). The limit of the salt wedge is defined as the upstream position of the salt-water intrusion on the bottom of a stratified estuary (Dyer 1973). The salt wedge may be steady or time-changing. It is steady if its form is conserved. This in general occurs when tidal and wind mixing is weak. The tides cause only a back and forth movement of the estuarine structure. In this case there exists an equilibrium between time-averaged horizontal forces: the longitudinal baroclinic pressure gradient due to the change in density, the longitudinal barotropic pressure gradient due to the river inflow, and the frictional forces (Geyer and Farmer 1989). An increase in freshwater outflow can increase the frictional forces between the layers and push the salt wedge downstream.

Compared with variations in the estuary, salinity in the bay, the seawater source, varies little

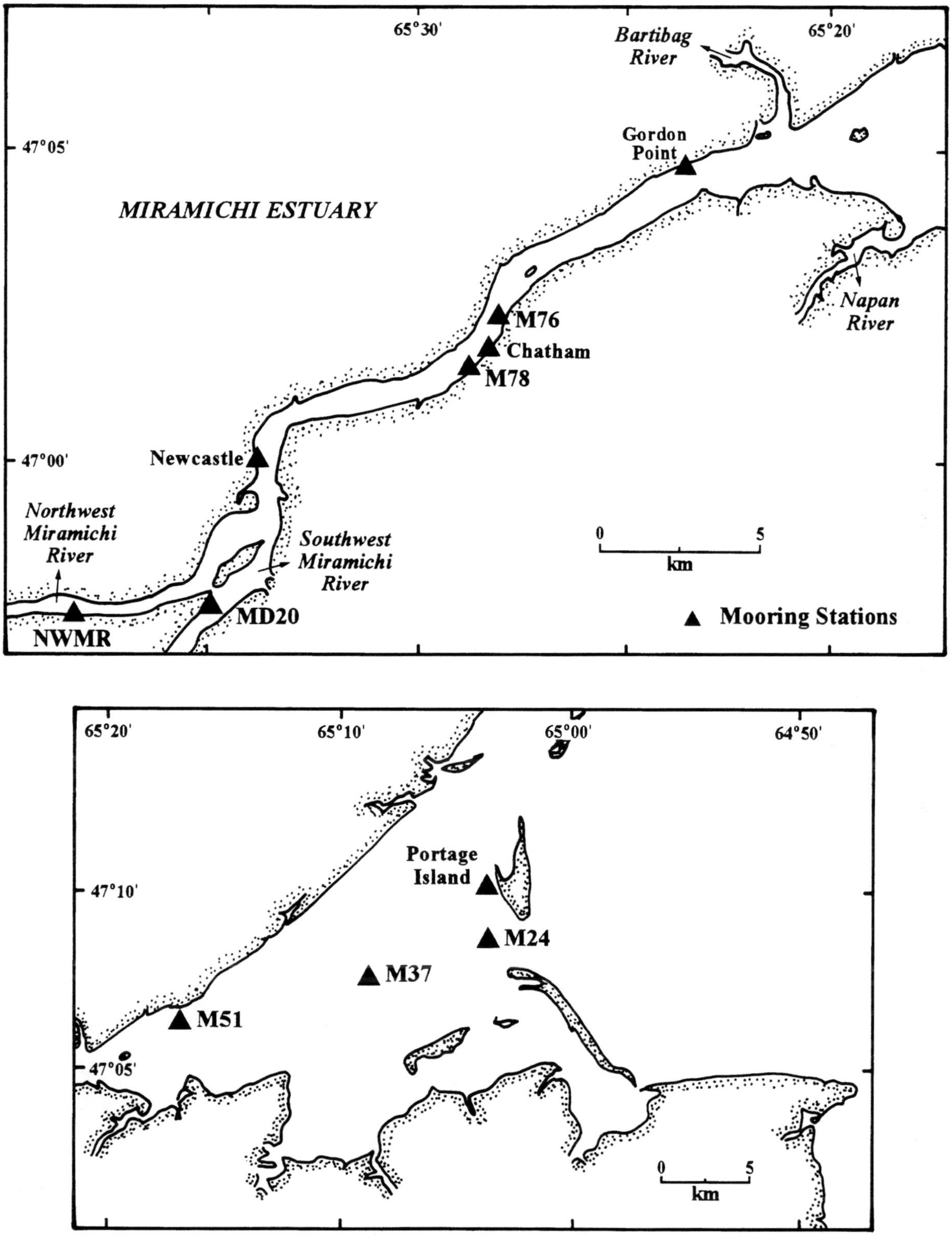

Fig. 1. Mooring positions in the Miramichi Estuary.

Table 1. Mooring description. All data are supplied by the Institut Maurice-Lamontagne except the measurements taken at stations marked by a star which are from Université du Québec à Rimouski. (s=summer; w=winter; a=autumn; S=salinity; T=temperature; C=current)

Mooring	Position	Time	Duration (month)	Instruments	Measurements S	T	C	Instrument depth (m)	Water depth (m)
Portage Island*	47° 10.3' N 65° 3.8' W	a92	1	Aanderaa RCM7	X	X	X	4.2 m above bottom	9.4
M24*	47° 8.6' N 65° 3.9' W	a92	1	IntreOcean S4	X	X	X	3.9 m above bottom	9.7
M37*	47° 7.6' N 65° 8.9' W	a92	1	InterOcean S4	X	X	X	2.1 m above bottom	5.9
M51	47° 6.40' N	s91	5	Aanderaa RCM7 #9101	X	X	X	1.0 m above bottom	6.5
	65° 16.96' W			Thermograph Hugrun #4341		X		0.4 to 0.5 m below surface	
M51	47° 6.47' N	s92	4	Aanderaa RCM7 #9023	X	X	X	1.0 m above bottom	6.5
	65° 16.63' W			Thermograph Hugrun #4338		X		0.4 to 0.5 m below surface	
M51	47° 6.26' N	s93	4	Aanderaa RCM7 #9023	X	X	X	1.0 m above bottom	6.0
	65° 15.57' W			Thermograph Hugrun #4338		X		0.4 to 0.5 m below surface	
Gordon Pt	47° 4.76' N	w91	6	Aanderaa RCM4 #9101	X	X		On the bottom	1.5
	65° 23.62' W	s92	6	Aanderaa RCM4 #5807	X	X		On the bottom	
		w92	7	Thermograph Vemco #8615		X		On the bottom	
Chatham	47° 1.79' N	w91	6	Aanderaa RCM4 #6013	X	X		On the bottom	2.5
	65° 28.38' W	s92	6	Aanderaa RCM4S #7703	X	X		On the bottom	
		w92	7	Tidal gauge WLR7 #1114	X	X		On the bottom	
M76*	47° 2.26' N 65° 28.08' W	a92	1	InterOcean S4	X	X	X	3.8 m above bottom	9.9
M78	47° 1.52' N	s93	1	Aanderaa RCM7 #5807	X	X	X	1.0 m above bottom	5.5
	65° 28.25' W			InterOcean S4 #647	X	X	X	1.0 m below surface	
Newcastle	46° 59.89' N	w91	6	Aanderaa RCM4 #6011	X	X		1.0 m above bottom	3.0
	65° 33.75' W	s92	6	Aanderaa RCM4S #7704	X	X		On the bottom	
		w92	4	Thermograph Vemco #8613		X		1.0 m above bottom	
MD20	46° 57.73' N	s91	3	Aanderaa RCM7 #9023	X	X	X	1.0 m above bottom	6.5
	65° 34.95' W			Thermograph Hugrun #4402		X		0.4 to 0.5 m below surface	
MD20	46° 57.45' N	s93	4	Aanderaa RCM4S #7703	X	X	X	2.5 m above bottom	7.0
	65° 34.55' W			Thermograph Hugrun #4401		X		0.4 to 0.5 m below surface	
NWMR	46° 57.42' N	s93	4	Aanderaa RCM7 #7704	X	X	X	1.0 m above bottom	7.5
	65° 38.11' W			InterOcean S4 #646	X	X	X	1.0 m below surface	

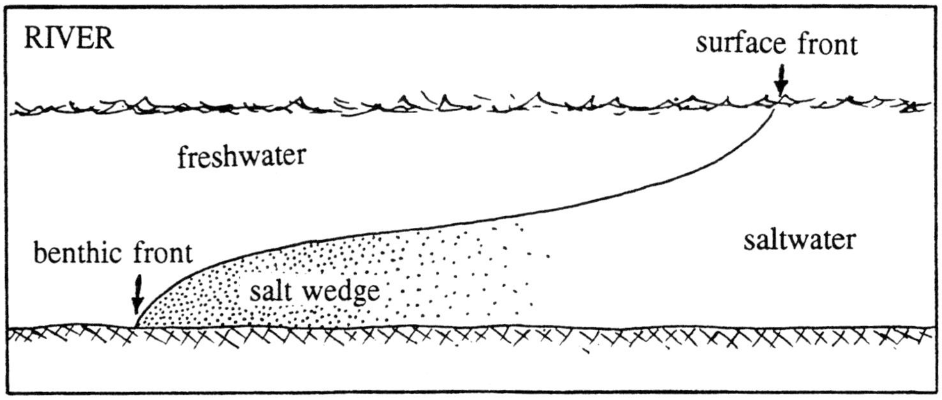

Fig. 2. Distribution of water masses in a highly stratified estuary.

Table 2. Details of STD profiles in the Miramichi Estuary.

Day	Station	Position	Time (h:min) (UTC)
497	M51	47° 6.5'N; 65° 16.0'W	18:18
(13 May 1992)	M56	47° 4.7'N; 65° 18.7'W	18:49
	M61	47° 5.0'N; 65° 21.8'W	19:09
	M65	47° 4.7'N; 65° 23.4'W	19:19
	M68	47° 3.9'N; 65° 25.1'W	19:32
	M71	47° 3.4'N; 65° 27.0'W	19:45
	M75	47° 3.0'N; 65° 27.8'W	19:55
	Chatham	47° 1.7'N; 65° 28.8'W	20:11
	M88	47° 1.0'N; 65° 31.0'W	20:26
	M95	47° 0.5'N; 65° 33.5'W	20:40
	Newcastle	47° 59.9'N; 65° 33.8'W	21:00
672	M51	47° 6.5'N; 65° 16.0'W	15:13
(4 Nov 1992)	M56	47° 4.7'N; 65° 18.7'W	15:38
	M61	47° 5.0'N; 65° 21.8'W	15:50
	M65	47° 4.7'N; 65° 23.4'W	16:00
	M68	47° 3.9'N; 65° 25.1'W	17:48
	M71	47° 3.5'N; 65° 27.1'W	18:00
	M75	47° 3.0'N; 65° 27.8'W	18:07
	Chatham	47° 1.7'N; 65° 28.8'W	18:21
	M88	47° 1.0'N; 65° 31.0'W	18:31
	Newcastle	47° 59.9'N; 65° 33.7'W	18:49
884	M51	47° 6.3'N; 65° 16.1'W	12:26
(3 June 1993)	M54	47° 5.1'N; 65° 17.5'W	12:46
	M56	47° 4.4'N; 65° 18.5'W	12:55
	M61	47° 4.6'N; 65° 21.5'W	13:08
	M65	47° 4.4'N; 65° 23.3'W	13:17
	M68	47° 3.6'N; 65° 25.0'W	13:28
	M71	47° 3.3'N; 65° 27.0'W	13:39
	M75	47° 3.0'N; 65° 27.5'W	13:47
	Chatham	47° 1.5'N; 65° 28.3'W	13:56

(Vilks and Krauel 1982) except during spring runoff. Hence, as the freshwater discharge increases and as the salt wedge is pushed downstream, the longitudinal salinity gradient should increase as a result of the continuous supply of salt from the bay (Bowman 1988), thus shortening the zone of fresh and salt water mixing. As the longitudinal gradient of salinity increases, the vertical salinity gradient also increases because isohalines are squashed together.

Tides and winds change the salinity distribution because of vertical mixing, which has the effect of bending isohalines towards the vertical. The distribution of water masses in the mixing zone is thus altered. The length of the mixing zone should, to a first approximation however, stay unaffected, at least until the shear readjusts to the new density distribution. The horizontal gradient of the

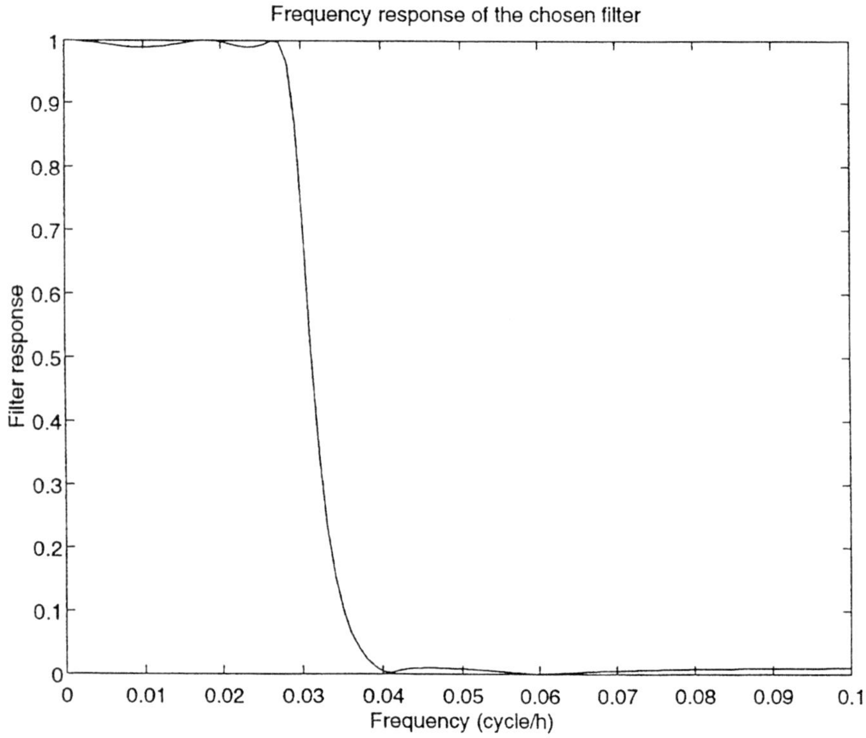

Fig. 3. Frequency response of the chosen low pass filter. The cutoff frequency is 0.04 cycle·h^{-1} equivalent to a period of 25 h.

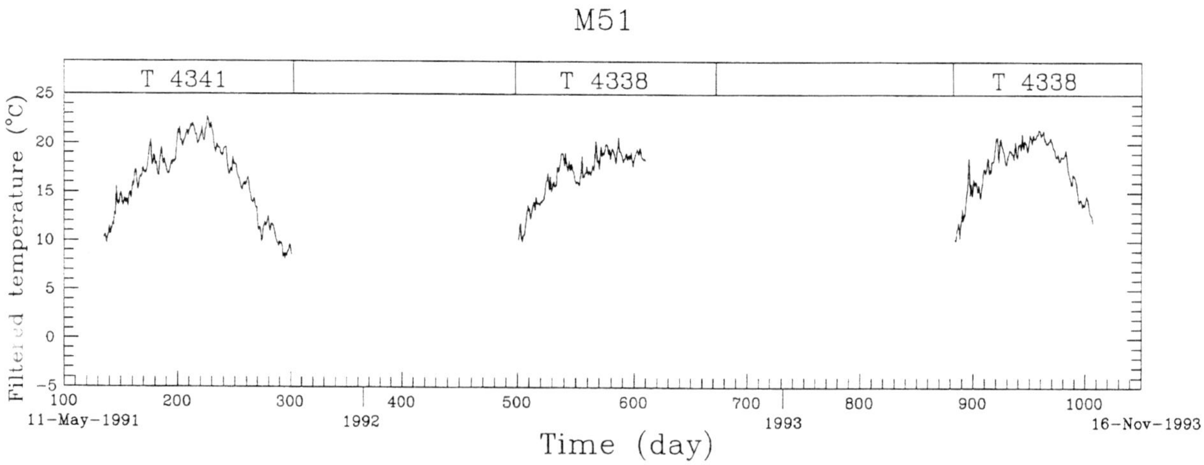

Fig. 4a. Filtered temperature series at 0.4 – 0.5 m below the water surface at buoy M51.

depth mean salinity is initially little affected by the vertical mixing.

This paper examines the processes that change the stratification and the form of the salt wedge in the Miramichi. We start by presenting temperature and salinity data collected in the Miramichi between 1991 and 1993. We then examine in particular the physical processes that control the distribution of water masses in the estuary.

Materials

Oceanographic data

Long-term series of oceanographic data were obtained during summer and winter between 1991 and 1993 using a variety of instruments: current-meters, thermographs, and tide gauges. These instruments were deployed at various places along the main channel of the estuary (Fig. 1). Details of moorings and sources of data are contained in Table 1. All summer instruments were moored on U-shaped moorings and were cleaned periodically. Winter instruments were directly attached to quays and were not visited for cleaning. Typical deployment durations lie between 1 and 6 months with sampling intervals of 15, 30, or 60 minutes. Apart from the meters deployed by Université du Québec à Rimouski (UQAR), all instruments were calibrated before deployment and after recovering. Mean calibration coefficients were used to convert original data to real values. In the case of the three rented *InterOcean S4*s deployed by UQAR, the factory calibration was used. The *Aanderaa* meter placed for 1 month near Portage Island was calibrated only before deployment.

Three grids of STD profiles were also completed at the time of mooring installation or recovery. The STD was an Applied Microsystems STD 12 equipped with a 70 dbar pressure sensor. The details of profiles, locations, and times of acquisitions are given in Table 2.

Supplementary data

Meteorological data for Point Escuminac were supplied by Environment Canada. They were composed of hourly records of wind speed and direction. Unfortunately, wind series were segmented by missing data.

River flow data from Environment Canada were obtained for four sites: station 01BO001 on the Southwest Miramichi river at Blackville, draining 5050 km^2; station 01BO004 on the Bartibog River 10 km upstream from its mouth, draining 316 km^2; station 01BP001 on the Little Southwest Miramichi river at Lyttleton draining 1340 km^2; and station 01BQ001 on the Northwest Miramichi river at Trout Brook, draining 948 km^2. The drainage area gauged by these stations constitutes 55% of the total drainage area of the Miramichi system. The stations operate only during the ice-free season. Major gaps in the wind and discharge data are indicated in the figures.

Finally, for tidal data, we calculated elevations for Chatham with tidal constituent amplitudes and phases supplied by the Canadian Hydrographic Service, Department of Fisheries and Oceans. These were obtained from a tidal height analysis of a 88 days tidal height record.

Methods

Filtering of oceanographic data

To highlight nontidal events, the time series were low-pass filtered to remove semidiurnal and diurnal frequencies. The filter used was an infinite-impulse response (IIR) filter (Parks and Burrus 1987) of the MATLAB library with a cutoff of 0.04 cycle·h^{-1} (Fig. 3). The filter was applied twice, once forwards and once backwards to avoid phase shifts.

Each salinity and temperature was filtered in turn. The sample frequency was then reduced to twice a day with samples at 12 and 24 h UTC irrespective of the original sampling interval.

Estimation of the wind mixing power

In shallow coastal waters, strong winds generate vertical mixing that reduces stratification and causes a variation on a short-time scale (i.e., a few days) in the salinity and temperature distributions. To quantify the effects of wind on the oceanographic conditions, we estimated the energy per unit of time used to mix the water column (P_w) at Chatham. To estimate P_w, we used the wind

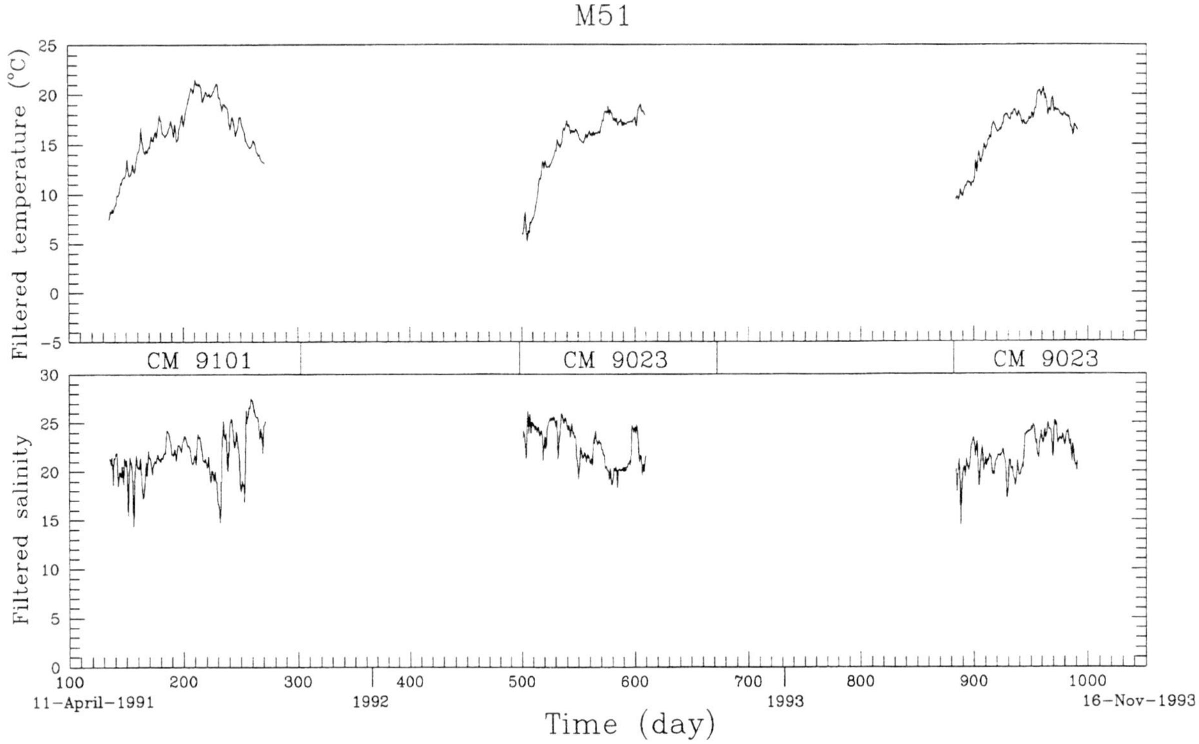

Fig. 4b. Filtered temperature and salinity series at 1 m above the bottom at buoy M51.

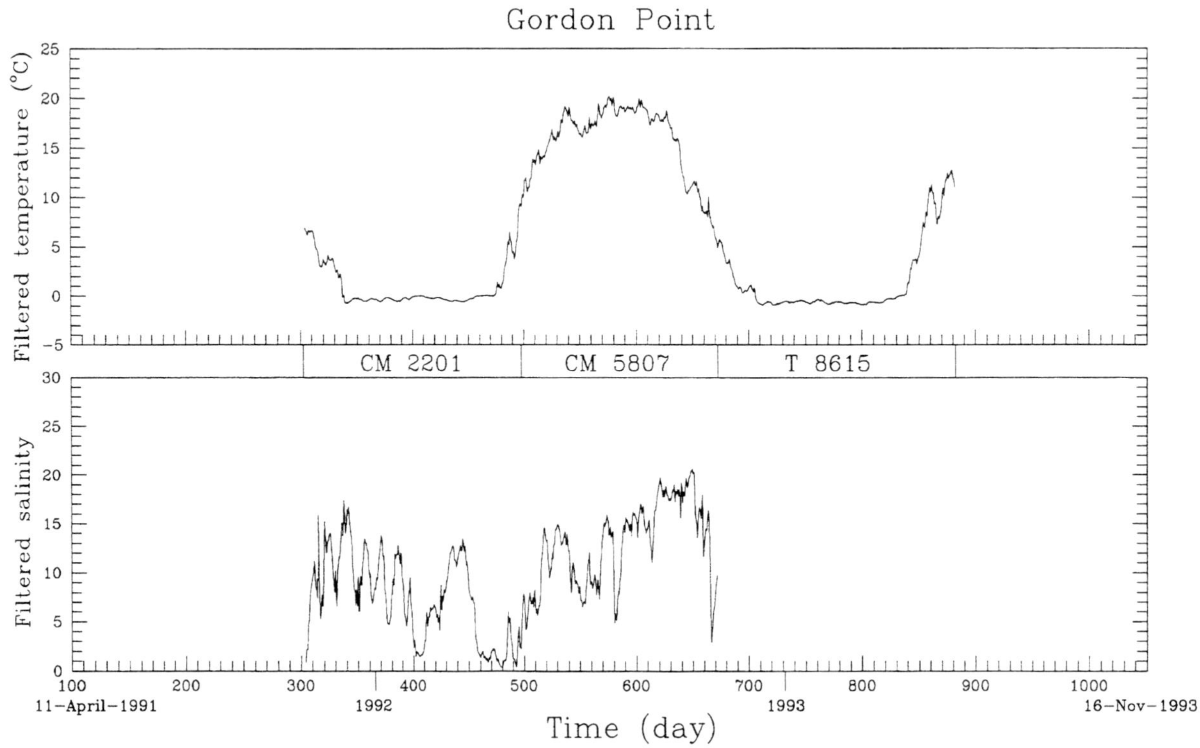

Fig. 4c. Filtered temperature and salinity series at about 1.5 m from the surface at Gordon Point.

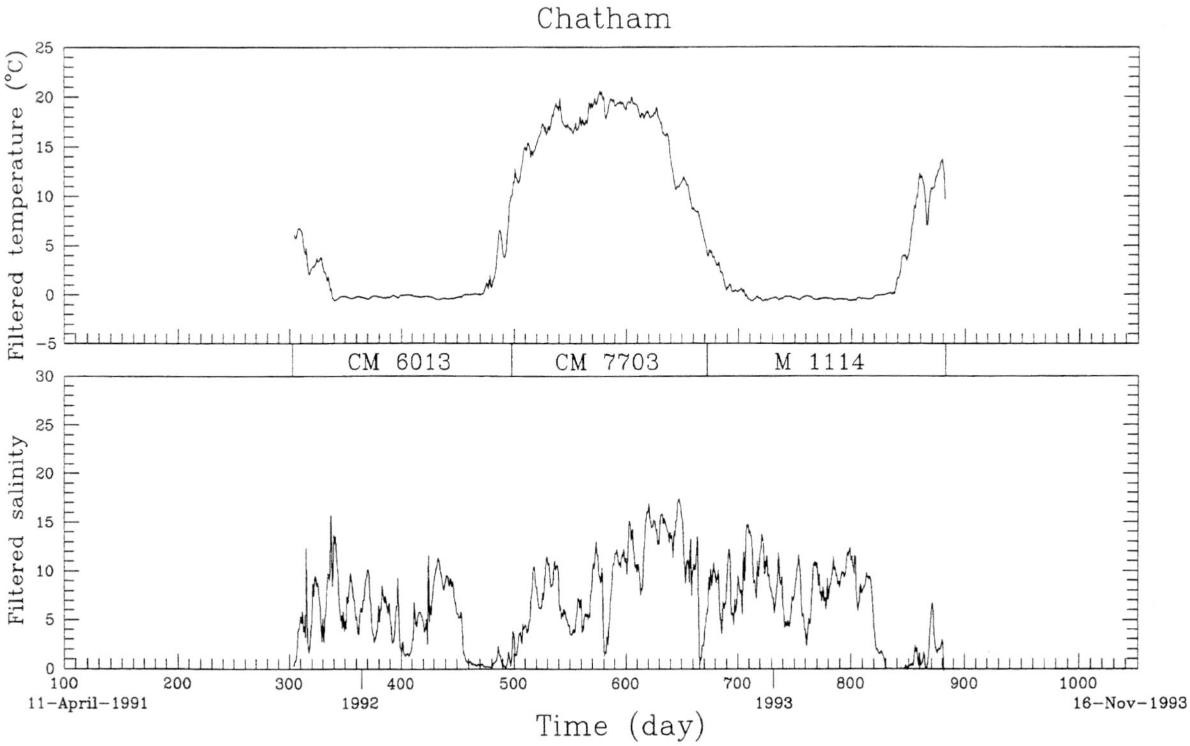

Fig. 4d. Filtered temperature and salinity series at about 2.5 m from the surface at Chatham.

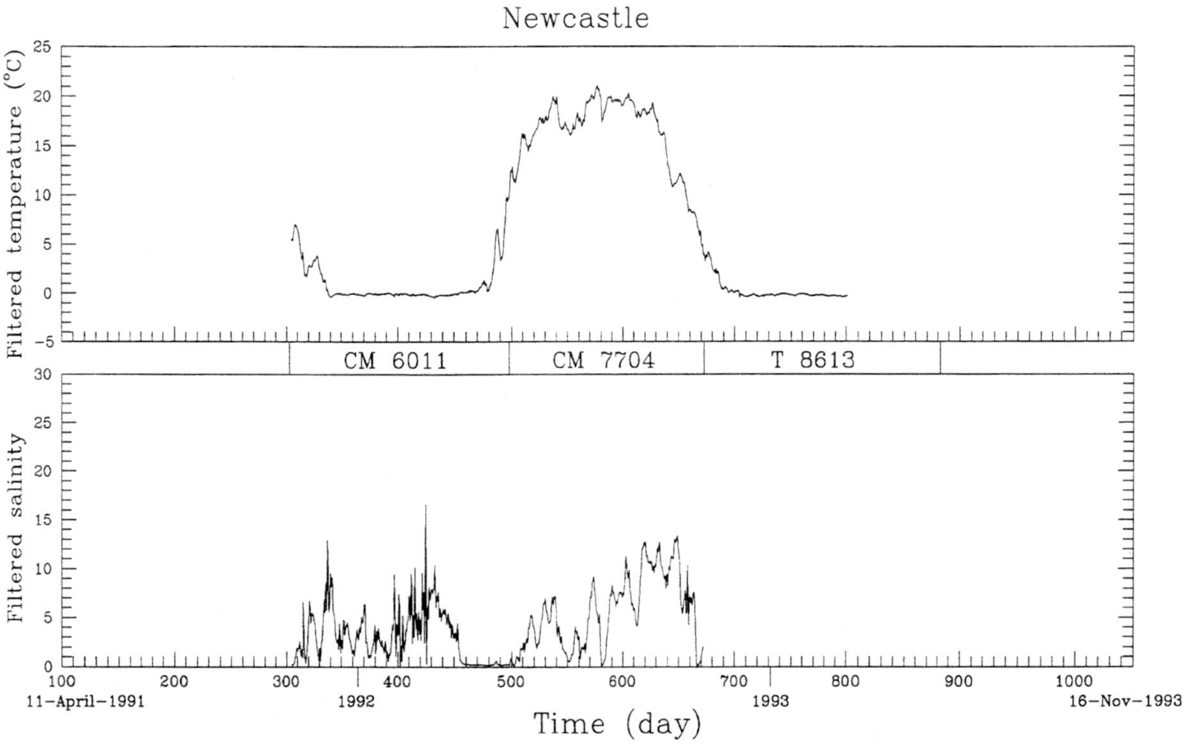

Fig. 4e. Filtered temperature and salinity series at about 3 m from the surface at Newcastle.

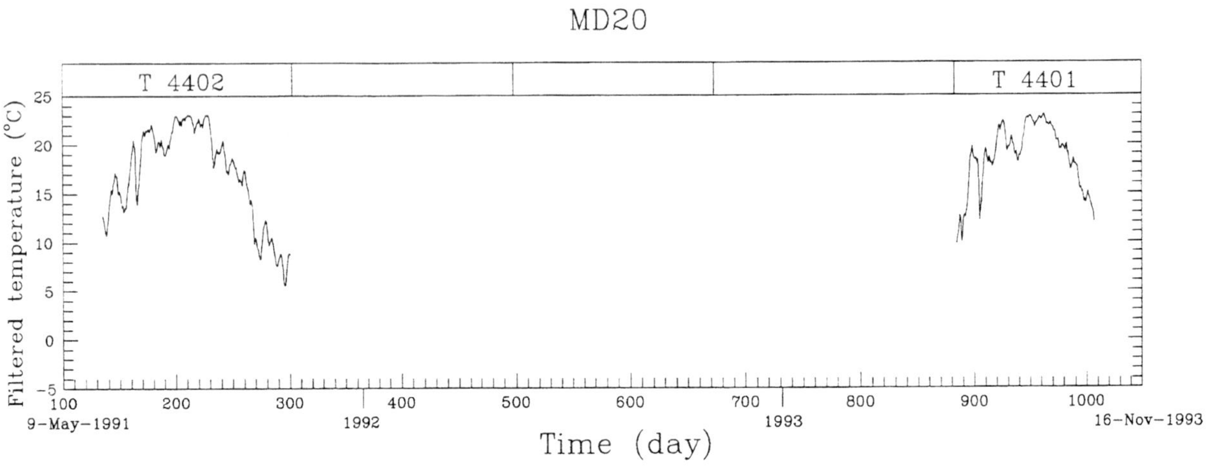

Fig. 4f. Filtered temperature series at 0.4 – 0.5 m from the water surface at buoy MD20.

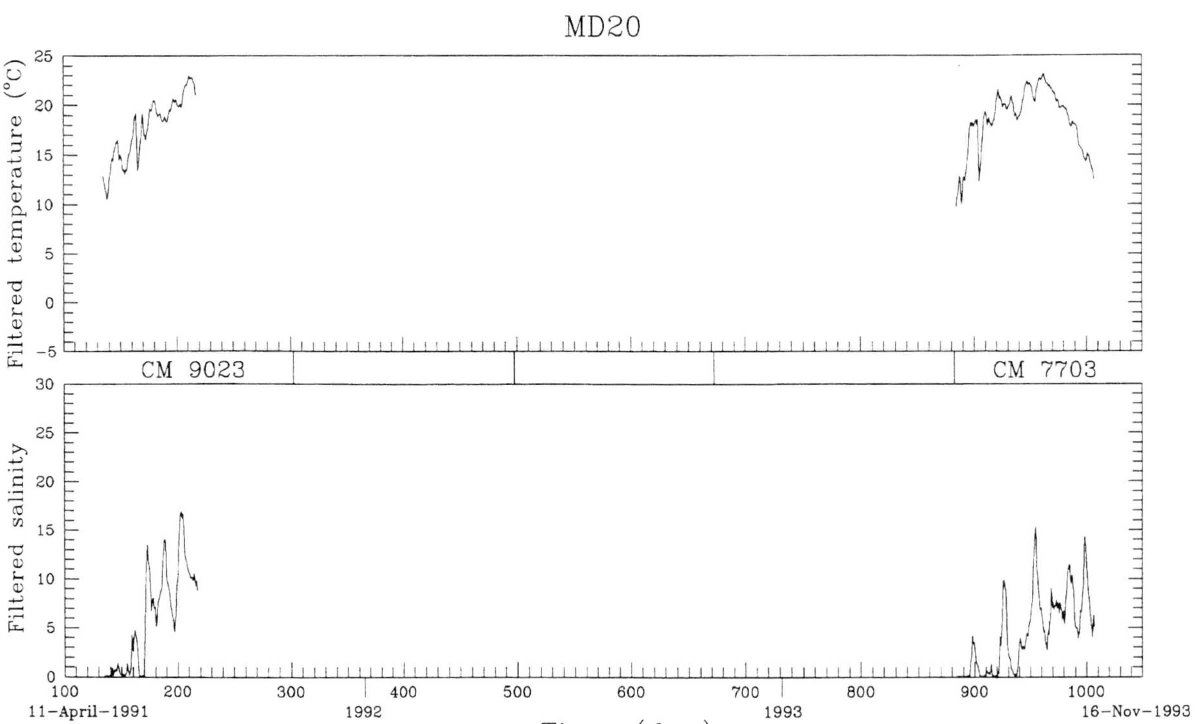

Fig. 4g. Filtered temperature and salinity series at 1 m in 1991 and 2.5 m in 1993 from the bottom at buoy MD20.

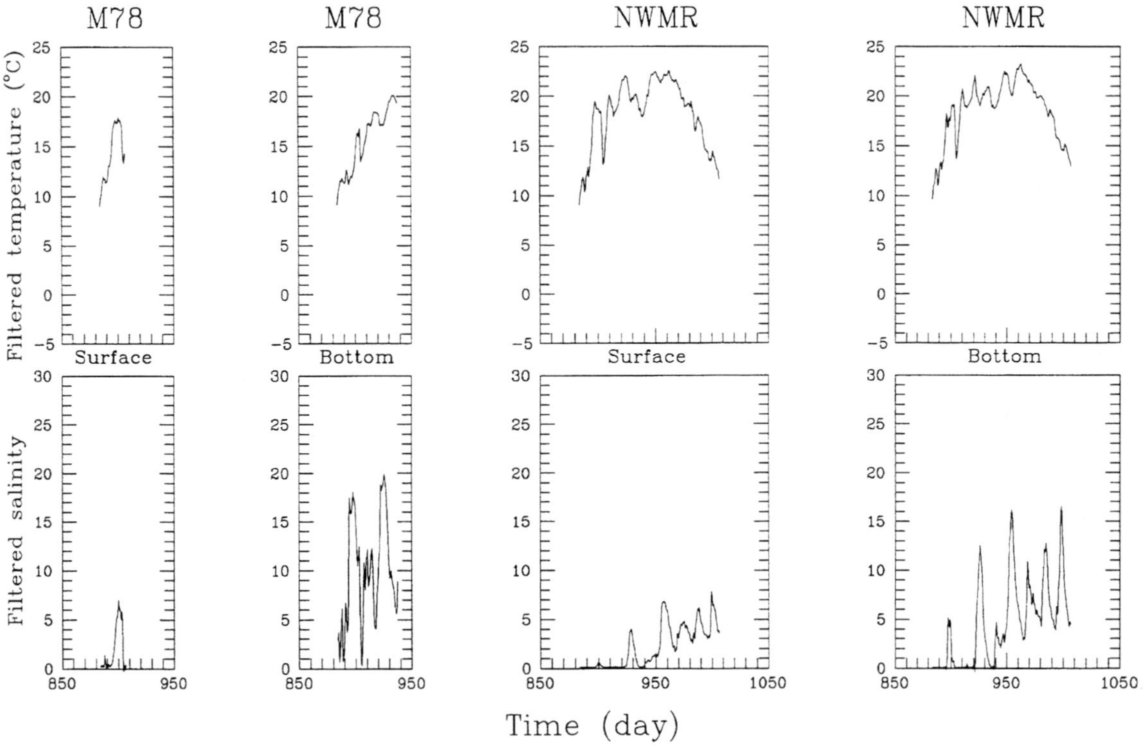

Fig. 4h. Filtered temperature and salinity series at 1 m below the surface and above the bottom at station M78 and NWMR.

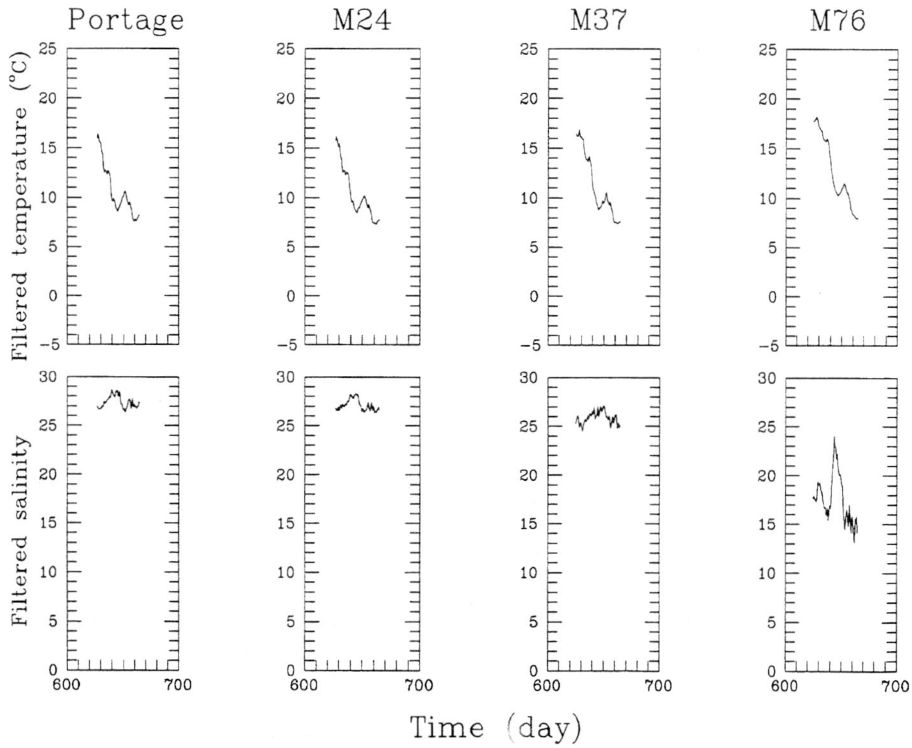

Fig. 4i. Filtered temperature and salinity series at 4.2 m from the bottom at Portage Island, at 3.9 m from the bottom at buoy M24, at 2.1 m from the bottom at buoy M37, and at 3.8 m from the bottom at buoy M76.

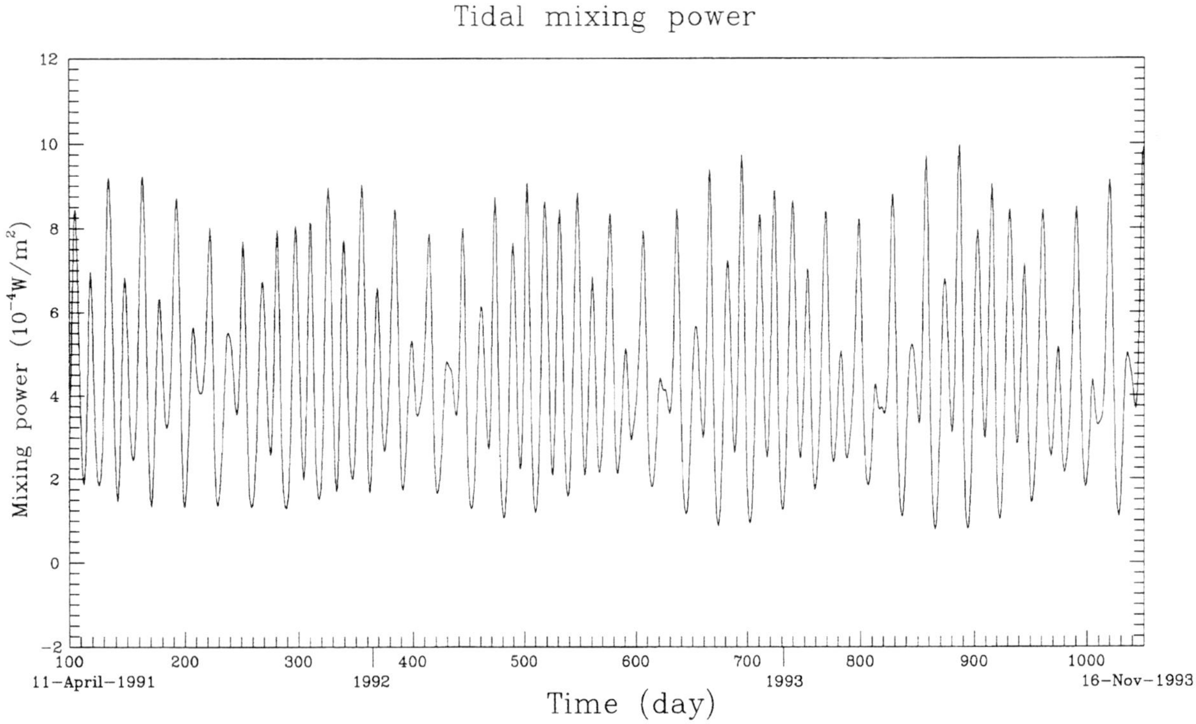

Fig. 5. Filtered tidal mixing power calculated using predicted currents for Chatham in equation (2).

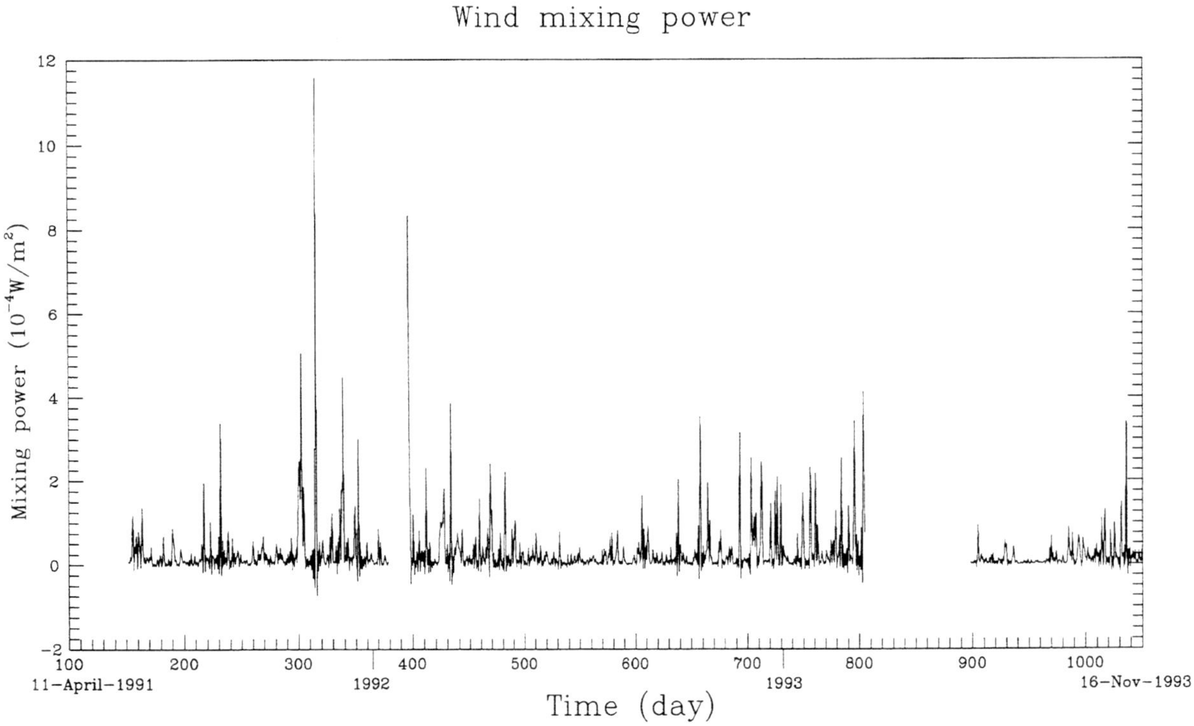

Fig. 6. Filtered wind mixing power calculated using the wind component parallel to the channel at Chatham in equation (1).

velocity parallel to the main channel (u_a) in the following expression (modified from Atkinson and Blanton 1986):

(1) $P_w = R_f \rho_w [(\rho_a/\rho_w) C_{ds}]^{3/2} u_a^3$

where R_f = Flux Richardson number, ρ_a = air density, ρ_w = water density and C_{ds} = water surface drag coefficient. The Flux Richardson number, which is equal to the fraction of turbulent energy used in vertical mixing (Dyer 1988), was given a standard value of 0.1 (Linden 1979). The water density was taken as the mean density at Chatham, 1008 kg·m^{-3}. The air density was put at 1.2 kg·m^{-3} and C_{ds} was put equal to 0.0013 (Schroeder et al. 1990). Series of hourly values of P_w were then filtered using the method described in "Filtering of Oceanographic Data" to reduce the sample frequency to twice a day.

Estimation of the tidal mixing power

Tidal currents also cause vertical mixing. The intensity of this mixing is a function of current speed and bottom roughness. The tidal currents cause cyclic variations in temperature and salinity. To obtain the mixing power due to tidal currents for 3 years, we used Foreman's (1977) program to predict tidal elevations. From these predicted tidal heights, mean horizontal currents (u_c) were obtained using the following:

(2) $u_c(\varsigma) = (S/A)\, \partial\varsigma/\partial t$

where u_c is the mean horizontal current over the cross section, S = the area of the tidal basin upstream from Chatham, A = the vertical area of the cross section at Chatham, and ς = the tidal elevation with respect to the lowest low water level. In equation (2), S was kept constant at a value of 31.58 km^2 (neglecting tidal flats since the sides of the estuary are quite steep) and A varied with ς according to following equation:

(3) $A(\varsigma) = A_o + \alpha \varsigma$

in which A_o = the transverse area below the lowest low-water level (3630 m^2) and α = the width of the estuary at Chatham (580 m).

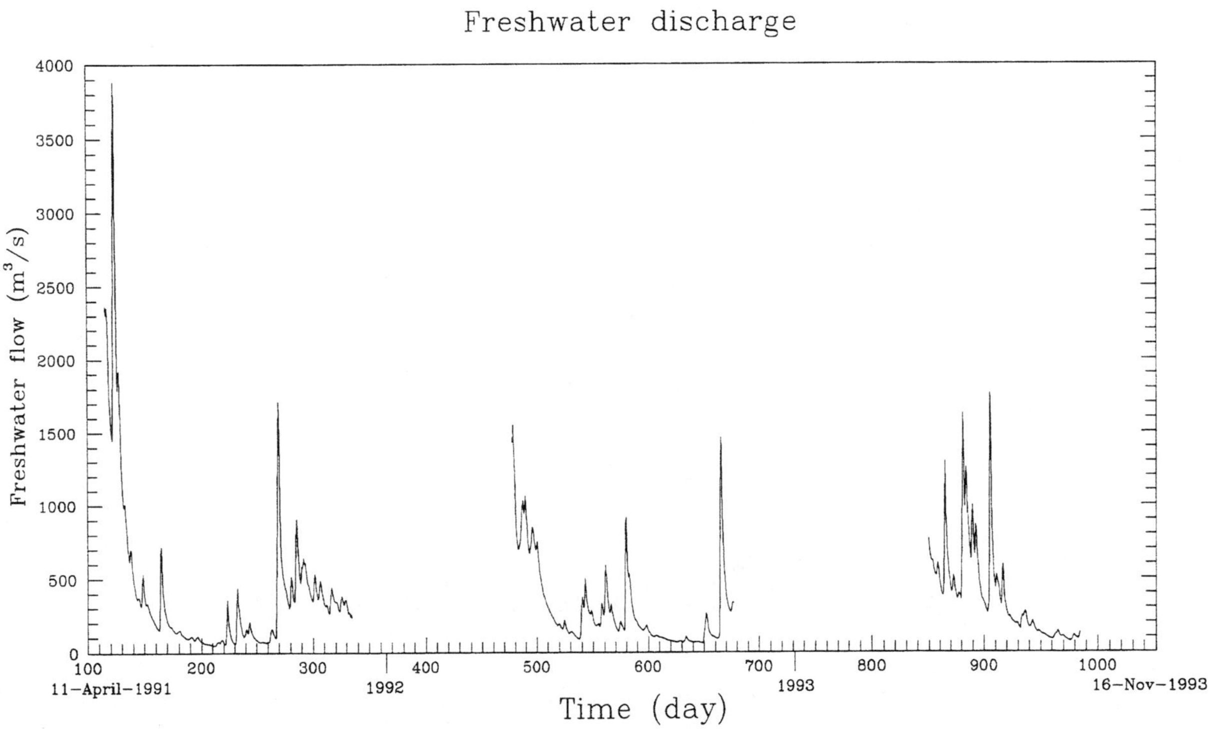

Fig. 7. Freshwater discharge of the Miramichi system.

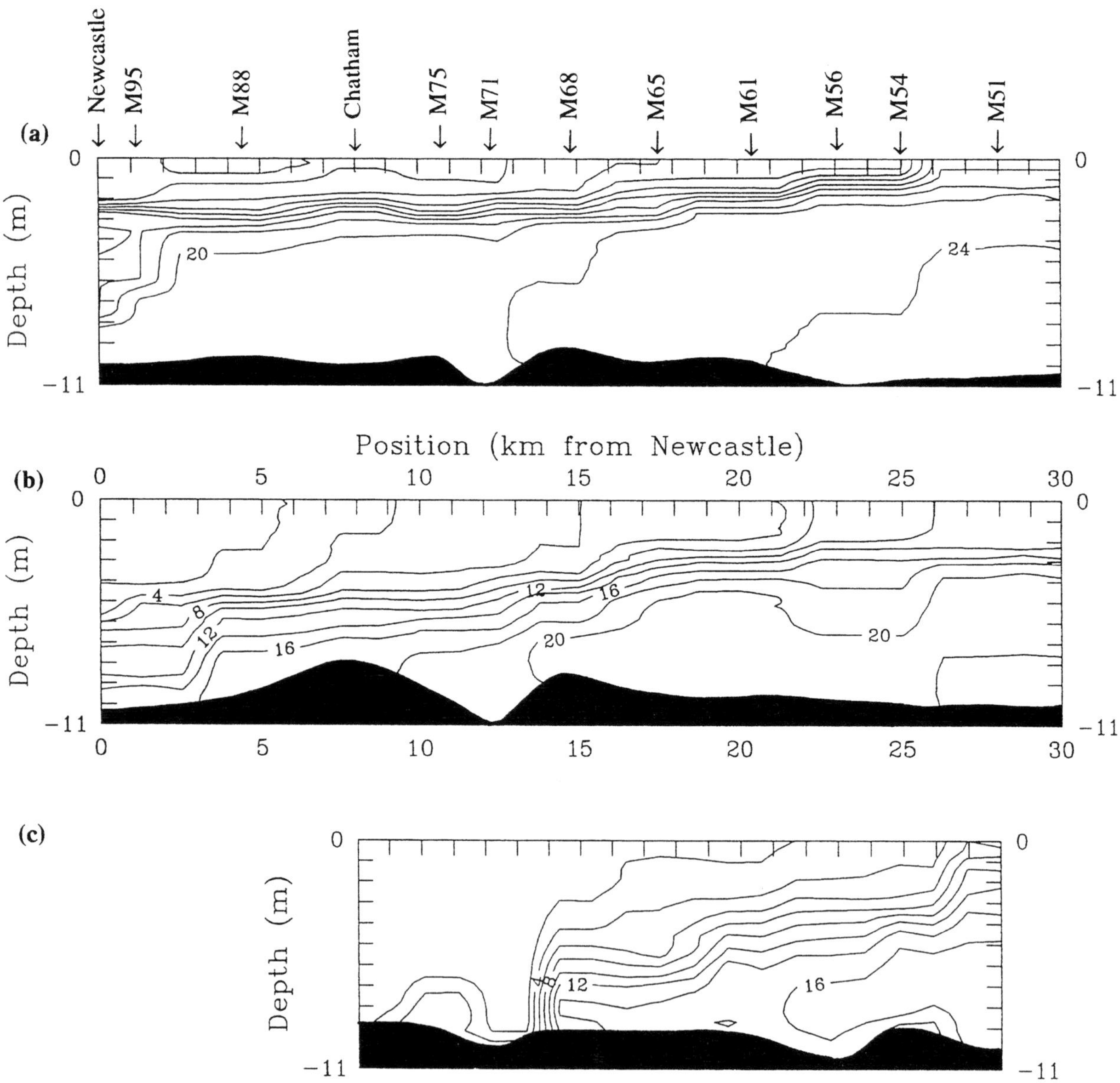

Fig. 8. Longitudinal salinity distribution from Newcastle to buoy M51: (a) around low water on day 672 (neap tide) with a freshwater discharge of 310 $m^3 \cdot s^{-1}$, (b) at the beginning of the flood on day 497 (neap tide) with a freshwater discharge of 740 $m^3 \cdot s^{-1}$, (c) during the flood on day 884 (spring tide) with a freshwater discharge of 1190 $m^3 \cdot s^{-1}$.
(Note that the distribution (c) is for a section downstream of Chatham.)

Table 3. Salinity and temperature semidiurnal variation range as function of the time of the year and the approximate position in the longitudinal salinity distribution. This table is based on the unfiltered series.

Season	Location	Salinity variations	Temperature variations (°C)
Summer	1. Near salt wedge limit	–	–
	2. In max. stratification	7 to 13	0 to 12
	3. Downstream from 2.	1 to 5	0 to 4
Autumn	1. Near salt wedge limit	–	–
	2. In max. stratification	7 to 17	<1
	3. Downstream from 2.	around 4	<1
Winter	1. Near salt wedge limit	–	–
	2. In max. stratification	0 to 23	~0
	3. Downstream from 2.	1 to 5	~0
Spring	1. Near salt wedge limit	2 to 7	1 to 5
	2. In max. stratification	1 to 11	1 to 3
	3. Downstream from 2.	1 to 2	<1

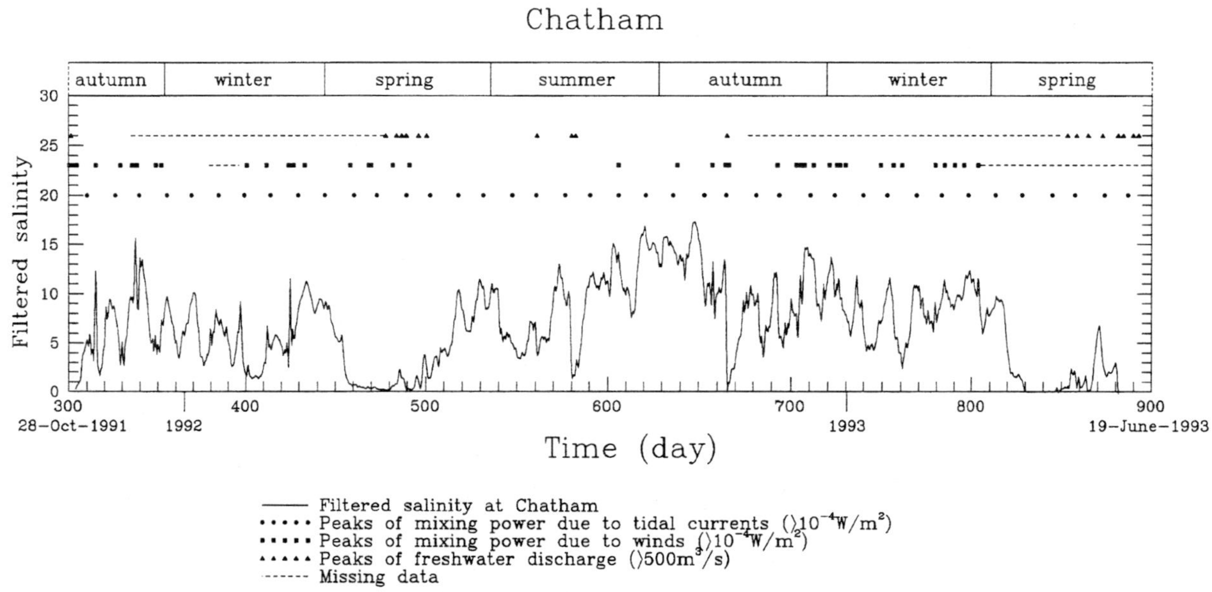

Fig. 9. Low-pass filtered salinity for water at 2 – 3 m below the surface at Chatham as a function of time. Events of mixing power and freshwater discharge are indicated by symbols.

The hourly predicted tidal currents (u_c) were then used to calculate the mixing power due to currents. The mixing power (P_c) was evaluated as for the wind mixing power with an equation modified from Atkinson and Blanton (1986):

$$(4) \quad P_c = R_f \, \rho_w \, C_{db}^{3/2} \, u_c^3$$

where C_{ds} = the bottom drag coefficient. As in equation (1), R_f was taken as 0.1 and ρ_w as 1008 kg.m^{-3}. The bottom drag coefficient was given a standard value of 0.002 (Schroeder et al. 1990). Hourly P_c series were filtered as above. Hence, only periods greater than 25 h are perceptible.

Calculation of freshwater discharge

The total freshwater discharge of the Miramichi was obtained by multiplying the total of four gauged-station discharges by a factor of 1.82 to take account of ungauged catchment areas. The sample frequency was then reduced to twice a day by taking mean values for periods of 13 h centred on noon and midnight UTC. Simple means were used instead of usual low-pass filter because of numerous gaps in the data.

A problem occurred with the gauge placed at station 01BP001 leaving gaps in the 1992 summer series. To fill these gaps, three stations were used to calculate the freshwater discharge for days 525 – 576 and for days 600 – 655 (days are given as number of days since January first, 1991). For these periods, a correcting factor of 2.21 was used to obtain the total freshwater discharge.

Results

Filtered temperature and salinity series

We consider first the filtered time series of temperature and salinity. Data for all positions between the rivers and Portage Channel are presented in Figs. 4a – 4i. Units of time used in the time-series plots are in days after 1 January 1991 (day 0). Hence, 1 January 1992 is day 365 and 1 January 1993 is day 731.

The variability removed by the filter is presented in Table 3. Short-time-scale variations in salinity depended on the season and on the location with respect to the salinity distribution. In general, temperature variations at the same time scale were small.

The most obvious variation in the filtered series, other than the seasonal one, was the 15 day cycle, a result of the neap and spring tidal cycle. Other variation periods were also present. We try to

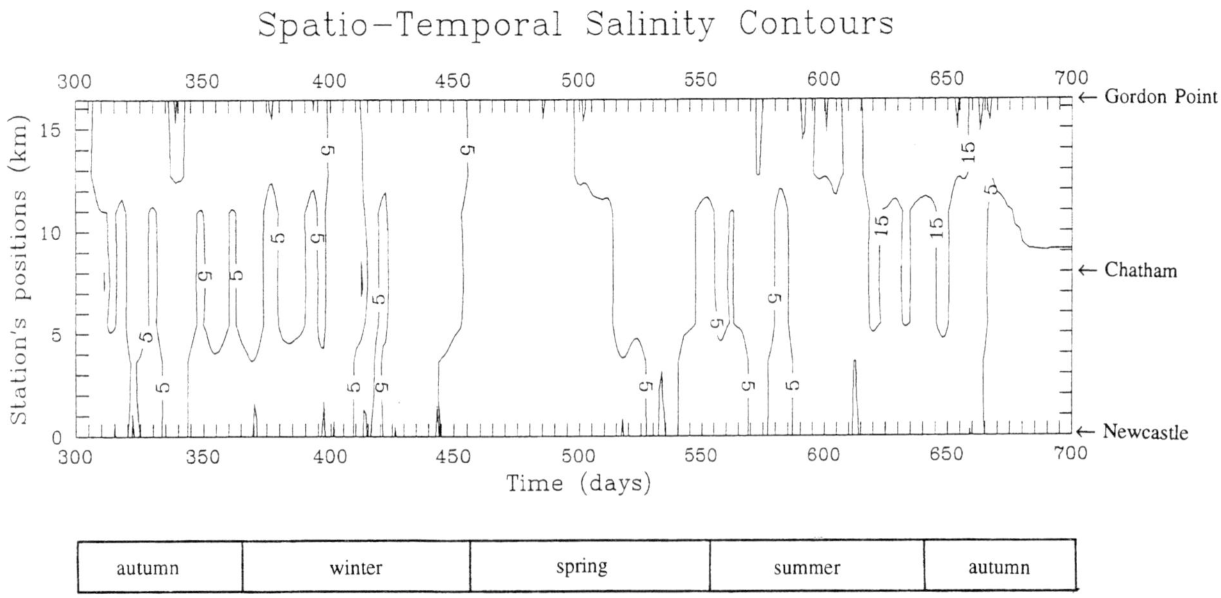

Fig. 10. Spatio-temporal salinity contours for the region between Newcastle to Gordon Point at 2 – 3 m from the water surface.

identify the origin of these variations by considering the mixing power and the freshwater discharge.

Mixing powers and freshwater discharge

Mixing power due to predicted tidal currents and to winds are presented in Figs. 5 and 6 with the same time axis as for the salinity and temperature series. Freshwater discharge is shown in Fig. 7. The mixing power of wind showed a succession of short-duration peaks of different intensities. On the other hand, the mixing power due to tidal currents had a more regular pattern with maxima at spring tides and minima at neap tides. Mixing power of the tides was in general greater than that of the winds. The freshwater-discharge series showed elevated runoff in spring, primarily due to snow melting, followed by a decrease to low summer runoff. Superimposed on this general pattern were short-duration peaks, probably associated with precipitation.

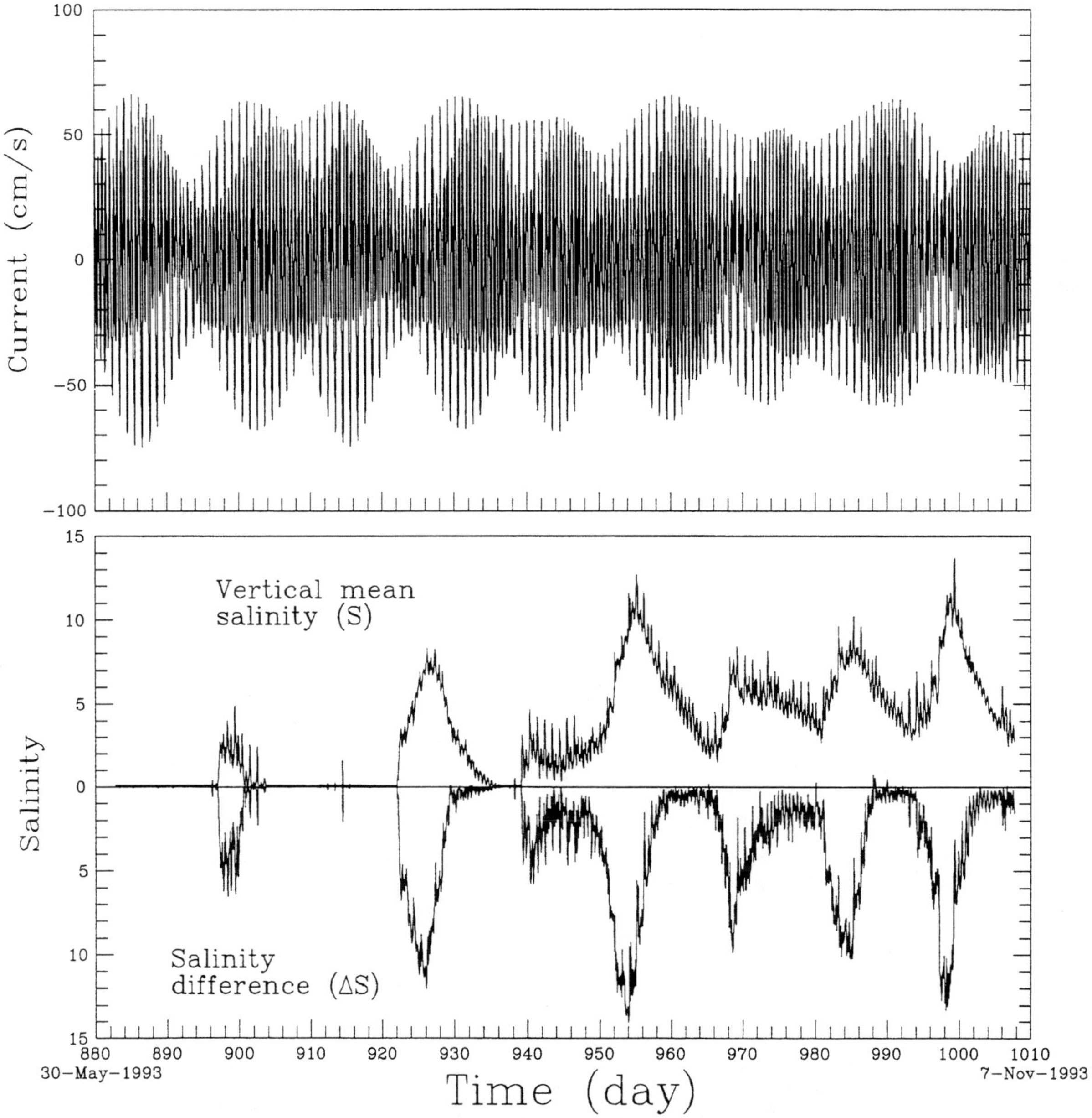

Fig. 11. Predicted tidal currents at Chatham, vertical mean salinity ($\bar{S}$) and salinity difference (ΔS) at station NWMR from June to November. (Note the inverted axis for ΔS).

STD profiles

The STD data give the longitudinal distribution of stratification in the estuary for a given time interval within the tidal cycle. Each series of STD profiles, spread along a 20 or 30 km stretch of the estuary, were made within a period of 1.5 – 3.5 h. Figure 8 illustrates the estuarine structure for three different freshwater conditions: 310, 740, and 1190 $m^3 \cdot s^{-1}$.

The freshwater discharge affected both longitudinal and vertical salinity distributions. With increased freshwater runoff, the halocline at a given location became thicker and the longitudinal salinity gradients between Chatham and buoy M51 strengthened at the bottom but weakened at the surface. The estuary between Newcastle and buoy M51 changed from a stratified one at low

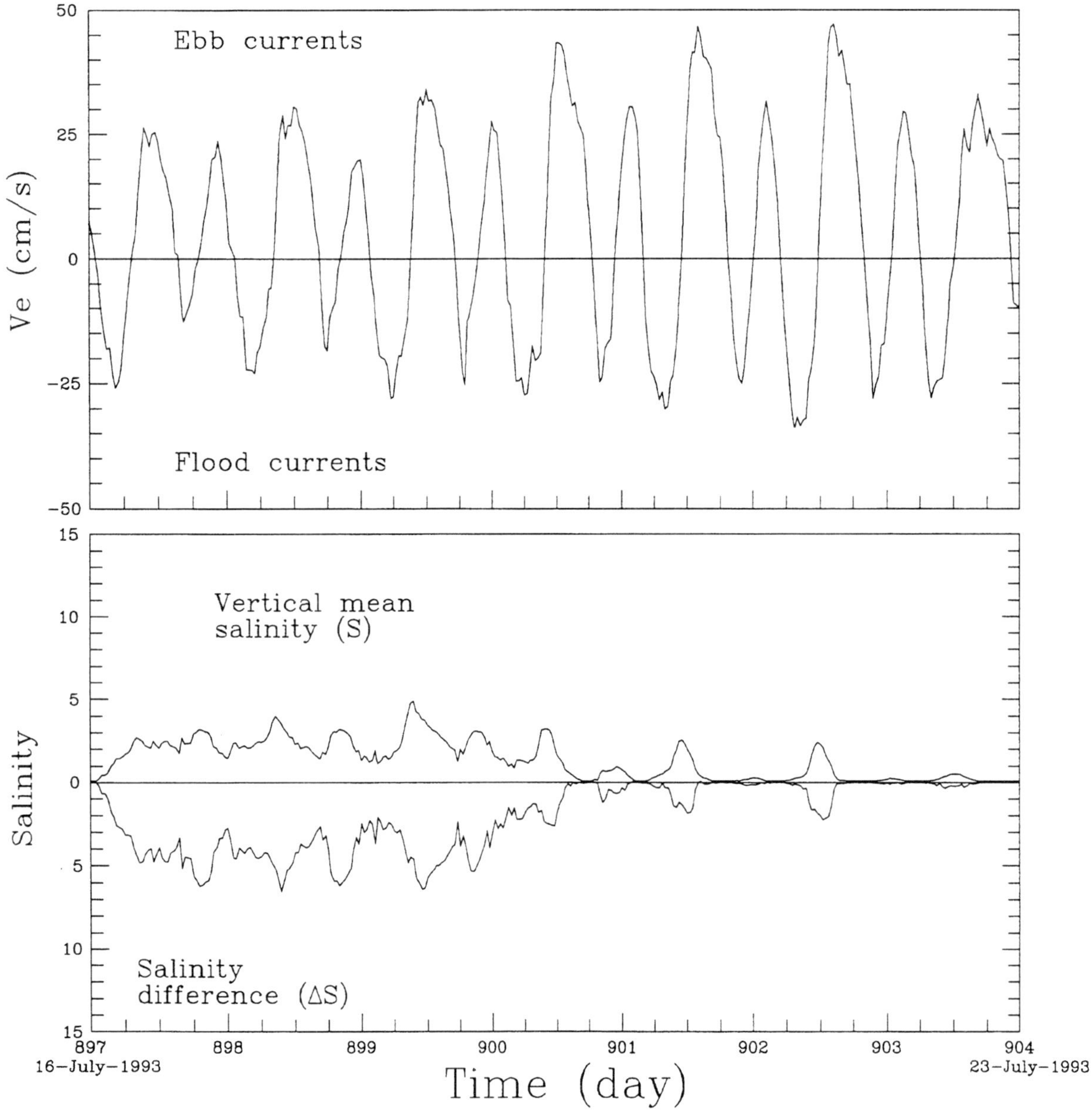

Fig. 12. Current velocity and salinity near the benthic front represented by the vertical mean velocity parallel to the channel ($Ve=[Ve_b+Ve_s]/2$ where Ve_b and Ve_s are the observed longitudinal currents near the bottom and the surface) and the semidiurnal variations of the vertical mean salinity ($\bar{S}$) and salinity difference (ΔS) for 7 days in June at station NWMR. (Note the inverted axis for ΔS.)

runoff to a partially mixed one at moderate runoff. This, however, cannot be generalized. Stratification depends on other factors other than just the discharge. Data for large discharge (Fig. 8c) were taken during spring tides in conditions of more intense vertical mixing.

These three contour graphs also indicate that the salt wedge is located around Chatham for freshwater discharge of about 1200 $m^3 \cdot s^{-1}$ and that it migrates upstream past Newcastle as the discharge diminishes.

Temperature and salinity variations

Temperature series at Gordon Point, Chatham, and Newcastle stations (Figs. 4c – 4e), clearly indicate the seasonal temperature variation. Temperatures close to 0°C were observed at

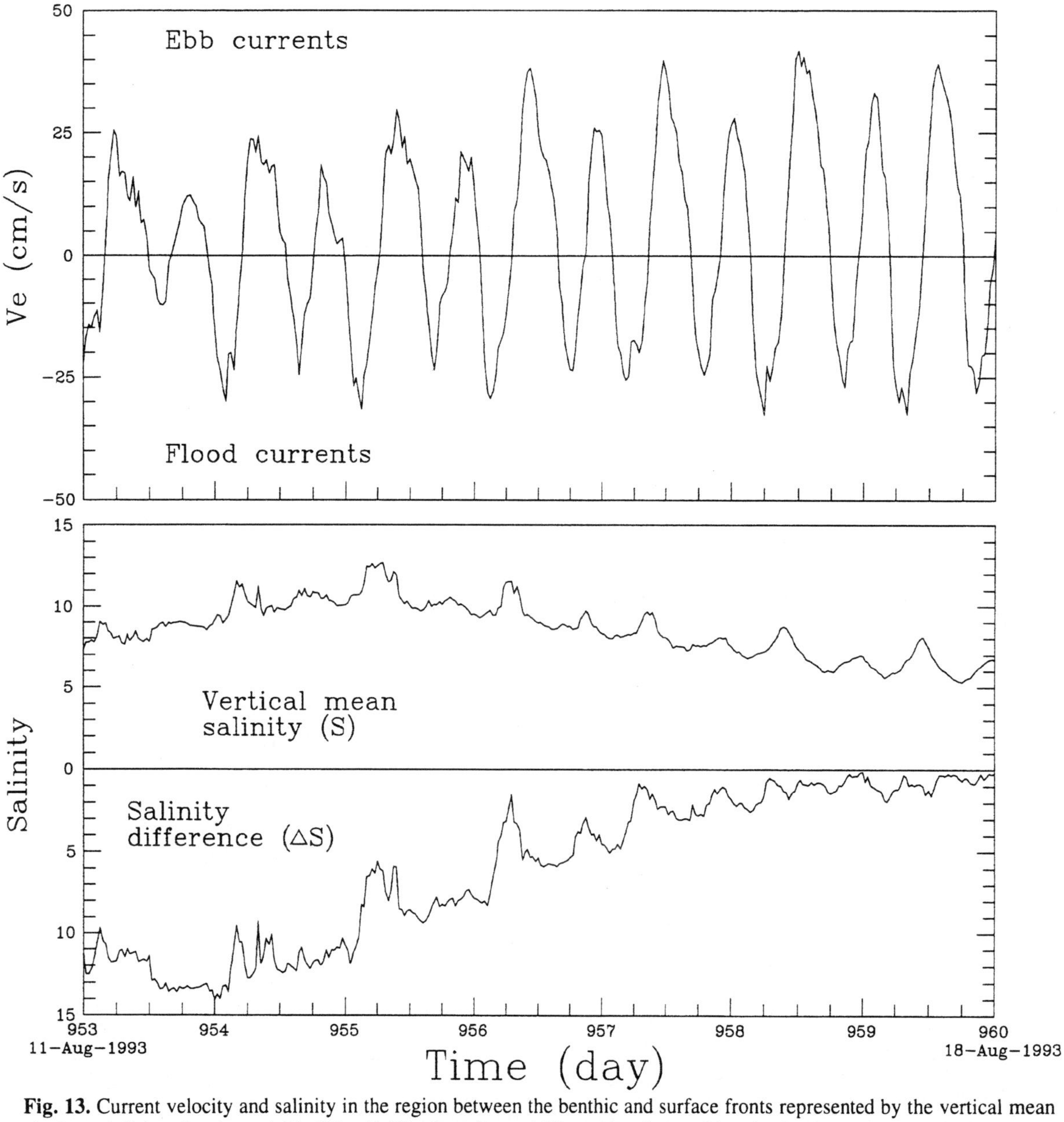

Fig. 13. Current velocity and salinity in the region between the benthic and surface fronts represented by the vertical mean velocity parallel to the channel ($Ve=[Ve_b+Ve_s]/2$ where Ve_b and Ve_s are the observed longitudinal currents near the bottom and the surface) and the semi-diurnal variations of the vertical mean salinity ($\bar{S}$) and salinity difference (ΔS) for 7 days in August at station NWMR. (Note the inverted axis for ΔS)

Newcastle and Chatham during winter while the temperatures at Gordon Point stayed close to –0.5°C. Maximum temperatures reached values around 19°C from mid-July to mid-August. The temperature series also demonstrated short-period fluctuations, particularly at the spring–neap period. The seasonal cycle dominated the series, illustrating the relative importance of heat exchange. To study the effects of tides, winds, and freshwater discharge on the water property distribution in the Miramichi Estuary, salinity, however, is the most useful variable.

Variations at a variety of frequencies were found in all the salinity series. The most evident variation is that of the neap–spring cycle. This variation was different in the surface and bottom layers. In the bottom layer (Figs. 4b, 4g, and 4h), the salinity was generally higher during neap and lower at spring tides. In the upper layer (Figs. 4c – 4e, and 4h), the salinity varied inversely with max-

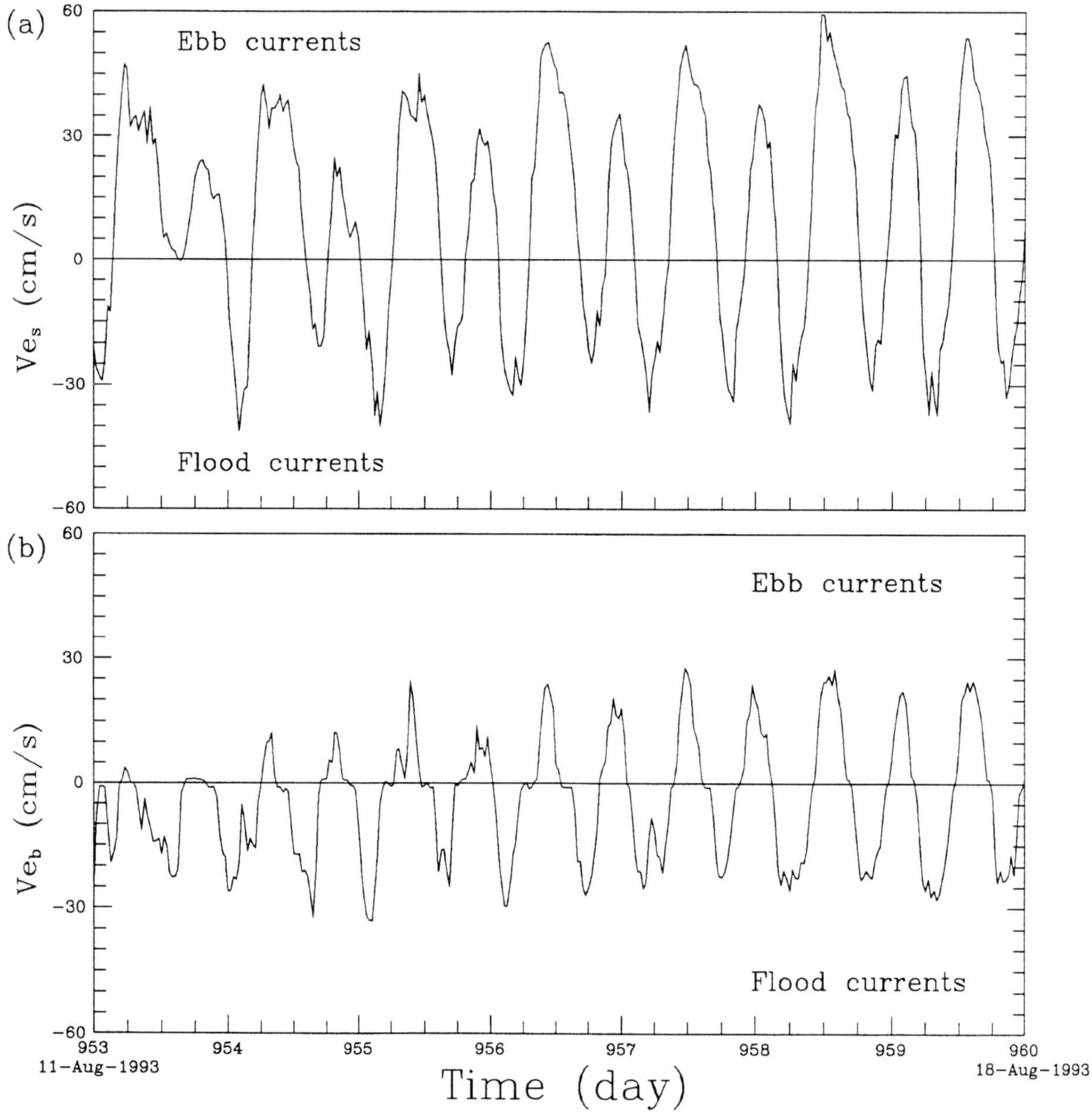

Fig. 14. Measured longitudinal currents (a) at 1 m below the surface and (b) at 1 m from the bottom for 7 days in August at the NWMR station.

ima at spring tides. This difference in behaviour between the two layers indicates that the spring–neap tides do not just advect the estuarine structure, but that vertical mixing is also an important process.

Along the estuary, the salinity increased downstream but the neap–spring oscillations in salinity did not. The greatest salinity variations, of 8 – 17, are seen at Chatham (station M78) probably because this station is situated in the region of greatest longitudinal salinity gradient. Station M51 has the lowest neap–spring salinity variations.

To illustrate the relation between variations in salinity and those of freshwater discharge and mixing by wind and tide, Fig. 9 is provided. In this figure of salinity at 2 m below the surface at Chatham, peaks in freshwater discharge and mixing power are indicated by points above the time

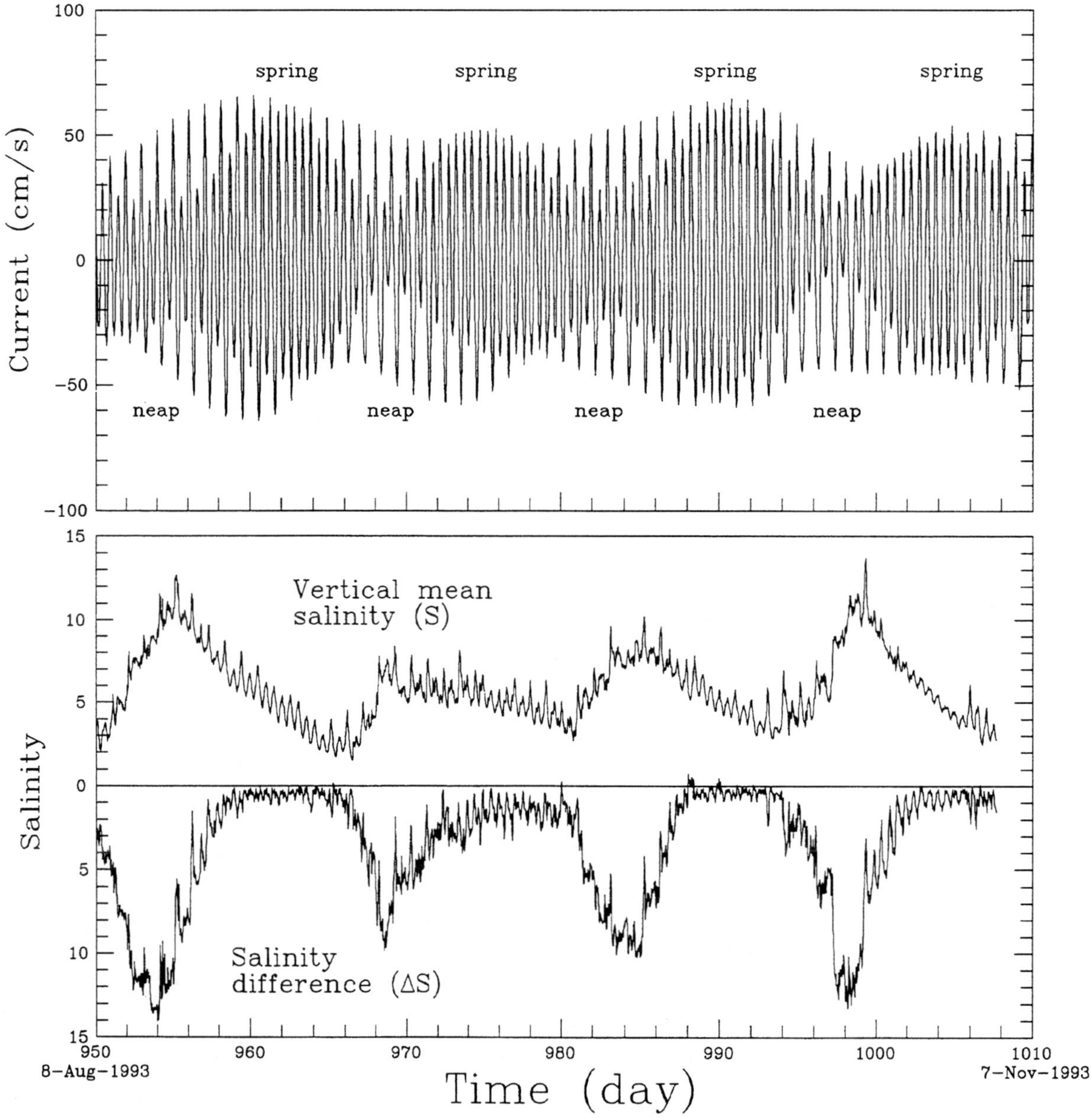

Fig. 15. Predicted tidal currents at Chatham indicating the succession of neap and spring tides, vertical mean salinity (S) and salinity difference (ΔS) at station NWMR from August to November, i.e., the latter part of Fig. 11. (Note the inverted axis for ΔS.)

axis. It summarizes the major information obtained in preceding figures and facilitates the comparison with the salinity series. Increases in freshwater discharge caused an abrupt decrease in salinity followed by a slower increase. Certain wind events, however, seem to have caused a thin salinity peak, probably a result of mixing between surface and bottom waters. Fortnightly variations in salinity appear to be associated with tides; high salinities occurred at peaks in tidal mixing. These variations are investigated below.

Salt wedge

In the following sections, we investigate the salt-wedge migrations using the bottom-salinity series, the spatio-temporal variations of salinity at 2 m below the surface, and the neap–spring variations in the mean vertical salinity.

Bottom-salinity series

Bottom salinity was recorded during the ice free months at 1 m from the bottom at buoys M51 and MD20, and in the North West Miramichi River (NWMR) (Figs. 4b, 4g, and 4h). Recorded bottom salinities at M51 were greater than 14 (Fig. 4b) whereas at Chatham, values of near zero were observed (Fig. 4d). These results as well as STD observations suggest that the salt-wedge limit migrated downstream to near Chatham. The salt wedge reached this position only in spring time (April) when the freshwater discharge was at its maximum. After spring runoff, the salt wedge migrated gradually upstream reaching buoy MD20 in the Southwest Miramichi (and NWMR) at the end of May or at the beginning of June when the freshwater discharge had decreased to a value of about 400 ± 100 $m^3 \cdot s^{-1}$. At low discharge, effects of tides became apparent. Spring–neap tidal oscillations seem to move the salt wedge back and forth. Even with this 15 day oscillation, however, the salt wedge continued to migrate further upstream during the low freshwater runoff of the summer, from mid-July to October. This was marked by a general increase of bottom salinity at mooring MD20 (Fig. 4g) and at mooring NWMR (Fig. 4h). Unfortunately, due to lack of data, the position of the salt wedge during winter is unknown.

Spatio-temporal salinity variations

The spatio-temporal salinity contour graph (Fig. 10) was put together with data recorded at about 2 m from the surface near Newcastle, Chatham, and Gordon Point. This graph illustrates the migration of the salt wedge. Interpretation is, however, constrained by the interpolation. Isohalines are not necessarily evenly spaced (see Vilks and Krauel 1982). Nevertheless, we consider the migration of the 2 m isohalines of 5 and 15. During winter, the isohaline 5 stayed around Chatham but travelled about 8 km each 15 day. When the spring freshwater discharge arrived at the end of March, isohaline 5 was pushed downstream past Gordon Point. After this event, winter-like conditions were restored in about 90 day. During summer time, the isohaline 15 gradually travelled up to Chatham, and also showed 15 day oscillations. At the same time, the isohaline 5 proceeded upstream past Newcastle. Autumn freshwater runoff restored the conditions found in winter. With lower water runoff during winter, one might expect the winter salinity near Chatham to be higher than 5 at this depth but this was not observed, probably the result of the ice cover. As ice formed, the interface between the water and the ice descended. The instrument, therefore, was closer to the water surface than the 2 m at the time of deployment and thus recorded lower salinities.

This contour graph is probably not an accurate representation of the salt-wedge displacement; vertical movements should also be considered. It does, however, illustrate the seasonal cycle caused by freshwater discharge and confirms the presence in winter of low-salinity water near Chatham.

In summary, it is clear that the salt wedge is subject to both seasonal migrations and 15 day longitudinal oscillations. The seasonal migration appears to be about 18 km (Fig. 10), more than twice the observed spring–neap oscillation. It is not clear, however, how the form of the salt wedge changes.

Salt wedge migration on the neap–spring tidal cycle

To further evaluate the spring–neap excursion, we consider the vertical mean salinity, approximated here by the mean of the surface and near-bottom values. The vertical mean salinity is unaffected by vertical mixing and thus illustrates the horizontal progression and regression of the salt wedge. Although the mean vertical salinity does not indicate the position of the salt wedge, it may be taken to evaluate the longitudinal change in salinity. Interpretation is, however, limited by vertical shear in the circulation.

Consider first that the vertical mean salinity is defined as the average of surface and bottom salinity ($\bar{S}=[S_s+S_b]/2$). This mean value may be expanded in a Taylor series of the first order as follows:

$$(5) \quad (\bar{S})_{t2} = (\bar{S})_{t1} + (\Delta\bar{S}/\Delta x)\, L$$

where $(\bar{S})_{t1}$ and $(\bar{S})_{t2}$ are the vertical mean salinities at times $t1$ and $t2$, between which the water mass has been advected as a block over a distance of L. If $t1$ is the spring tide time and $t2$ is the neap tide time, the distance L approximates the spring–neap tidal excursion of the mean vertical salinity.

The longitudinal salinity gradient ($\Delta\bar{S}/\Delta x$) was calculated by considering tidal variations in salinity at the surface and at the bottom, ΔS_s and ΔS_b. The mean of ΔS_s and ΔS_b was associated with the horizontal excursion given by the following:

Table 4. Results of the spring–neap salt-wedge excursion estimations by the parameter L for stations NWMR and M78.

Station	Periods used for estimation (Julian days)	Excursion (km)
NWMR	949.5 to 957.5	16.6
	957.5 to 965.0	18.0
	965.0 to 973.0	6.7
	973.0 to 981.0	4.1
	981.0 to 988.0	6.0
	988.0 to 995.0	8.8
	995.0 to 1003.0	8.6
M78	890.0 to 897.5	7.9
	897.5 to 905.0	14.0

$$(6) \quad \Delta x = \int_0^{T/2} v \sin(\omega t)\, dt = T\, v / \pi$$

where T = the tidal period, $\omega = 2\pi/T$, and v = the maximum value of the tidal current averaged over the cross section. Values of $\Delta\bar{S}/\Delta x$, averaged over half a spring–neap cycle were used to estimate the mean longitudinal salinity gradient in equation (5).

Calculations of L were performed with data obtained on the Northwest Miramichi river station and at station M78. We calculated L for 7 different half spring–neap periods in the former and 2 for the latter. For NWMR, the spring–neap salt-wedge excursion had a mean value of 10 km with a standard deviation of 5 km (Table 4). For M78, the average of the two estimates gave a value of 11 ± 4 km. These estimations are limited by the simplicity of the process envisaged and should be considered as approximate.

Time variations of the stratification

The estuarine structure in the Miramichi is variable. The tide alone can have both stratifying and destratifying effects. Correlations between the oscillatory tidal mixing power and the salinity series at Chatham indicated that the 15 day oscillations are not totally explained by tides, at least in a linear manner. These correlations, although inconclusive and difficult to interpret in terms of physical processes, demonstrated how complex the salinity variation pattern may be in the Miramichi Estuary.

Using a simpler approach, we attempt to relate tidal oscillations with the salinity time series. To separate vertical and horizontal processes, we consider the salinity difference between surface and bottom, $\Delta S=(S_b-S_s)$, and the mean water column salinity, $\bar{S}=[S_s+S_b]/2$, at the NWMR station from June to November (Fig. 11). The water column at this station was completely fresh ($\bar{S}$=0) for the first half of the month of June before the first arrival of the salt wedge. Then, an increased freshwater discharge pushed the salt wedge further downstream again until mid-July (around day 922). Thereafter, $\bar{S}$ clearly showed a gradual migration upstream with a back and forth movement of the salt wedge related to spring–neap tides.

Both the salinity difference and the vertical mean salinity varied at a daily scale. We examine these high-frequency variations first close to the benthic front, (i.e., close to the upper limit of the salt wedge), and then in the salt wedge itself. At the upper limit of the salt wedge (Fig. 12), maximum $\bar{S}$ and ΔS were observed at high water and minimum at low water. They vary at a semidiurnal frequency. In the salt wedge (Fig. 13), between the surface and the benthic fronts, the behaviour is different. The mean depth salinity, $\bar{S}$, was again maximum at high water and minimum at low water, but the stratification, given by ΔS, was weaker at high water and stronger at low water. Furthermore, the tidal variation of ΔS was more evident at neap tides than at spring tides and was largely diurnal (Figs. 13 and 15). During neap tides, variations up to 5 units of salinity occurred. During spring tides, these variations were reduced to around 1 and were not clearly identified with the tidal oscillations. It is important here to note that currents near the water surface were generally higher than those at 1 m above the bottom by a factor varying between 1.5 and 2 (Fig. 14). It is also important to note that, at salinities less than 5, semidiurnal variations in the unfiltered salinity series were generally larger at the bottom than at the surface.

The neap–spring tidal cycle imposed even stronger changes in salinity than did the semidiurnal cycle (Fig. 15). During a 15 day period, $\bar{S}$ tended to be maximum just after neap tides and minimum about 5 days before. Variations were of 5 – 10 units of salinity. On the other hand, ΔS was minimum for 7 days during spring tides and reached a maximum at neap tides. Variations in ΔS were of between 8 and 13. These variations suggest that vertical mixing, which tends to destroy stratification, was pronounced during periods of strong current. Variations in $\bar{S}$ indicated a rapid progression of the water structure upstream during a few days before neaps followed by a slower retreat for the rest of the 15 day cycle.

Freshwater events

We examine the effect of a sudden increase of 1500 $m^3{\cdot}s^{-1}$ in freshwater discharge (day 904) on the salinity of the bottom layer between buoy MD20 (Fig. 4g) and buoy M51 (Fig. 4b). At MD20, the salinity stayed at 0; at Chatham (station M78, Fig. 4h), the salinity dropped from 10 to 0.5 within 1 day; at M51, the salinity decreased slightly from 20 to 18 also within 1 day. The salinity distribution responded to this increase in freshwater runoff in a delay of less than 1 day with a salt wedge retreat of about 10 km. As mentioned in the introduction, an increase of freshwater discharge may increase the longitudinal salinity gradient if the downstream boundary conditions remain almost constant. This is confirmed by this example. Before the increase of the river flow, the longitudinal gradient between Chatham and M51 was 0.5 salinity units$\cdot$km^{-1} at 1 m from the bottom. After the abrupt increase of 1500 $m^3{\cdot}s^{-1}$, the gradient was 0.9 salinity units$\cdot$km^{-1}.

Discussion

The following discussion examines the migration of the salt wedge and the stratification changes at various temporal scales, from seasonal to semidiurnal. At the seasonal scale, variations in the estuarine structure are principally determined by river discharge and heat flux. Tides influence the structure at scales of between 0.5 day and 2 weeks. Wind effects are found at the time scale of days.

Seasonal variations

Freshwater discharge changes seasonally and causes large variations in the estuarine structure of the Miramichi. Runoff is low in winter, but it is subject to an enormous increase in spring due to snowmelt. In the summer and autumn, the freshwater discharge is determined principally by precipitation. The salt wedge, which is subject to the incoming flow of fresh water, is pushed down the estuary to a location between Chatham and Gordon Point in spring. It then migrates upstream with the decrease in freshwater inflow to a position situated above stations MD20 and NWMR. The temperature variations are dominated by the seasonal cycle, the winter temperature being determined by the freezing point of salt water. Thus the seasonal variation of the estuarine structure is determined mainly by river discharge and heat fluxes.

Weekly and monthly variations

Variations in the estuarine structure at this time scale are dominated by the half month tidal oscillations. During the few days at neap tides, the salt wedge migrates inland and the stratification is enhanced. As the currents increase with the oncoming spring tides, vertical mixing increases (Fig. 5). The vertical mixing weakens the stratification and the estuarine circulation. This weakening seems to cause a slow retreat of the salt wedge (Fig. 11). Then with neap tides, the stratification strengthens and the salt wedge progresses upstream again. It is of interest to note the relatively slow retreat of the salt wedge during spring tides compared to the upstream migration during neap tides. This may simply be a result of increased bottom friction.

Variations within a few days

Changes in salinity distribution on scales of a few days are due to freshwater discharge and wind events. Mixing enhanced by winds is generally weaker than that caused by tides (Fig. 5 and 6) although certain wind events briefly provide more mixing power than the tides. It should be noted, however, that we took the longitudinal component of the wind rather than its total speed to calculate the power. If the local hills deviate the wind rather than block the transverse component, our estimates of the wind power may be underestimated.

Superimposed on the fortnightly migration, precipitation events impose rapid retreats of the salt wedge over a distance varying with the quantity of water input to the estuarine system. In the freshwater event examined above, the increase in freshwater discharge enhanced the longitudinal salinity gradient. This necessarily changed the estuarine circulation. The final result was a retreat of the salt wedge of about 10 km. After such an event, the salt wedge migrates upstream again. The time to restore initial conditions is larger than that taken for the perturbation (Fig. 9).

As mentioned in the introduction, we can expect an increase in the vertical salinity gradient to be associated with an increase in the longitudinal salinity gradient. STD profiles (Fig. 8) seem to indicate rather a weakening of the vertical salinity gradient. They do not, however, display the same salt wedge. Both wind and tidal conditions differed from one STD grid to another. Hence, the observed change in vertical salinity gradients may be a result of mixing rather than shear. Nevertheless, the increase in the longitudinal salinity gradient in response to an increase in freshwater inflow suggests that the mixing zone between fresh and salt waters acts as a buffer reducing the impact of an event as it moves down the estuary. Vilks and Krauel (1982) have also observed this process.

The estuarine structure is also affected by winds. In the filtered salinity series, only wind events with effects longer than one day are seen. Wind events are difficult to quantify since they are variable in strength and duration. In general, winds do not immediately cause a migration of the salt wedge but rather modify the stratification of the estuary. Vertical mixing increases with wind but is modified by a decrease in efficiency due to stratification. Wind events are relatively short with observed effects on the salinity distribution usually lasting less than three days. Initial stratification is restored more rapidly after a wind event than after a freshwater event (Fig. 9).

Diurnal and semidiurnal variations

Tides in the Miramichi are mixed. Both diurnal and semidiurnal constituents are evident. During spring tides, stratification at the NWMR station was almost inexistant for the whole sampling period (Fig. 11). Tidal variations in ΔS were small (Fig. 13). The stratification seemed to be reduced during peak currents suggesting that the change in the vertical structure is controlled by mixing. During neap tides, however, salinity changes differ according to location. Near the upper limit of the salt wedge, where the depth mean salinities, $\bar{S}$, were less than 5, both $\bar{S}$ and the stratification, represented by ΔS, showed a semidiurnal signal, with maximum $\bar{S}$ and maximum ΔS occurring at high water (Fig. 12). Further downstream, $\bar{S}$ was also maximum at high water but ΔS rather showed an inverse variation with maxima at low water (Fig. 13). Furthermore, the signal here was diurnal rather than semidiurnal.

Vertical mixing by semidiurnal tides alone

cannot explain these results. Stratification would in this case be reduced four times a day during peak currents. Near the benthic front, the observed variations in both $\bar{S}$ and ΔS can be explained by simple advection of the estuarine structure, in which the maximum vertical gradients are downstream of the mooring. Maximum $\bar{S}$ and ΔS occur at high water as a result of the salt wedge migrating back and forth with the tide.

In an attempt to quantify variations in stratification near the benthic front caused by tidal advection, we use the estimate of the semidiurnal tidal excursion for the NWMR station which is of the order of 3 – 5 km. This estimate is smaller than the 12 km excursion calculated by St-Hilaire (1993) for a section of the estuary situated between buoys M88 and M51. There the currents are stronger than in the upper part of the estuary: using the predicted currents at Chatham for exemple, we get an excursion of about 8 km. To continue, we also use the only salinity data available for the upper limit of the salt wedge during similar freshwater discharge conditions. These are presented in Fig. 12 of Vilks and Krauel's report (1982). Their results showed an increase in the strength of the stratification with the increasing distance downstream from the upper limit of the salt wedge with a change in ΔS of 6.5 units of salinity over 5.3 km. The vertical mean salinity also increased by 3 over the same distance. Although these data are for ice-covered conditions, they indicate that a simple tidal advection of 3 – 5 km of the salt wedge may indeed cause variations in ΔS near the benthic front equivalent to those observed (Fig. 12).

In the region downstream of the benthic front, where depth mean salinities are about 10 (Fig. 13) and where ΔS changes little with the horizontal distance, tidal advection is clearly not sufficient to explain the observed variations. Minimum stratification during neap tides occurred at high water and in particular after every second flood tide. Although the peak flood tide currents at the bottom varied little from one cycle to another, the shear from top to bottom did (Fig. 14). Weak stratification occurred after flood currents with the stronger vertical shear. We can expect relatively salty water to be advected upstream at the surface during the flood tides thereby reducing the stratification. Thus vertical shear in the tidal flow may be the cause of the observed diurnal weakening in stratification. The ebb flow, with its strong vertical shear, should likewise restore the stratification.

To advance further, we need to evaluate the horizontal density gradients in both the top and the bottom layers (Nunes Vaz 1990; Simpson et al. 1991). An attempt to measure these gradients during the summer of 1993 failed when a mooring broke away soon after deployment. It is clear that, for future studies, the vertical profile of horizontal gradients cannot be neglected. Quantifying the process discussed is the next stage to be undertaken.

Conclusion

The estuarine structure of the Miramichi is subject to changes, at a variety of time scales. At the seasonal scale, freshwater discharge and heat exchange are the dominant processes to be considered. At the monthly scale, vertical mixing by spring tides and subsequent restratification during neap tides determine the variations in the structure. There is also a spring–neap migration of the salt wedge. At neap tides, the salt wedge is several kilometres upstream of its position at spring tides. At the daily scale, simple advection explains, at least in part, the observed variations in both stratification and vertical mean salinity close to the benthic front. Within the salt wedge, however, vertical shear in the tidal currents and maybe the diurnal tidal constituent must be included to explain the observed variations. At this scale, both wind and discharge events also affect the estuarine structure.

Acknowledgements

Financial support for this work was provided by Public Works Canada through the Department of Fisheries and Oceans. We thank all who contributed to the project and especially the technicians at the Institut Maurice-Lamontagne, Sylvain Cantin, Roger Pigeon, Remi Desmarais and Robert Bélanger, who prepared and deployed most of the instruments. We also thank Vernon Goodfellow who dived to clean the instruments during the summer of 1993 and Stéphane Lapierre who helped deploy 4 moorings in 1992. Finally, we thank the referees for some constructive comments.

References

Atkinson, L.P., and Blanton, J.O. 1986. Processes that affect stratification in shelf waters, p. 117–130. *In* C.N.K. Mooers [*editor*]. Baroclinic processes on continental shelves. American Geophysical Union, Washington DC.

Bowman, M.J. 1988. Estuarine fronts, p. 85–132. *In* Björn Kjerfve [*editor*.] Hydrodynamics of Estuaries. CRC Press, FL.

Dyer, K.R. 1973. Estuaries: a physical introduction. John Wiley and Sons, New York, NY. 140 p.

Dyer, K.R. 1988. Tidally generated estuarine mixing processes, p. 41–57. *In* Björn Kjerfve [*editor*]. Hydrodynamics of Estuaries. CRC Press, FL.

Foreman, M.G.G. 1977. Manual for tidal heights analysis and prediction. Pac. Mar. Sci. Rep. 77–10: 101 p.

Geyer, W.R., and Farmer, D.M. 1989. Tide-induced variation of the dynamics of a salt wedge. J. Phys. Oceanogr. 19: 1060–1072.

Linden, P.F. 1979. Mixing in stratified fluids. Geophys. Astronom. Fluids Dynamics 13: 3–24.

Nunes Vaz, R.A. 1990. Periodic stratification in coastal waters, p. 69–105. *In* A.M. Davies [*editor*]. Modeling marine system, Vol II. CRC Press, FL.

Parks, T.W., and Burrus, C.S. 1987. Digital filter design. John Wiley and Sons inc., New York, 342 p.

Schroeder, W.W., Dinnel, S.P., and Wiswman, W.J. Jr. 1990. Salinity stratification in a river-dominated estuary. Estuaries 13(2): 145–154.

Simpson, J.H., Sharples, J., and Roppeth, T.P. 1991. A prescriptive model of stratification induced by freshwater runoff. Estuarine Coastal Shelf Sci. 33: 23–35.

St-Hilaire, A. 1993. Étude hydrodynamique de l'Estuaire de la Miramichi durant la saison sans glace — 1991. Master Thesis, Université de Moncton, Moncton, N.-B. 153 p.

Vilks, G., and Krauel, D.P. 1982. Environmental geology of the Miramichi Estuary: physical oceanography. Geol. Surv. Can. 81–24: 53 p.

Miramichi Pulp and Paper mill is located near the mouth of the Northwest Miramichi River; it is both the major employer and the main source of suspended organic matter on the Miramichi River.

CHAPTER 5

Is the Miramichi a Stratifed Estuary?

A. St-Hilaire

*INRS-EAU, 2800, rue Einstein, suite 105, C.P. 7500,
Ste-Foy, Québec, Canada G1V 4C7*

D. Booth

*Oceanography Department, Université du Québec à Rimouski, 2100, route des Ursulines,
Rimouski, Canada*

C. Bettignies

*School of Engineering, Université de Moncton,
Moncton, New Brunswick, Canada E1A 3E9*

E.M.P. Chadwick and S.C. Courtenay

*Department of Fisheries and Oceans, Science Branch, Gulf Region , P.O. Box 5030,
Moncton, New Brunswick, Canada E1C 9B6*

St-Hilaire, A., Booth, D., Bettignies, C., Chadwick, E.M.P., and Courtenay, S.C. 1995. Is the Miramichi a stratified estuary?, p. 73-82. *In* E.M.P. Chadwick [*editor*]. Water, science, and the public: the Miramichi ecosystem. Can. Spec. Publ. Fish. Aquat. Sci. 123.

Abstract

Stratification in estuaries affect the productivity and the distribution of fish. It is therefore important to quantify it. Stratification of the Miramichi Estuary was studied during the ice-free season. Based on salinity, wind, freshwater discharge and current velocity measurements, the mixing power of wind and tidal currents are compared to the power needed to mix the narrow portion of the Miramichi Estuary during the ice-free season. For typical values of density, salinity, velocity and depth, calculations predict that the estuary should be stratified during the ice-free season. Calculations also show a predominance of wind mixing power. Salinity profiles taken during the ice-free season show that there were stratifiaction during both high and low fresh water discharge but there were also some weak salinity gradients upstream caused by flushing.

Résumé

Dans les estuaires, la stratification a un effet sur la productivité et la distribution des poissons; il est donc important de la quantifier. Nous avons étudié la stratification de l'estuaire de la Miramichi pendant la saison libre de glace. À partir des mesures de la salinité, du vent, de l'apport d'eau douce et de la vitesse des courants, nous comparons la puissance de mélange du vent et des courants de marée à la puissance nécessaire pour mélanger les eaux de la portion étroite de l'estuaire pendant la saison libre de glace. Pour des valeurs typiques de densité, de salinité, de vitesse et de profondeur, les calculs prédisent que l'estuaire devrait être stratifié pendant la saison libre de glace. Les calculs montrent aussi une prédominance de la puissance de mélange du vent. Les profils de salinité pris pendant la saison libre de glace montrent qu'il existait une stratification quand l'apport d'eau douce était haut comme quand il était bas, mais on observait aussi quelques faibles gradients de salinité en amont à cause de l'effet de chasse.

Introduction

Stratification of the water column in estuaries affects the productivity (Malone 1977). It also sets the geographical and temporal limits of various fish species. In shallow waters, the action of the wind at the water surface and turbulence created by tidal currents near the bottom are often the most important factors among the various physical phenomena that prevent stratification (Simpson et al. 1990).

The main objective of this study is to quantify the mixing power of the wind and tides and compare them to the total energy budget of the estuary. Measurements of salinity, wind, fresh water discharge and current velocity were made during the ice-free season of 1991 to provide an insight into the variation of stratification in the estuary.

Study area

The drainage area of the Miramichi basin covers approximately 14000 km^2. For the purpose of this study, the estuary can be divided into two parts: a narrow portion and a triangular shaped bay. The narrow portion is less than 1.5 km in width from the Enclosure Park to Loggieville. The bay is separated from the Gulf of St Lawrence by a chain of islands. A 60 km channel has been dredged to a minimum depth of 7.6 m. Depths in the channel exceed 10 m in certain areas. Outside the channel, the estuary is shallow with depths seldom exceeding 7 m.

Tides in the estuary are mixed, with a range between 0.2 to 1.2 m. Tidal range in the narrow section is 20 % larger than at the mouth of the bay. Tidal period vary between spring tides which are almost diurnal, and neap tides showing mostly semi-diurnal characteristics (Willis 1990). Freshwater flow originates mostly from the two main tributaries: the Northwest Miramichi and the Southwest Miramichi. Vilks and Krauel (1982) estimated the mean annual discharge for the entire estuary to be 305 m^3 s^{-1}. Spring floods can exceed 1600 m^3 s^{-1} (Philpott 1978).

Methods

Figure 1 shows a simplified mixing model of an estuary. The processes creating turbulent kinetic energy near the surface are friction from wind and wave action. Wind transfers potential energy to the surface of the water column. At the bottom, tidal currents generate turbulence which in turn provides energy to mix the water column. By moving denser water up in the water column or, conversely, by moving less dense water down in the water column, mixing increases the potential energy in the model.

Expressions for the mixing power provided by wind and tidal action have been derived by many authors and are based on the expression for the Reynolds stress (i.e., stress caused by the friction between two fluids of different densities) at a fluid interface (e.g., Atkinson and Blanton 1986). The equation used for wind mixing power in this work is from Kullenberg (1976):

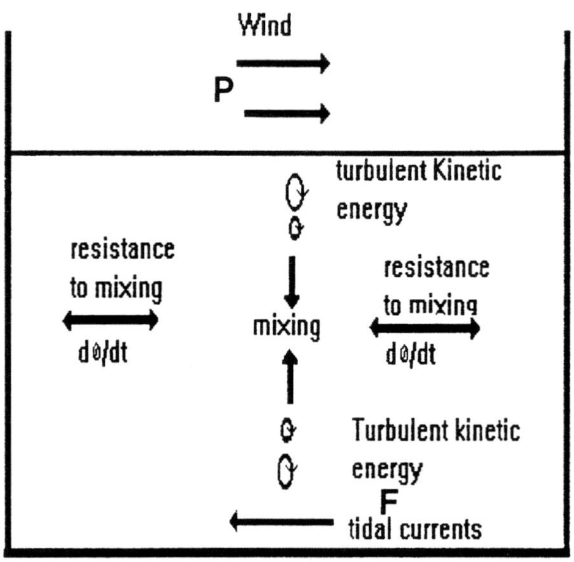

Fig. 1. General mixing model of an estuary.
P: Mixing power from the wind
F:Mixing power from tidal action
∂φ/∂t:Energy input form inflow and outflow

$$P = R_1\, \rho_a\, C_d\, U_w^3 \quad \text{(W m}^{-2}\text{)} \qquad (1)$$

where P:
wind mixing power per unit area of the estuary (W m^{-2})

R_1 :
empirical constant related to the efficiency of energy transfer from the wind kinetic energy to potential energy of the water column (0.0018).

ρ_a : air density (1200 g m^{-3})
C_d : friction coefficient (0.0017)
U_w : wind speed 10 m above the surface (m s^{-1})

Wind data were obtained from the Environment Canada meteorological station in Chatham (Environment Canada 1991). Available data covered most of the sampling period (from 1 May to 14 September 1991). Winds used in our calculations were 30 minute averages. Figure 2 shows the values of P calculated from the wind data.

Similar equations exist for tidal current

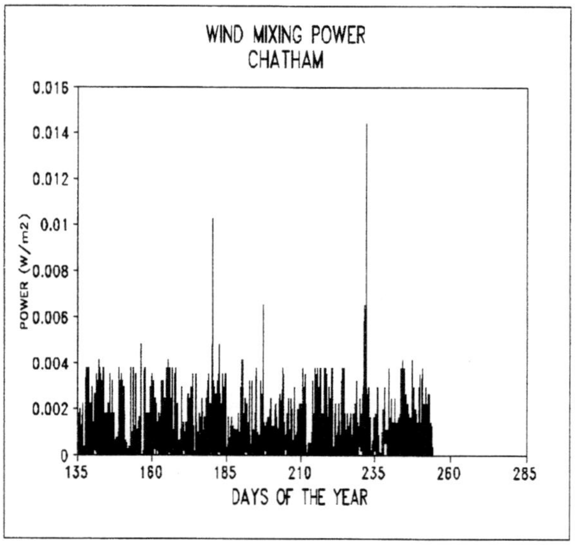

Fig. 2. Wind mixing power calculated from wind speed time series. From May 15 (day 135) to October 12 (Day 285).

mixing power. In this study, the equation provided by Simpson and Bowers (1981) was used:

$$F = \epsilon\, C_f\, \rho\, U_f^3 \quad \text{(W m}^{-2}\text{)} \qquad (2)$$

where F:
tidal mixing power per unit area of the estuary (W m^{-2})

ϵ : empirical constant related to the efficiency of the energy transfer between the kinetic energy of the turbulence caused by the current and the potential energy of the water column (0.0037).

C_f: friction coefficient (0.0025)
ρ : density of water (1000 kg m^{-3})
U_f: tidal current velocity (m s^{-1})

A current meter was installed 1 m above the bottom at each of two stations (Fig. 3). The total depth at both stations was 8 m. The meters were of the type Aandera RCM7, equipped with a rotor and vane for current speed and direction, and sampling every 30 minutes. The moorings were deployed on 14 May 1991 and recovered on 29 October 1991. Although current meters were

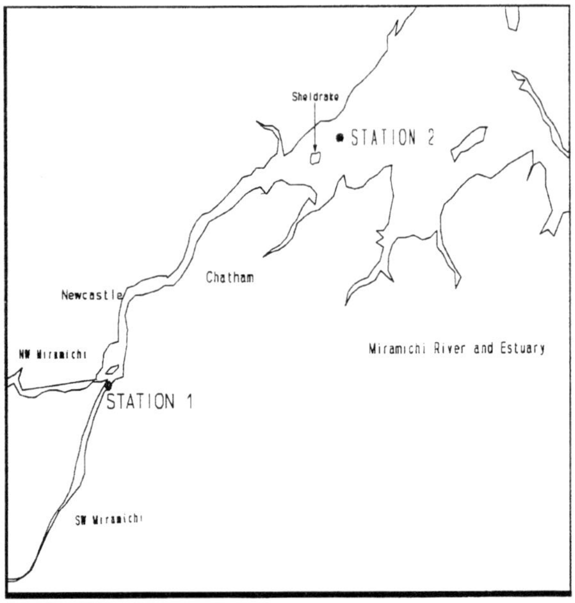

Fig. 3. General Area and Station Locations.

cleaned every 2 weeks by a diver, the instrument at station 1 stopped recording on August 7, while the one at station 2 stopped on October 2. It is believed that, in spite of regular clean ups, the premature end of sampling may have been caused by algae or woody debris. Calibration was performed on the meters before and after deployment at the Institut Maurice Lamontagne of the Department of Fisheries and Oceans, Canada. Averaged calibration coefficients were used to produce the data.

Using equantion (2), F was calculated from current meter data at stations 1 and 2. Results are shown in Figs. 4 and 5.

Before we can compare the contributions of the wind and tidal action to the mixing of the estuary, we need to estimate the resistance of water of different densities to mix. The energy flux, in the form of buoyancy variations caused by the freshwater inflow and outflow, quantifies the power needed to mix an estuary. The approach taken in this case is similar to that of Simpson et al. (1990) who defined the flux for the water column (i.e., the density ρ_m is the average density of the water column). However, we define pm as the mean density for the entire estuary. The potential energy, for the estuarine model shown in Fig. 2, can be defined as follows:

(3) $PE = (LH)^{-1} \left[\int_0^L \int_{-H}^0 (\Delta\rho)\, g\, z\, dx\, dz \right]$

where PE:
difference between potential energy of stratified and mixed states (J m^{-3}).

L: length of the estuary (m, positive eastward)

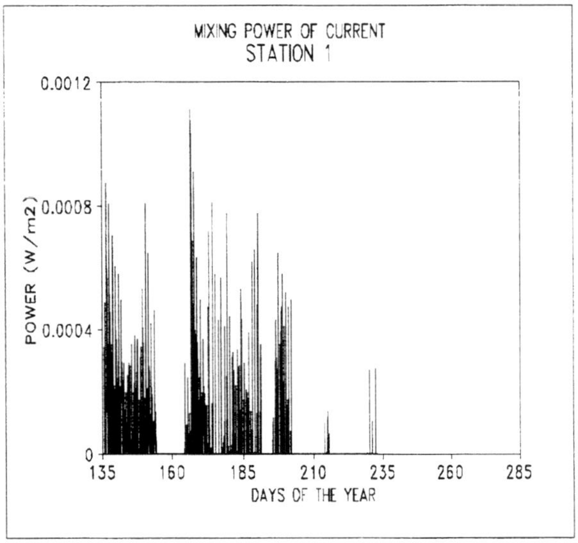

Fig. 4. Current mixing power, upstream station, calculated from current time-series.

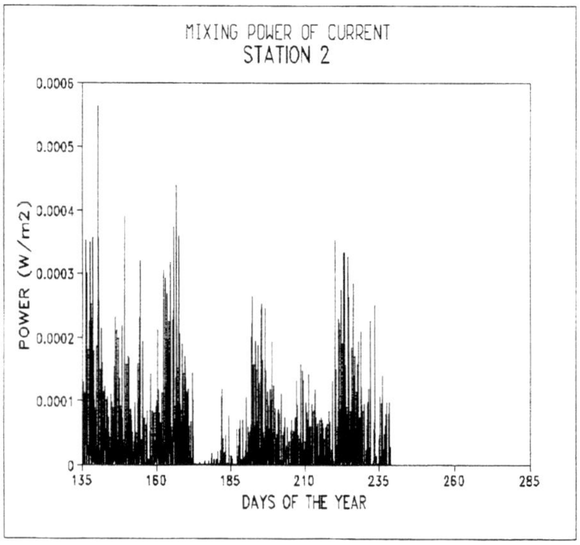

Fig. 5. Current mixing power, downstream station, calculated from current time-series.

H: mean depth (m)
ρ: density (kg m^{-3})
g: gravitational constant
z: depth of the estuary (m, positive upwards)
$\Delta\rho$: $\rho-\rho_m$ difference in density (kg m^{-3})

The energy needed to mix the estuary ϕ, is :

(4) $\phi = -H\,\mathrm{PE}$ (J m^{-2})

The power needed per unit area to mix the estuary is a measure of the resistance to mixing. It can be defined as the time derivative of ϕ:

$(\partial\phi/\partial t) = -H\,\partial\mathrm{PE}/\partial t$ which can be approximated by

(5) $(\partial\phi/\partial t) = -g\,(L)^{-1}\,[\,{}_{-H}\!\int^{0} u\,(\rho-\rho_m)\,z\,dz\,|_{x=0}$

$-\,{}_{-H}\!\int^{0} u\,(\rho-\rho_m)\,z\,dz\,|_{x=L}]$

where
ρ_m: average density for the entire estuary (kg m^{-3})
ρ_0: density of freshwater (1000 kg m^{-3})
u: current velocity (m s^{-1})

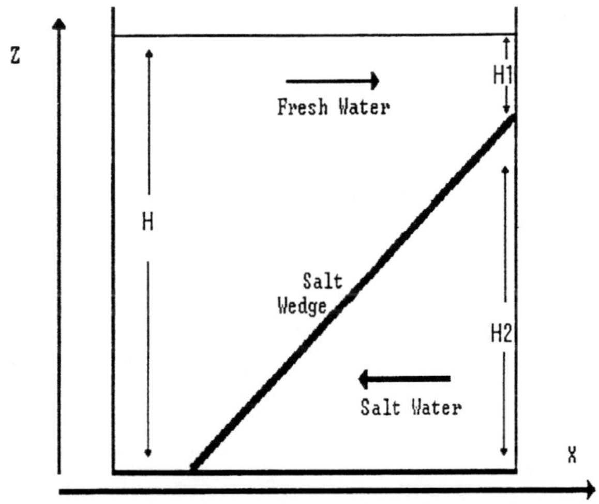

Fig. 6. Estuarine two-layer Model.
X:Horizontal coordinate, positive towards the east.
Z:Vertical coordinate, positive upwards.
H: Total depth, upstream.
H_1:Thickness of the top layer.
H_2:Thickness of the bottom layer.

and where $\delta\rho/\delta t$ is taken as zero (i.e., density is considered constant on time scales less than one month).

For a stratified estuary, equation (3), i.e., the potential energy, is negative while the power per unit area is positive.

A first approximation is that of a two-layered model. This first approach is often used to approximate water exchange and mixing processes (e.g., Stigebrandt 1988; Bowden and Hamilton 1975; and Tee, 1988). A two-layered model has been used to describe the narrow portion of the estuary in previous studies (Bousfield 1955). Vilks and Krauel (1982) describe the presence of a halocline which separates the saline bottom water from fresher surface water in the narrow section of the estuary.

Using equation (5) for the specific example of a two-layered model such as the one shown in Fig. 6 implies that the following limits can be applied to the integral of equation (5) as follows:

(6) $(\partial\phi/\partial t) = -g\,(L)^{-1}\,[\,{}_{-H}\!\int^{0} u\,(\rho_0-\rho_m)\,z\,dz\,|_{x=0}\; -$

${}_{-H^1}\!\int^{0} u\,(\rho_1-\rho_m)\,z\,dz\,|_{x=L}$
$-\,{}_{-H}\!\int^{H^1} u\,(\rho_2-\rho_m)\,z\,dz\,|_{x=L}]$

where
ρ_0: density of freshwater (1000 kg m^{-3})
ρ_1: density of the top layer (kg m^{-3})
ρ_2: density of the bottom layer (kg m^{-3})
H_1: thickness of the top layer (m)

When the integration of equation (6) is performed, the velocity is treated as constant for each of the two layers. These velocities u_1 and u_2 are obtained through the mass conservation equations:

(7) $0 = u_1H_1S_1 + u_2H_2S_2$ and
$u_0H = u_1H_1 + u_2H_2$

where $H = H_1 + H_2$
and the state of the estuary is taken as steady

The depth and density for each layer were also treated as constant. The velocity at the upstream station was calculated from the fresh water flow.

Freshwater flows were obtained from 3

hydrometric stations of the Environment Canada network. The stations were located on the Northwest Miramichi, Southwest Miramichi, and Bartibog rivers (Environment Canada 1990). Discharge values from these stations were extrapolated to the entire drainage area using the ratio of the gauged area to that of the entire drainage basin. Fig. 7 shows calculated hourly discharges for the ice-free season of 1991.

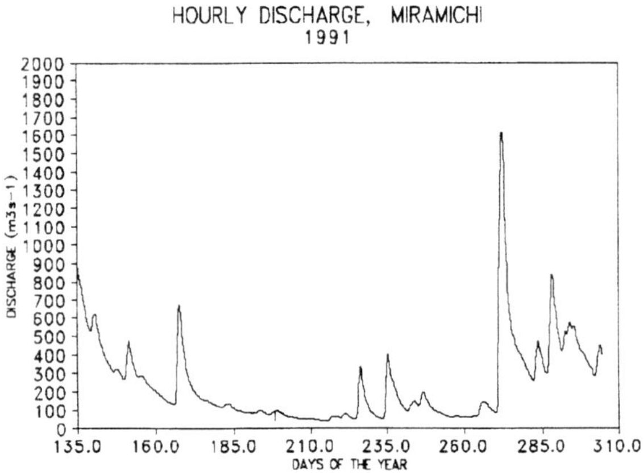

Fig. 7. Hourly Discharge, Miramichi Estuary.

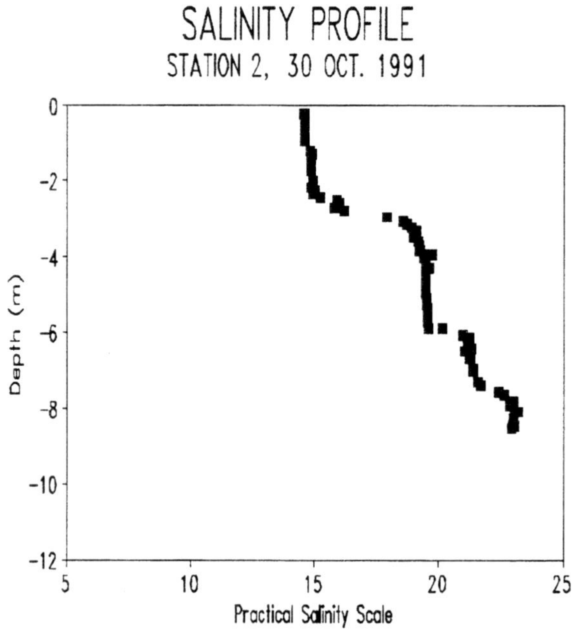

Fig. 8. Salinity Profile, 30 October 91, Station 2.

Results

Figure 2 shows the time series of the wind mixing power (*P*) using calculated with equation 1. The maximum value (0.014 W m^{-2}) was obtained on August 17 (day 229) for a wind speed of 56 km h^{-1}. Monthly averages of *P* are shown below.

	P (10^{-4} W m^{-2})	Standard Error (10^{-4} W m^{-2})
May	7.6	10
June	7.5	9.8
July	7.2	9.7
August	6.7	9.9
September	8.2	9.8

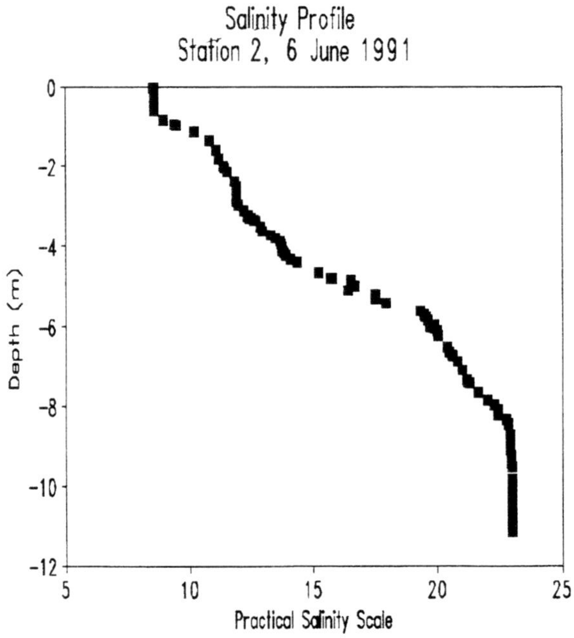

Fig. 9. Salinity Profile, 6 June 1992, Station 2.

	F Station 1	Standard Error	*F* Station 2	Standard Error
Month	(10^{-4}W m^{-2})	(10^{-4}W m^{-2})	(10^{-4}W m^{-2})	(10^{-4}W m^{-2})
May	1.1	1.7	0.6	0.87
June	0.9	1.5	0.4	0.89
July	0.3	0.88	0.3	0.49
August	0.7	1.74	0.4	0.65
September	—	—	0.2	0.39

Figures 4 and 5 show the time series of current mixing power *F* calculated with equation 2. These calculations were done using the vector sum of the two orthogonal (east and north) components. The monthly averages of *F* at both stations are shown above.

Table 1 summarizes monthly averages of discharge and mixing power. Discharge data used are monthly means of hourly flow calculated from the data shown in Fig. 7. The CTD data of St-Hilaire (1993) were used to calculate typical density values. Examples of CTD profiles from station 2 are given in Figs. 8 and 9. To represent the two-layered model, the top layer was defined as the first 3 m (above the haloclines of Fig. 8 and 9) while the bottom layer was established to be 2 m thick based on typical CTD profiles for the area. Values of ρ_1 were taken as the average of the density of the top layer from all the casts taken during that month at station 2, and ρ_2 is caculated as the average density of the bottom layer from CTD casts at the same site. The average densities (ρ_m) were provided by averaging all CTD casts in a given month, whereas the densities of the two layers at the downstream station were found by doing the monthly average of CTD density data. CTD casts were taken every two weeks during the ice-free season. Variations in density values were greater in months of high discharge (may and June) and during spring tides. In spite of this variations, montlhy means of density appear to be a good representation of the typical conditions in the estuary. The progressive increase of surface density values (ρ_1) from May to August is indicative (which can be visualised by

Table 1. Values of monthly discharge, salinity, density, velocity and $\partial\phi/\partial t$.

Month	May	June	July	Aug.	Sept.
Discharge ($m^3 s^{-1}$)	850	235	81	118	189
S_1	11	13	21	22	23
ρ_1 (kg m^{-3})	1007	1011	1014.5	1016	1015.5
S_2	21	20	22	25	24
ρ_2 (kg m^{-3})	1015	1013	1015	1018	1017
ρ_m (kg/m^{-3})	1008	1007	1012	1011	1011
U_1 (m s^{-1})	0.397	0.148	0.394	0.218	1.028
U_2 (m s^{-1})	-0.312	-0.144	-0.565	-0.288	-1.48
$\partial\phi/\partial t$	.0026	.0005	.0025	.0029	.0152

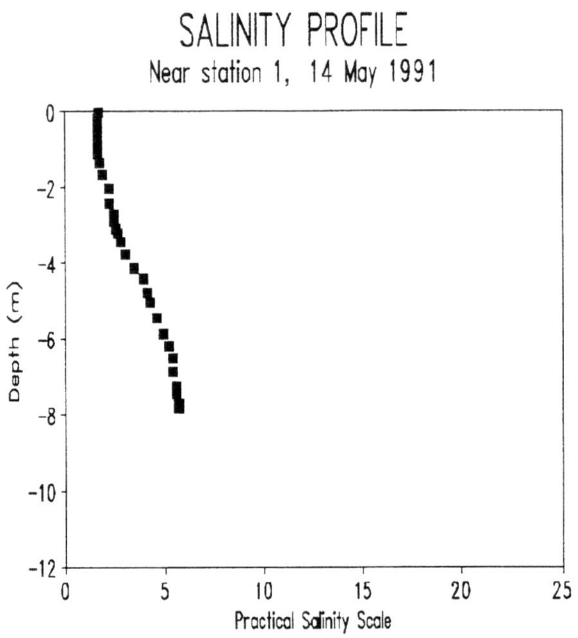

Fig. 10. Salinity profile, 14 May 1991, Station 1.

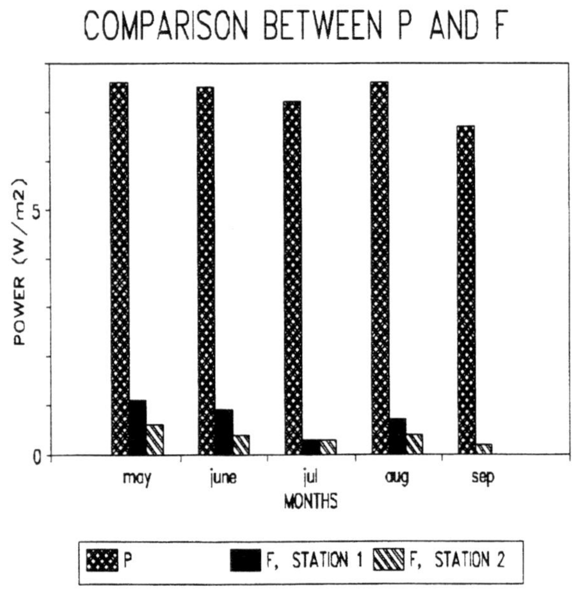

Fig. 11. Comparison between the mixing powers of wind (*P*)and tidal currents (*F*)

comparing values of salinity in the top layer of figs. 8 and 9) of an upstream migration of the salt wedge which is expected when discharge decreases.

The estuary appeared stratified in all months. Values of $\partial\phi/\partial t$ were positive for all months. It should be noted that the calculated surface velocity for the month of september was exceptionally high. Current meter measurements near the bottom seldom exceeded 0.35 m s^{-1} and it is unlikely that the calculated velocity for September is realistic. Orders of magnitude of $\partial\phi/\partial t$ varied between 10^{-2} and 10^{-4}. The minimum was in June. Maximum stratification was in September.

Discussion

The best way to interpret the values for the Miramichi is to compare them with another well-studied shallow estuary of similar size. Mobile Bay, on the south coast of the United States was studied by Shroeder et al. (1990). For winds between 18 and 54 km h^{-1}, mixing power of the wind was between 10^{-5} and 10^{-4} W m^{-2} for Mobile Bay. It should be noted that instantaneous wind speed measurements were used to calculate values of *P* in the Miramichi estuary but not in Mobile Bay . If the vector component of the wind velocity parallel to the main channel was used, values of *P* would have generally been smaller by a factor of 3 to 5.

There was a twofold difference in the magnitude of the tidal mixing power (*F*) for the two stations. This difference was because currents measured at station 1, located in a narrow section of the estuary, were generally greater than those measured at station 2. Schroeder et al. (1990) calculated the mixing power of the fresh water inflow separately. The effect of the fresh water discharge in our model is thought to be mostly to input potential energy at the upstream boundary of the estuary, thereby contributing to the stratification.

By comparison, calculations of *F* in Mobile Bay (Schroeder et al. 1990), were on the order of 10^{-6} to 10^{-5} W m^{-2}, 100 fold less than in Miramichi Bay. These values can only be compared if one takes into account the difference between the root mean square (rms) current values used in the Mobile Bay calculations and the monthly means of *P* (using current vector amplitudes) used in this work. Because root mean square of a pure sinusoidal wave is defined as roughly 70 % of the amplitude, if rms values had been used in our case, values of *F* would be smaller by a factor of 0.8.

A comparison of the mixing power of the wind (*P*) and the tide (*F*) is shown in Fig. 11. It can be seen that the power from the wind-induced mixing is generally 5 to 25 times greater than the tidal-induced power. Again, this is consistent with other shallow estuaries with small tidal amplitudes. Schroeder et al. (1990) also found that wind mixing power typically exceeded tidal mixing power by one order of magnitude. By contrast, on the much deeper continental shelf, winds are important sources of energy only during storm events (Atkinson and Blanton 1986).

The resistance to mixing, $\partial\phi/\partial t$ remained positive throughout the study. A comparison of the magnitudes of the sum of *P* an *F* (Table 2) with $\partial\phi/\partial t$ allows one to see that, for the typical values calculated the estuary tends to get more stratified during the ice-free season. Even if the initial state would be a completely mixed estuary, the fact that $\partial\phi/\partial t$ exceeds the sum of *P* and *F*, for the values shown here implies that the estuary becomes and remains stratified. The mixing by both wind and tides are never sufficient to mix the estuary.

Uncertainties in the wind mixing power may be caused by uncertainties in the friction coefficient and the efficiency factor. Further uncertainties may arise from varying wind fetch, a cross wind being less effective at mixing than a longitudinal wind. This latter effect would reduce the wind mixing power. Topographic guiding of the wind, however, should limit this uncertainty as winds will be steered along the river valley along a northeast-southwest axis (Philpott 1978).

Uncertainties in the tidal mixing power also arise from uncertainties in the friction coefficient (cf) and efficeincy ($\in$) factor, the latter of which depends on stratification through the flux Richardson number (number which quantifies the transfer of kinetic energy to potential energy). Although is is important to choose a friction coefficient appropriate for the height above the bottom, uncertainties in the efficiency factor are probably the greater source of error since they are taken directly from the litterature (Simpson et Bowers 1981).

Uncertainties in $\partial\phi/\partial t$ are more difficult to evaluate. The calculations were done under assumption of steady state conditions, implying that the total mixing power was equal to stratifying power. Although short term events do occur in the estuarine structure (St-Hilaire 1993), this simplification was considered adequate for monthly values. A more serious limitation is probably the inadequacy of the CTD data to estimate density ρ_m and salinity at downstream boundary.

Despite uncertainties, the results suggest that, at the scale of one month, the stratifying power is considerably greater than the mixing power. This is substantiated by the field measurements of Vilks and Krauel (1982) and St-Hilaire (1993). The former study showed that, on the time scale of 1 month, the estuary was always stratified during the ice-free season. Our work shows that even for weak freshwater flows (e.g., mid summer), the stratifying effects are always greater than the mixing power of winds and tidal currents.

These typical values were not necessarily

Table 2. Comparison of mixing power (*P* and *F*) with $\partial\phi/\partial t$.

Month	*P+F* STATION 1 (10^{-4} W/m^2)	*P+F* STATION 2 (10^{-4} W/m^2)	$\partial\phi/\partial t$ (10^{-4} W/m^2)
May	8.7	8.2	26
June	8.4	7.9	50
July	7.5	7.5	25
August	7.4	7.5	29
September	—	8.4	152

representative of the conditions prevailing throughout the ice free season. In fact, surface and the bottom temperatures at stations 1 and 2 were often identical during spring tides, which was indicative of a mixed water column (St-Hilaire et al. 1992). Moreover, salinity data gathered through CTD profiles during the ice-free season of 1991 showed that in the upper portion of the estuary (near station 1) there were occurrences of a very weak salinity gradient (as shown in Fig. 10) mainly due to flushing (St-Hilaire 1993), especially during the spring flood events. Isohaline plots taken at various periods show important migrations of the salt wedge. This confirms that brief events in mixing or sudden migrations of the salt wedge can temporally change the estuarine structure. These short term changes, however, were not included in our energy budget.

In conclusion, the Miramichi estuary is essentially stratified on a monthly time scale, during the ice-free season. Wind is the dominant source of mixing power in the estuary. Stratification of the estuary is not always detrimental to productivity. Although mixing of the water column is an important source of nutrients, Corell (1978) explained that the presence of a salt wedge and its associated counter current may act as a nutrient trap and prevent downstream flushing of phytoplancton and nutrients. A stratified water column may also keep phytoplankton in the photic zone for a longer period. In terms of water quality, there is a lesser chance that sediment-trapped pollution will be recirculated in the water column of a stratified estuary.

Acknowledgements

The authors are endebted to Mr. Bernard Pettigrew and Mr. Roger Pigeon for their work in the field and calibration of current meters. The authors also wish to thank Mr. Dave Moore and Ms. Emily Chin for their help during the sampling season.

References

Atkinson, L., and Blanton, J. 1986. Processes that affect stratification in Shelf Waters. *In* Baroclinic Processes on continental shelves. Am. Geophys. Un.: 117-130.

Bousfield, E.L. 1955. Some physical features of the Miramichi estuary. J. Fish. Res. Board. Can. 12, 3: 342-361.

Bowden, K.F., and Hamilton, P. 1975. Some Experiments with a Numerical Model of Circulation and Mixing in a Tidal Estuary. Estuarine, Coastal and Shelf Science 3: 281-301.

Corell, D.L. 1978. Estuarine Productivity. BioScience 28, 10: 646-650.

Environment Canada. 1991. Monthly records, metorological observations in Eastern Canada, A.E.S., Environment Canada, monthly publications.

Environment Canada. 1990. Historical streamflow Summary: Atlantic Provinces. Inland Waters Directorate, Water Resources Branch, Ottawa, 294p.

Malone, T. 1977. Environmental Regulation of Phytoplankton Productivity in the Lower Hudson Estuary. Estuarine, Coastal Mar. Sci. 5: 157-171.

Philpott, K.L. 1978. Miramichi Channel Study. Public Works Canada, 284 p.

Simpson, J.H., Brown, J., Matthews, J., and Allen, G., 1990. Tidal Straining, Density currents and Stirring Control of Estuarine stratification. Estuaries 13(2): 125-132.

Schoreder, W., Dinnel, S., and Wiseman, W. 1990. Salinity Stratification in a river-dominated estuary. Estuaries, 13(2): 145-154.

St-Hilaire, A. 1993. Étude Hydrodynamique de l'estuaire de la rivière Miramichi durant la saison sans glace-1991. Masters of Applied Sciences Thesis, Université de Moncton, Moncton (N.-B.), 167 p.

St-Hilaire, A., Bettignies, C., Booth, D., and Chadwick, E.M.P. 1992. Tidal Stratification in the Miramichi Estuary, Proceedings of the ASCE Hydraulic conference, Water Forum 92, Baltimore: 953-958.

Stigebrandt, A. 1988. Dynamic Control by Topography in Estuaries. *In* Hydrodynamics of Estuaries, Bjorn Kjerfve, editor: 17-26.

Tee, K.T. 1988. Modeling of Tidally induced Residual Currents. *In* Hydrodynamics of Estuaries, Bjorn Kjerfve, editor: 133-148.

Vilks, G., and Krauel, D. 1982. Environmental Geology of the Miramichi Estuary: Physical oceanography. Geol. Surv. Can. Pap. 81-24: 53 p.

Willis, D. 1990. Miramichi Estuary-Physical Oceanography. *In* Review of impacts of channel dredging on biological resources of the Miramichi estuary. Acadia Center for Estuarine Research: 1-9.

CHAPTER 6

Hydrology of the Miramichi River Drainage Basin

D. Caissie

Department of Fisheries and Oceans, P.O. Box 5030,
Moncton, New Brunswick, Canada E1C 9B6

and N. El-Jabi

Université de Moncton, Moncton, New Brunswick, Canada E1A 3E9

Caissie, D., and El-Jabi, N. 1995. Hydrology of the Miramichi River drainage basin, p. 83-93. *In* E.M.P Chadwick [*editor*]. Water, science, and the public: the Miramichi ecosystem. Can. Spec. Publ. Fish. Aquat. Sci. 123.

Abstract

Hydrological data are important for the management of aquatic and water resources as the hydrologic cycle affects not only the stream biota but also the design and operation of water resource infrastructures. The present study provides some hydrological characteristics of the Miramichi River basin including data on air temperatures, precipitation, and river discharge. Also, possible ecological implications of streamflow variation and extreme events in the Miramichi River basin are presented in the discussion.

The Miramichi River drainage basin is coldest in January with a mean monthly temperature of -11.8°C and warmest in July with a mean monthly temperature of 18.8°C. Between these two extremes, gradual changes in monthly temperatures occur. This region of New Brunswick can receive between 860 mm and 1365 mm of annual precipitation with a long term average annual value of 1130 mm. The precipitation is evenly distributed throughout the year at approximately 100 mm of precipitation monthly.

Discharge (runoff) characteristics of the Miramichi River can be studied using five existing gauged drainage basins within the Miramichi system (Barnaby, Renous, Northwest Miramichi, Little Southwest Miramichi, and Southwest Miramichi River). The average annual runoff (or discharge per drainage) was estimated at 714 mm which represents 63% of total precipitation (1130 mm) and therefore a water consumption by plant or storage in aquifers (groundwater) of 416 mm or 37%. The median flow, which is the flow equalled or exceeded 50% of the time on the annual flow duration curve, was estimated at approximately 1/2 of the mean annual runoff (e.g., 714/2 = 357 mm) for the region. High flows (floods) for the Miramichi drainage basin were estimated at 16.3 mm for a 2-year flood and 46.4 mm for a 100-year flood. The value of 16.3 mm represents approximately 18.9 m^3/s for a drainage basin of 100 km^2 (or 0.189 m^3/s per km^2 of drainage basin area). Higher return floods were expressed as a factor of 16.3 mm (2-year flood) by the index flood method. For instance the 5, 25, 50 and 100-year flood were approximately 1.5, 2.0, 2.5, and 3.0 times the 2-year flood, respectively. Low flows were estimated at 0.24 mm for the Miramichi basin or 0.28 m^3/s for a basin of 100 km^2. The higher return low flows can also be expressed as a ratio of the 2-year low flow however basin size was noted to be an important factor. The calculated values are 1/2, 1/3, and 1/5 of the 2-year low flow for a 10, 25, and 50-year low flow, respectively.

Résumé

Les données hydrologiques jouent un rôle important dans la gestion des ressources aquatiques et hydriques, car le cycle de l'eau affecte non seulement la biologie d'un cours d'eau mais aussi la conception et le fonctionnement des infrastructures d'exploitation des ressources en eau. La présente étude décrit certaines caractéristiques hydrologiques du bassin de la Miramichi, avec notamment des données sur la température de l'air, les précipitations et le débit. Nous présentons aussi dans l'analyse les incidences écologiques possibles des variations du débit et des événements à caractère extrême qui se produiraient dans le bassin de la Miramichi.

C'est en janvier que la température dans le bassin hydrographique de la Miramichi est la plus basse, avec une moyenne mensuelle de −11,8 °C, et en juillet qu'elle est la plus haute avec une moyenne mensuelle de 18,8 °C. Entre ces deux extrêmes, on observe des changements graduels de la température d'un mois à l'autre. Cette région du Nouveau-Brunswick peut recevoir entre 860 et 1 365 mm de précipitations annuelles, avec une moyenne annuelle à long terme de 1 130 mm. Les précipitations se répartissent également tout au long de l'année, avec une moyenne d'environ 100 mm par mois.

On peut étudier les caractéristiques de l'écoulement (débit) de la Miramichi grâce aux cinq bassins instrumentés qui se trouvent dans son réseau (Barnaby, Renous, Northwest Miramichi, Little Southwest Miramichi et Miramichi Southwest). On a estimé le débit annuel moyen (ou l'écoulement par bassin) à 714 mm, ce qui représente 63 % des précipitations totales (1 130 mm) et donc une consommation d'eau par les plantes ou par stockage dans les nappes souterraines de 416 mm, soit 37 %. L'écoulement médian, c'est-à-dire celui qui est atteint ou dépassé 50 % du temps sur la courbe annuelle des débits classés, a été estimé à environ la moitié du débit annuel moyen (714/2 = 357 mm) pour la région. On a estimé l'écoulement de crue pour le bassin de la Miramichi à 16,3 mm pour une crue bisannuelle et à 46,4 mm pour une crue centenaire. Cette valeur de 16,3 mm représente environ 18,9 m^3/s pour un bassin de drainage de 100 km^2 (ou 0,189 m^3/s par km^2 de surface de bassin). Les crues dont la période de récurrence est supérieure à deux ans ont été exprimées sous la forme d'un facteur de 16,3 mm (crue bisannuelle) par la méthode indice crue. Par exemple, les crues de 5, 25, 50 et 100 ans correspondaient respectivement à environ 1,5, 2,0, 2,5 et 3,0 fois la crue bisannuelle. On a estimé l'écoulement d'étiage à 0,24 mm pour le bassin de la Miramichi, soit 0,28 m^3/s pour un bassin de 100 km2. Les débits d'étiage dont la récurrence est supérieure à deux ans peuvent aussi être exprimés par rapport au débit d'étiage bisannuel, mais on note que la taille du bassin est un facteur important. Les valeurs obtenues correspondent à 1/2, 1/3 et 1/5 du débit d'étiage bisannuel, respectivement, pour les débits d'étiage de 10, 25 et 50 ans.

Introduction

Environmental factors, including hydrological processes, have major influences on our fisheries and aquatic resources. Hydrological processes such as mean flows, rainfall events, baseflow periods, etc., not only affect stream biota but also the effective operation of the water supply. Extreme events also play an important role in hydrological processes, in the design of water resource infrastructures, and for the management of fisheries resources.

From a fisheries management perspective, hydrological events such as floods have often been identified as having important impacts on fish (Elwood and Waters 1969; Erman et al. 1988). Similarly, low flows can affect fish movement and stream water temperature (Cunjak et al. 1993; Edwards et al. 1979). In order to increase our understanding of water resources in the Miramichi River, we need to better understand the hydrology of its drainage basin.

A regional hydrological analysis was carried out using meteorological and hydrometric data as the data base. The hydrometric data came from gauged streams and rivers in the Miramichi River basin. Water resources in the region were studied by calculating mean annual runoff of rivers (average flow divided by drainage area) and flow duration data. The flow duration analysis provides information on the percentage of time of the year that a certain amount of water (or discharge) is present or equalled in the watercourse. Flood flows and low flows were each fitted to a statistical distribution function in a frequency analysis to estimate the T-year events (Kite 1978). For instance, the 25-year ($T = 25$) low flow is a low flow which occurs on the average every 25 years so that 4 such events could have occurred in the last 100 years. Calculated flows (mean, high and low) are not only important for gauged rivers, but also to establish regional trends for the estimation of discharge of ungauged basins.

The objective of the present study is to provide regional hydrological information on the Miramichi River drainage basin for aquatic and water resource management. The specific objectives are to: (a) provide air temperature and precipitation information for the region; (b) determine the mean annual runoff of gauged hydrometric stations; (c) carry out a flood and low flow frequency analyses; and d) study the streamflow availability by undertaking a flow duration analysis.

Study Region

The Miramichi River drainage basin covers approximately 14 000 km^2 (Fig. 1). The most central meteorological station in the Miramichi River basin is situated at the Department of Natural Resources and Energy ranger station near McGraw Brook. Therefore, this station was used for temperature and precipitation data analysis. Data available from this station covered the period from 1969 to 1991.

Mean monthly air temperature varied at McGraw Brook between –11.8°C in January to 18.8°C in July. Between these two extreme months the mean air temperature changed gradually from month to month with 7 months of the year having above freezing mean air temperature (April to October; Fig. 2). The difference between the mean monthly minimum and the mean monthly maximum was found to be approximately 13°C for winter and summer monthly extremes (coldest and warmest months) while during the fall and spring the difference was smaller at approximately 10°C in the fall and 11°C in the spring (Fig. 2).

The annual precipitation recorded at McGraw Brook ranged from 860 to 1365 mm

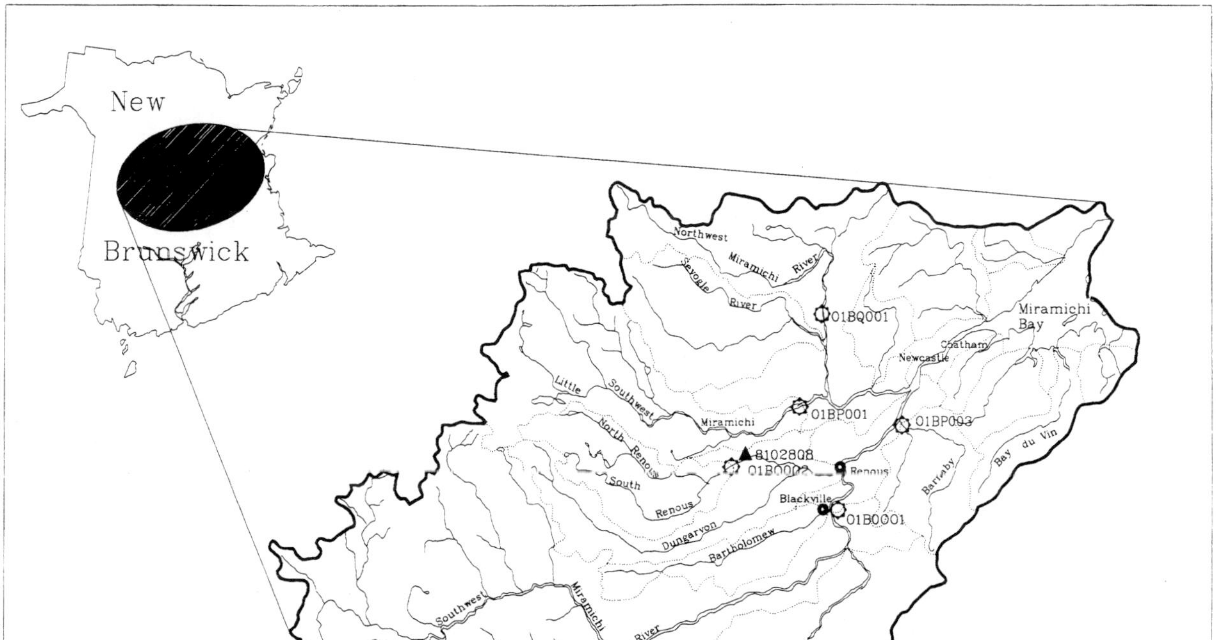

Fig. 1. Miramichi River drainage basin showing the location of the hydrometric and meteorological gauges.

since 1969 with a long-term average of 1130 mm for the period of record. On a monthly basis the precipitation was calculated at approximately 100 mm per month (Fig. 3). February showed lower precipitation at 71.7 mm compared with other months but this is partially explained by this being a short month. For the period of record (1969–1991), the observed monthly extremes were January 1970 at only 7.1 mm for the minimum and August 1991 at 245 mm for the maximum. During August of 1991, over 100 mm of rain fell in 14 hours coincident with the passing of Hurricane Bob in the Miramichi River region as reported by Cunjak et al. (1993).

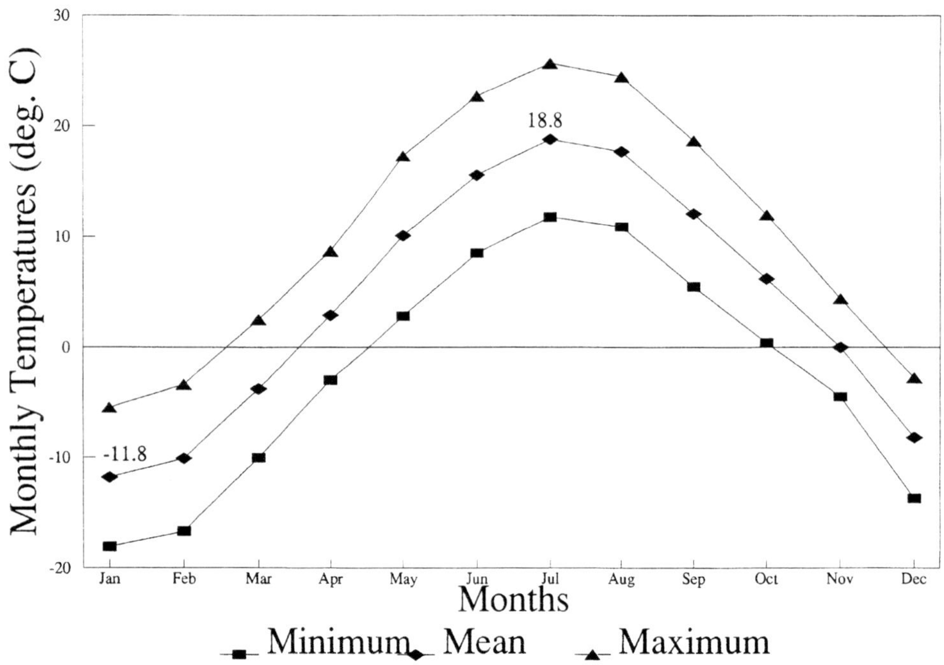

Fig. 2. Monthly air temperature (deg. C) in the Miramichi River drainage basin.

Table 1. List of hydrometric and meteorological stations used in for the hydrological analysis of the Miramichi River drainage basin. (*N* = number of years of record).

Hydrometric Stations

River	Station	Lat.	Long.	Area (km^2)	*N*
Barnaby	01BO003	45° 53' 19"	65° 35' 44"	484	19
Renous	01BO002	46° 49' 17"	66° 06' 53"	611	26
Northwest	01BQ001	47° 05' 41"	65° 50' 14"	948	30
L. Southwest	01BP001	46° 56' 09"	65° 54' 26"	1340	40
Southwest	01BO001	46° 44' 10"	65° 49' 36"	5050	44

Meteorological Station

River	Station	Lat.	Long.	Altitude (m)	*N*
McGraw Brook	8102808	46° 49'	66° 07'	53	23

Mean Annual Runoff

To establish flow conditions in the Miramichi River basin, five hydrometric gauges were used (Table 1) which are part of the Environment Canada stream gauging network (Environment Canada 1990). These gauged basins were on Barnaby River, Renous River, Northwest Miramichi River, Little Southwest Miramichi River and Southwest Miramichi River (Fig. 1). The drainage areas of the analyzed gauged rivers ranged from 484 km^2 for the Barnaby River to 5 050 km^2 for the Southwest Miramichi River (Table 1).

Figure 4 presents the runoff characteristics by rivers including the mean annual runoff and the median runoff. The mean annual runoff provides information on the discharge per unit area expressed in mm as opposed to a volume per unit of time (e.g., m^3/s). Runoff characteristics in mm provide a good method for comparing between

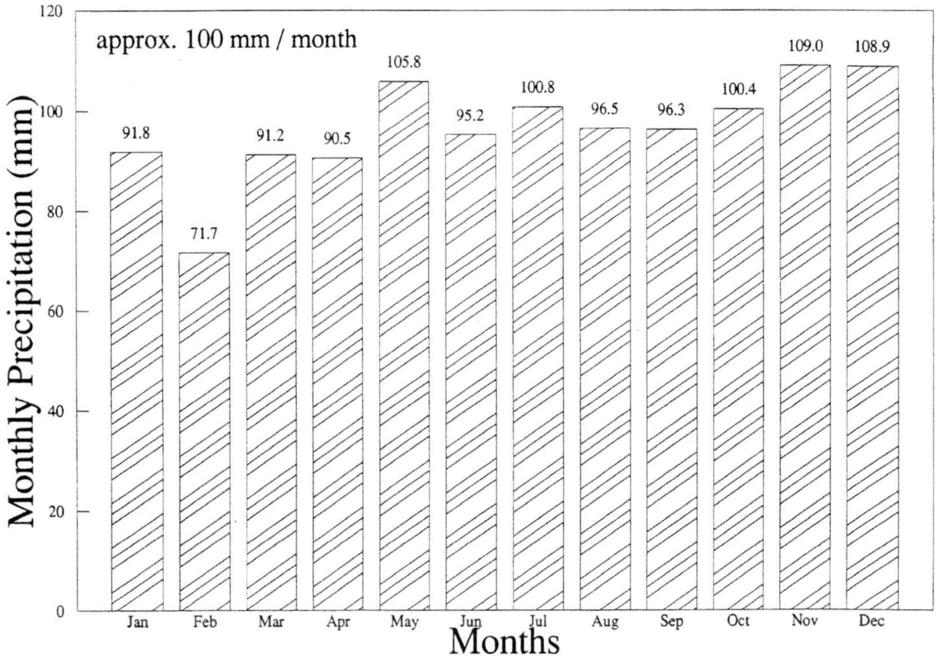

Fig. 3. Monthly precipitation (mm) for the Miramichi River drainage basin.

Table 2. Results of frequency analysis of annual maximum daily runoff (mm) for the Miramichi River drainage basin for different recurrence intervals (*T*) in years. Index of flood represents a ratio of Q_T/Q_2 .

	Recurrence intervals (*T*) in years					
River	*T* = 2	5	10	25	50	100
Barnaby	17.2 mm	23.8	27.9	31.7	36.4	39.9
Renous	18.7	26.8	32.5	38.1	45.6	51.5
Northwest	16.6	24.9	30.5	36.0	43.2	48.7
L. Southwest	15.1	24.2	30.8	37.7	47.2	54.8
Southwest	13.9	19.7	23.7	27.7	32.9	37.0
Mean	16.3 mm	23.9	29.1	34.2	41.1	46.4
Index of flood	1.0	1.5	1.8	2.1	2.5	2.8

basins of different sizes. The mean annual runoff in the studied area ranged from 631 mm (Barnaby River) to 763 mm (Little Southwest Miramichi River). The average annual runoff was calculated at 714 mm which represents approximately 63% of the annual precipitation (1130 mm). The difference between precipitation and runoff (i.e., 416 mm) represents water used by evapotranspiration (evaporation and transpiration, i.e., consumption by plants) with the assumption that changes in groundwater storage over the years is negligible.

Also calculated was the median runoff, which is the river flow available or exceeded 50% of time on the flow duration curve. The flow duration analysis consist of ranking the flows and calculating a probability that such flow will be exceeded. The discharge showing a 50% chance of being exceeded is referred to the median flow. This discharge was calculated using a flow duration analysis software package FLODUR (Caissie 1991). The median runoff in the Miramichi River basin ranged from 248 mm to 400 mm (Fig. 4). It was observed that the median runoff in this region of New Brunswick represents approximately half of the mean annual runoff. As with the mean annual runoff, the Little Southwest Miramichi showed the highest median flow per unit of area; the Barnaby river showed the lowest median runoff. The results showed that more precipitation from the Little Southwest Miramichi River basin makes its way to the stream with a higher runoff. This could be explained by topography and/or type of forest cover. For the whole Miramichi River basin (at approximately 14 000 km^2) the mean annual discharge was estimated at 317 m^3/s with a median flow of approximately 158 m^3/s based on the above calculated mean and median average runoff for the studied region.

Flood Frequency Analysis

A flood frequency analysis was carried out for the Miramichi River basin using the above five hydrometric or gauged stations. For this analysis, the 3-parameter lognormal distribution function fitted to annual maximum daily runoff was used (Kite 1978) and the results are shown in Table 2. For comparative purposes on a regional basis, discharges were expressed in runoff (mm). The larger drainage basin showed a lower flood runoff with 13.9 mm for the Southwest Miramichi River compared with 18.7 mm for the Renous River. This was expected as smaller basins have a faster response to rainfall events and also less channel

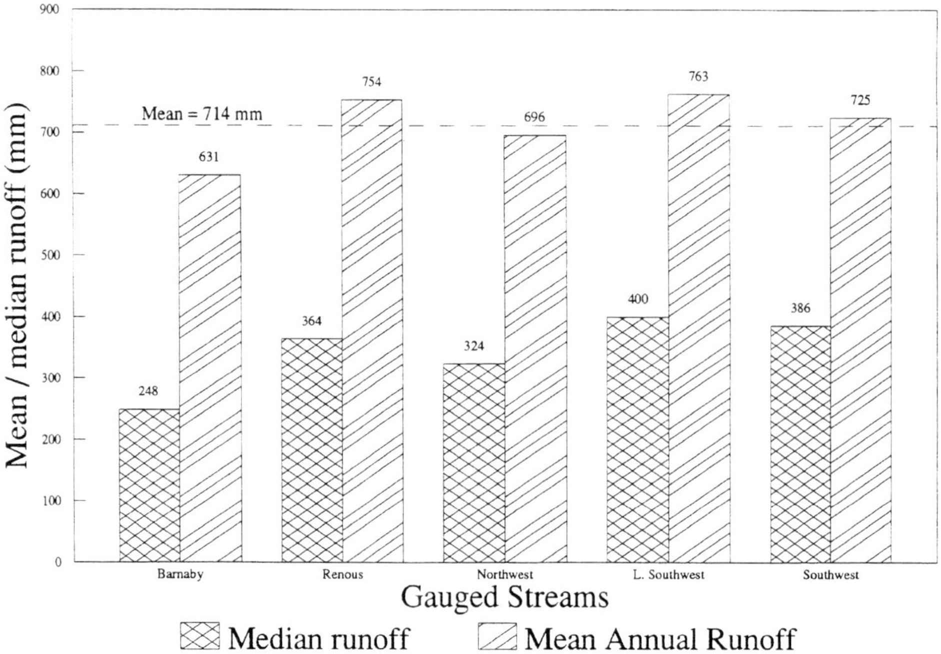

Fig. 4. Mean annual runoff (mm) and median runoff (mm) of studied hydrometric stations.

storage than bigger river systems. In contrast, the smaller Barnaby River showed a different behaviour with high flow values less than the larger Renous River. This could be due to the location of the Barnaby River which is closer to the coast and also the different topography within the respective basins. It was noted that for higher return floods (50 and 100 years), the Little Southwest Miramichi River showed the highest flood runoff at 47.2 mm and 54.8 mm for a 50-year and 100-year flood, respectively (Table 2). The Southwest Miramichi River showed the lowest flood runoff of all studied basins at 37.0 mm. Therefore, for a given rainfall and/or snowmelt event and subsequent flood event, comparatively more water will become streamflow in the Little Southwest Miramichi River than would be the case in all of the other gauged streams in the Miramichi River.

On a regional basis, for the Miramichi River area, the mean 2-year flood was estimated at approximately 16.3 mm of daily runoff (Table 2). A flood runoff of 16.3 mm in a day represents approximately a discharge of 18.9 m^3/s for a drainage basin of 100 km^2. Therefore, flood discharge of ungauged basins of different sizes can be calculated using the mean values of flood runoff in Table 2. The 50-year and 100-year floods for the region were estimated at 41 mm and 46 mm respectively (Table 2).

Low return floods such as the 2 to 5-year floods are more accurately estimated by the flood frequency analysis as these events occur more often. Although estimates of higher return flood are made, these events do not occur frequently and therefore the error associated with their estimate is greater than for lower return floods. This is especially true for estimates of high return floods (such as 100-year) with fewer than 30 years of record on a particular stream.

Often in flood frequency analysis and on a regional basis, floods are not only expressed as flood runoff but also as index of floods. The index of flood is calculated as the ratio of the T-year flood event to the 2-year flood (Q_T/Q_2; Caissie and El-Jabi 1991). Values of index of floods for the Miramichi River were calculated and are shown in Table 2. Given the 2-year flood estimate of any rivers in the region, the flood of a recurrence interval of 25 years, for instance, is 2.1 times the 2-year flood (i.e., $Q_{25} = Q_2 * 2.1$; Table 2). As a rule of thumb for the region, the index of flood factor for the 5, 25, 50, and 100-year floods can be estimated at 1.5, 2.0, 2.5 and 3.0.

Low-flow Frequency Analysis

A low flow frequency analysis was also carried out for the Miramichi River basin using gauged rivers. The minimum daily discharges on an annual basis were used for the analysis and fitted to a Type III extremal distribution function (Kite 1978). The estimated 2-year low flow runoff for the studied area is presented in Table 3. Results show that the 2-year low flow increased directly

Table 3. Results of low flow frequency analysis for the Miramichi River drainage basin for different recurrence intervals (T) in years. Discharges are expressed in daily runoff (mm).

	Recurrence intervals (T) in years				
River	$T = 2$	5	10	25	50
Barnaby	0.102 mm	0.057	0.039	0.026	0.014
Renous	0.191	0.110	0.075	0.050	0.028
Northwest	0.260	0.183	0.137	0.096	0.049
L. Southwest	0.324	0.214	0.160	0.118	0.076
Southwest	0.337	0.238	0.187	0.147	0.104
Mean	0.24 mm	0.16	0.12	0.087	0.054

with drainage basin area. Although Barnaby River showed a different flooding response to other basins in the region, low flow conditions were similar and were used in the comparative analysis. In fact, Barnaby River showed the lowest 2-year low flow at 0.102 mm while Southwest Miramichi River, the biggest gauged drainage basin, showed a value of 0.337 mm (3 times higher; Table 3). The higher low flow runoff for larger river systems is mainly due to the greater channel storage and longer response time. The variability in low flow runoff between rivers of different drainage sizes was noticed to be higher than during high flows. The variability is also more important for higher return (e.g., T = 50 years) low flows especially when comparing Barnaby River (0.026 mm) with Southwest Miramichi River (0.147 mm), which was 5.7 times higher for Southwest Miramichi River (Table 2).

The mean low flow runoff for the region was estimated at 0.24 mm for a recurrence interval of 2 years (or approximately 0.28 m^3/s for a 100 km^2 drainage basin). For the 50-year low flow, the mean value was calculated at 0.054 mm although the variability was significant. In the case of the low flow analysis, the drainage basin area is an important factor and observed low flows can differ from the mean low flow runoff value depending on basin size as shown in Table 3. As a rule of thumb for low flow analyses in the Miramichi River basin, the 10-year low flow is approximately 1/2 of the 2-year low flow while the 25-year and 50-year are approximately 1/3 and 1/5 of 2-year low flow respectively.

Discussion

The mean annual runoff for the Miramichi River basin was found to be comparable to that for similar size basins elsewhere in the province. This was consistent for flood runoff in most New Brunswick (NB) rivers except for those located in the southern part of the province or the Bay of Fundy region (Acres Consulting Services Ltd. 1977) which can be twice that observed in the Miramichi River area. For instance, the 100-year flood runoff for Lepreau River was estimated at 100 mm compared to 55 mm for Little Southwest Miramichi River, the highest runoff for the Miramichi River Basin. The corresponding 100-year index flood ratio for the Little Southwest Miramichi was computed (i.e., Q_{100}/Q_2) at 5.4, the provincial maximum (Acres Consulting Services Ltd. 1977). The flood index calculated in the present study was 3.6 and similar to the value of 3.5 calculated by New Brunswick Department of Municipal Affairs and Environment (1987). This value, 3.6, is still significant for the region given that the second highest value was 2.9 for Northwest Miramichi River (Table 2). The discrepancy between low and high return floods for Little Southwest Miramichi River could be attributed to a modified topography or a different snowmelt-runoff process.

The results of the low flow frequency analysis for the Miramichi River basin were similar to those observed for other regions of the province. Low flow in rivers is highly dependant on factors such as swamp areas, lakes, groundwater discharges, and other physical characteristics which are somewhat variable within a particular region. More variability is thus found in the estimation of low flows than for flood flows. This was true for the Miramichi River basin with a coefficient of variation of 0.67 (67%) for a 50-year low flow (Table 3). However, smaller river systems generally show more severe low flows (lower runoff) as observed in the Miramichi River basin.

Ecological implication of streamflow and streamflow variability on aquatic resources is of importance (Poff and Ward 1989). Extreme event conditions such as high and low flows seem to have a greater impact on these resources than do average flow conditions. For instance, there is some evidence that winter floods reduced young-of-the-year (YOY) fish as noted by corpses and significantly decreased summer electrofishing success for that age class (Erman et al. 1988). Although YOY fish are most affected by flooding, the population of larger fish can also be reduced. For instance, Jowett and Richardson (1989) found that the abundance of small brown trout (Salmo trutta) was reduced by 90–100% in a comparison of 7 rivers. Population of medium size fish was reduced by 62–87% while large fish experienced a reduction of approximately 26–57%. Factors implicated ranged from reduced visibility due to higher suspended sediments, lack of low velocity refuge sites and others. Erman et al. (1988) found that the dominant factor in fish mortality during floods was the large-scale bed load movement.

Other studies have shown that changes in water quality parameters associated with floods can have adverse physical effects on fish and on fish population. These changes include an increase in aluminum concentration that occurs during the spring pH depression associated with high snowmelt runoff, as reported by Campbell et al. (1992). Monitoring has shown that during spring high flows, pH can reach lows of 5.3 pH units in the Little Southwest Miramichi River (Cunjak et al. 1993).

Also, during a particular summer rainfall event, a decrease of 0.7 pH unit (7.5 – 6.8) in one day was monitored at Catamaran Brook (a small stream within the Miramichi River Basin) at a time of year when snowmelt runoff was no longer a contributing factor (Caissie et al. 1994). Suspended sediment concentrations can also be a factor affecting fish during spring floods and breakups. Few studies have looked at the effect of high flows and suspended sediment concentrations on aquatic biota, however, the effect of suspended sediment concentration on fish and aquatic resources has been studied both in the field (McCrimmon 1954; Everest et al. 1987; Redding et al. 1987) and in the laboratory (Dakins 1965; Wickett 1954; Cooper 1965). According to Hynes (1973), the upper tolerable level of suspended sediment is 80 mg/L. Any amount greater than 80 mg/L is " . . . bound to have adverse biological consequences . . . ". Suspended sediment concentration is also well studied during ice breakup and other high flow events. Summer high flow events have been monitored for Catamaran Brook and results show that suspended sediments can reach 37 mg/L during such events (Cunjak et al. 1993). It is expected that higher concentrations would be measured during spring floods as other studies have shown. In fact, during the spring breakup of 1994 in the Saint John river system, suspended sediment concentrations of between 100 and 150 mg/L were monitored (S. Beltaos, pers. comm.). He (S. Beltaos) also monitored brief concentration spikes of between 300 and 800 mg/L during surges caused by ice jam releases. Ice jams can often be associated with massive sediment transport as was noted for the Little Southwest Miramichi River during the breakup of 1994 (D. Caissie, pers. obser.) In this instance an ice jam created high flows on a road adjacent to the Little Southwest Miramichi River and resulted in considerable erosion. Studies have also looked at the suspended sediment concentration during high flow or ice breakup (Prowse 1993; Williams 1989). It is believed that during thermal ice breakup (breakup without much potential for ice jams) the suspended sediment is about twice that experienced during similar open-water flow conditions (Milburn and Prowse 1994). It is also estimated that during a dynamic or mechanical breakup (ice broken by the force of water which often results in ice jams) the concentration of suspended sediment can be three times the value observed during a thermal breakup. This was clearly illustrated when Milburn and Prowse (1994) monitored suspended sediment concentrations of 1067 mg/L during such an event. Increases in trace metal concentration, which are often associated with these events, can have a direct detrimental effect on the aquatic resources (Milburn and Prowse 1994). They found a strongly correlated level of aluminium (Al), Ion (Fe), Zinc (Zn), and copper (Cu) with suspended sediment concentrations.

Low flow conditions can also effect fisheries and aquatic resources directly. In the Miramichi River basin, such conditions can occur during winter. The Little Southwest Miramichi River recorded the occurrence of 50% of its annual minimum daily discharge between January and March. The other low flow season in the Miramichi River basin is during August and September with the latter having a higher frequency of low flows.

Frenette et al. (1984) found that late winter low flows explained well the variation in the population of juvenile Atlantic salmon (*Salmo salar*). Low flow conditions in combination with cold winter have presumably affected salmonid population as reported by Chadwick (1982) and by Cunjak and

Randall (1993). Egg mortality could have resulted due to freezing of redds and low water may have limited movement of fish. Although many studies have implied low water conditions to potential problems for winter survival, physical conditions and their biological implication remains a complex issue (Cunjak 1994).

Summer low flows can certainly affect fish populations as the low water conditions are often associated with high stream water temperatures. Low concentration of dissolved oxygen (DO) is the most commonly encountered problem during periods of high stream water temperature. Stream water temperatures in the range of 23–25°C have been observed to affect salmonid mortality directly (Lee and Rinne 1980; Bjornn and Reiser 1991). During these high stream water temperatures, instream fish movements into colder small tributary streams have been observed (Cunjak et al. 1993). This is most problematic in the Miramichi River basin during late July and early August. This time of year typically experiences the highest stream water temperatures and greatest daily variation (D. Caissie, unpubl. data). For instance, stream water temperatures can vary by as much as 7°C during one day. This variation was observed both in small streams (e.g., Catamaran Brook) as well as larger rivers (e.g., Little Southwest Miramichi River).

Low flow conditions in the fall tend to cool larger and more exposed rivers such as Little Southwest Miramichi more rapidly than smaller streams. It is only at this time of year that those larger river systems experience colder stream water temperature than smaller streams. Another potential impact factor of low water conditions during the fall is for fish migration. Migration of spawners can be affected during fall by culverts, beaver dam, natural falls, and other obstacles that could prevent upstream fish migration at low flow. For instance, many spawners at Catamaran Brook during 1994, were limited to the first km of a 20 km stream as a result of a beaver dam and low water conditions on the main stem of the brook.

Acknowledgements

The authors would like to thank R.A. Cunjak for reviewing the manuscript and J.H. Conlon for his graphic assistance during the preparation of the paper. Many thanks are also offered for the provision of data by Environment Canada.

References

Acres Consulting Services Ltd. 1977. Regional flood frequency analysis, Canada–New Brunswick Flood Damage Reduction Program, 141p.

Bjornn, J.R., and Reiser, D.W. 1991. Habitat requirements of salmonids in streams, pp. 83–138. *In* Influences of Forest and Rangeland Management on Salmonid Fishes and their Habitats. Amer. Fish. Soc. Spec. Publ. 19.

Caissie, D. 1991. A computer software package for instream flow analysis by the flow duration method. Can. Tech. Rep. Fish. Aquat. Sci. 1812: 21p.

Caissie, D., and El-Jabi, N. 1991. A stochastic study of floods in Canada: frequency analysis and regionlization. Can. J. Civil Eng. 18: 225–236.

Caissie, D., Pollock, T.L, and Cunjak, R.A. 1994. Variation in stream water chemistry and hydrograph separation in a small drainage basin. J. Hydrol. 39p. (submitted).

Campbell, P.G.C., Hansen, H.J., Dubreuil, B., and Nelson, W.O. 1992. Geochemistry of Quebec North Shore salmon rivers during snowmelt: organic acid pulse and aluminum mobilization. Can. J. Fish. Aquat. Sci. 49: 1938–1952.

Chadwick, E.M.P. 1982. Stock-recruitment relationship for Atlantic salmon (*Salmo salar*) in Newfoundland rivers. Can. J. Fish. Aquat. Sci. 39: 1496–1501.

Cooper, A.C. 1965. The effect of transported stream sediment on the survival of Sockeye and Pink salmon eggs and alevins. Int. Pac. Salmon Fish. Comm. Bull. 18: 71p.

Cunjak, R.A., Caissie, D., El-Jabi, N., Hardie, P., Conlon, J.H., Pollock, T.L., Giberson, D.J., and Komadina-Douthwright, S.. 1993. The Catamaran Brook (New Brunswick) Habitat Research Project: Biological, Physical and Chemical Conditions (1990-1992). Can. Tech. Rep. Fish. Aquat. Sci. 1914: 81p.

Cunjak, R.A., and Randall, R.G. 1993. In-stream movements of young Atlantic salmon (*Salmo salar*) during winter and early spring, p.43–51. *In* R.J. Gibson and R.E. Cutting [*editors*]. Production of juvenile Atlantic salmon, Salmo salar, in natural waters. Can. Spec. Publ. Fish. Aquat. Sci. 118.

Cunjak, R.A. 1994. Winter ecology of juvenile Atlantic salmon (*Salmo salar* L.) in running waters: A review.

In R.A. Kazakov [*editor*]. The Atlantic Salmon, St. Petersburg, Russia (accepted).

Daykin, P.N. 1965. Application of mass transfer theory to the problem of respiration of fish eggs. J. Fish. Res. Board Can. 22(1): 159-171.

Edwards, R.W., Densen, J.W., and Russell, P.A. 1979. An assessment of the importance of temperature as a factor controlling the growth rate of brown trout in streams. J. Anim. Ecol. 48: 501–507.

Elwood, J.W., and Waters, T.F. 1969. Effects of flood on food consumption and production rates of a stream brook trout population. Trans. Amer. Fish. Soc. 98: 253–262.

Environment Canada. 1990. Historical streamflow summary: Atlantic Provinces. Inland Waters Directorate, Water Resources Branch, Ottawa, Ont. 294p.

Erman, D.C., Andrews, E.D., and Yoder-Williams, M. 1988. Effects of winter floods on fishes in the Sierra Nevada. Can. J. Fish. Aquat. Sci. 45: 2195–2200.

Everest, F.H., Beschta, R.L, Scrivener, J.C., Koski, K.V., Sedell, J.R., and Cederholm, C.J. 1987. Fine sediment and salmonid production: a paradox. *In* E.O. Salo and T.W. Cundy [*editors*]. Streamside management: forestry and fishery interactions. University of Washington, Seattle, Washington, Contribution No. 57: 98-142.

Frenette, M., Caron, M., Julien, P., and Gibson, R.J. 1984. Interaction entre le débit et les populations de tacons (*Salmo salar*) de la rivière Matamec, Québec. Can. J. Fish. Aquat. Sci. 41: 954–963.

Hynes, H.B.N. 1973. The effects of the sediment on the biota in running waters, pp. 652–663. *In* Proceedings of Hydrology Symposium on Fluvial Processes and Sedimentation, University of Alberta, Edmonton, Alberta, Canada.

Jowett, I.G., and Richardson, J. 1989. Effects of a severe flood on instream habitat and trout populations in seven New Zealand rivers. N.Z. J. Mar. Freshwater Res. 23: 11–17.

Kite, G.W. 1978. Frequency and risk analysis in hydrology. Water Resources Publications, Fort Collins, CO. 224p.

Lee, R.M., and Rinne, J.N. 1980. Critical thermal maxima of five trout species in the southwestern United States. Trans. Amer. Fish. Soc. 109(6): 632–635.

McCrimmon, H.R. 1954. Stream studies on planted Atlantic salmon, J. Fish. Res. Board Can. 11(4): 362–403.

Milburn, D., and Prowse, T.D. 1994. Observation on sediment–chemistry interactions during northern river breakup. Proceedings of the Workshop on Environmental Aspects of River Ice, Saskatoon, Sask., Aug. 18–20, 1993, pp. 21–41.

New Brunswick Department of Municipal Affairs and Environment. 1987. Flood frequency analyses — New Brunswick. A guide to the estimation of flood flows for New Brunswick rivers and streams. 49p.

Prowse, T.D. 1993. Suspended sediment concentrations during river ice breakup. Can. J. Civil Eng. 20: 872–875.

Poff, N.L., and Ward, J.V. 1989. Implication of streamflow variability and predictability for lotic community structure: A regional analysis of streamflow patterns. Can. J. Fish. Aquat. Sci. 46: 1805–1818.

Redding, J.W., Schreck, C.B, and Everest, F.H. 1987. Physiological effects of exposure to suspended solids in steelhead trout and coho salmon. Trans. Amer. Fish. Soc. 116: 737–744.

Wickett, W.P. 1954. The oxygen supply to salmon eggs in spawning beds. J. Fish. Res. Board Can. 11(6): 933–953.

Williams, G.P. 1989. Sediment concentration versus water discharge during single hydrologic events in rivers. J. Hydrol. 111: 89–106.

THE BIOLOGICAL ENVIRONMENT

"The Miramichi region has the most valuable fishery for smelts in the world."

Dr. A.J. Huntsman, 1918
quoted in The Commercial and the World, January 18, 1945

Fish traps like this one operated by the Department of Fisheries and Oceans near the mouth of the Southwest Miramichi River have been used since 1954 to estimate the abundance of migratory anadromous fish, principally Atlantic salmon.

CHAPTER 7

Ichthyoplankton and Invertebrate Zooplankton of the Miramichi Estuary: 1918-1993

A. Locke and S.C. Courtenay

Department of Fisheries and Oceans, Science Branch, P.O. Box 5030, Moncton, New Brunswick, Canada E1C 9B6

Locke, A., and Courtenay, S.C. 1995. Ichthyoplankton and invertebrate zooplankton of the Miramichi Estuary: 1918-1993, p. 97-120. *In* E.M.P. Chadwick [*editor*]. Water, science, and the public: the Miramichi ecosystem. Can. Spec. Publ. Fish. Aquat. Sci. 123.

Abstract

This document summarizes the results of studies of ichthyoplankton (1918, 1948-55, 1979-80, 1981, 1992-93) and invertebrate zooplankton (1951, 1992-93) in the Miramichi Estuary. Species diversity of the Miramichi is similar to that recorded for other Atlantic estuaries, with 33 taxa of larval fish and 75 invertebrate taxa recorded. Numerically dominant ichthyoplankton taxa are anadromous – rainbow smelt (*Osmerus mordax*) and two species of gaspereau (*Alosa aestivalis, A. pseudoharengus*). The location of spawning and early larval development of these species in the upper estuary near Newcastle places them at risk from industrial effluents especially those from wood processing plants. Atlantic tomcod (*Microgadus tomcod*) and striped bass (*Morone saxatilis*) also utilize this area as nursery habitat. Marine species for which the Inner and/or Outer Bay are probable nursery areas include Atlantic herring (*Clupea harengus*), winter (*Pleuronectes americanus*) and smooth (*Pleuronectes putnami*) flounders, sand lance (*Ammodytes* sp.), sculpin (*Myoxocephalus* sp.), Atlantic mackerel (*Scomber scombrus*) and fourbearded rockling (*Enchelyopus cimbrius*).

The invertebrate fauna is usually dominated by the estuarine calanoid copepod *Eurytemora affinis* and by copepod nauplius larvae, and these are an important food source for larval smelt, gaspereau and tomcod. Barnacle nauplius larvae, sporadically the most abundant zooplankters, are not found in the diet of these larvae because they rarely occur together. As is typical of other estuaries, the Miramichi contains planktonic organisms of both freshwater and marine origin and those with intermediate salinity preferences. The species assemblage varies along the salinity gradient. The two-layer circulation system is instrumental in determining the position of organisms in the estuary. In order to take advantage of this form of transportation, organisms apparently utilize tidal, rather than diel, vertical migration.

Résumé

Nous résumons ici les résultats d'études de l'ichtyoplancton (1918, 1948-1955, 1979-1980, 1981, 1992-1993) et des invertébrés zooplanctoniques (1951, 1992-1993) dans l'estuaire de la Miramichi. La diversité spécifique de la Miramichi est similaire à celle qu'on a observée dans d'autres estuaires de l'Atlantique, avec 33 taxons de poissons au stade larvaire et 75 taxons d'invertébrés. Les taxons qui dominent numériquement l'ichtyoplancton sont anadromes – éperlan arc-en-ciel (*Osmerus mordax*) et gaspareau (*Alosa pseudoharengus* et *A. aestivalis*). Le fait que la ponte et le début du développement larvaire de ces espèces se passent au fond de l'estuaire, près de Newcastle, expose ces poissons aux effluents industriels, particulièrement ceux des usines de traitement du bois. Le poulamon (*Microgadus tomcod*) et le bar rayé (*Morone saxatilis*) utilisent aussi cette zone comme nourricerie. Certaines espèces marines en font probablement de même dans les parties intérieures et extérieures de la baie : le hareng (*Clupea harengus*), la plie rouge (*Pleuronectes americanus*) et la plie lisse (*Pleuronectes putnami*), les lançons (*Ammodytes* sp.), les chabots (*Myoxocephalus* sp.), le maquereau (*Scomber scombrus*) et la motelle à quatre barbillons (*Enchelyopus cimbrius*).

La faune invertébrée est généralement dominée par le copépode calanoïde estuarien *Eurytemora affinis* et par des larves nauplius de copépodes, qui constituent une importante source de nourriture pour les larves d'éperlan, de gaspareau et de poulamon. Les larves nauplius de balanes, qui sont sporadiquement les zooplanctontes les plus abondants, ne se retrouvent pas dans l'alimentation de ces larves de poissons parce qu'elles sont rarement présentes ensemble. Comme bien d'autres estuaires, celui de la Miramichi abrite des organismes planctoniques d'origine dulcicole et marine, ainsi que d'autres qui ont des préférences intermédiaires en matière de salinité. L'assemblage spécifique varie le long du gradient de salinité. Le système de circulation à deux couches régit la position des organismes dans l'estuaire. Pour profiter de cette forme de transport, il semble que les organismes aient recours à une migration maréale plutôt que nycthémérale.

Introduction

Zooplankton play a critical role in pelagic food webs at intermediate trophic levels which link primary producers with top consumers such as pelagic fish. A wide range of fish and invertebrate taxa are represented in the zooplankton. Some species are planktonic throughout their entire life cycle while others are planktonic only in the egg and/or larval stage(s). Most fishes begin life as members of the zooplankton community, usually feeding upon smaller zooplankters. Advection of planktonic fish eggs and larvae by ocean currents serves as an effective dispersal mechanism for a number of fish species. Zooplankton community dynamics and distribution affect fisheries production by determining egg and larval dispersal, food web structure and energy flow, and recruitment variability.

Knowledge of the zooplankton of the southern Gulf of St. Lawrence is largely restricted to offshore habitats (e.g., Kohler et al. 1975, 1976, 1977; Faber 1976; de Lafontaine et al. 1984; de Lafontaine 1990), despite the importance of estuaries for early development of up to 70% of economically important Atlantic fish species (Clark 1967; McHugh 1966, 1967). The Miramichi Estuary is the largest in the southern Gulf, and one of the largest in Atlantic Canada, but surprisingly few studies have been conducted on its ichthyoplankton and invertebrate zooplankton (Table 1). None of the existing material on Miramichi zooplankton was included in a literature review of Gulf of St. Lawrence plankton research (de Lafontaine et al. 1991).

The purpose of this document is to review the data available on zooplankton of the Miramichi Estuary. This will be done by (1) summarizing the major findings of research conducted prior to the 1990's; (2) presenting the preliminary findings of the authors' research conducted since 1992, and comparing these to previous studies where possible; (3) identifying gaps in current knowledge of Miramichi zooplankton and suggesting directions for future research. Before discussing the biota of the estuary, we will begin by defining its parts: the

Table 1. Scope in time, space and subject matter of studies of Miramichi zooplankton and ichthyoplankton.

Dates sampled	No. of stations	Location of stations	Data collected	Plankton collecting gear	Published results
May–Nov. 1918, monthly	at least 13	Red Bank to 50 km outside barrier islands	temperature salinity phytoplankton zooplankton ichthyoplankton fish	plankton net 0.6 mm mesh 0.76 m diameter	Huntsman 1945 (general newspaper article) Rogers 1940 (larval smelt distribution)
Apr.– July, 1948-55, daily	1	Morrissey Bridge at Newcastle	smelt egg and larval abundance	plankton net 0.6 mm mesh 0.3 m diameter oblique and bottom tows from bridge	Table VIII in MacKenzie 1964
June-Aug. 1951, weekly	11	Red Bank to 14 km outside barrier islands	temperature salinity zooplankton	water pumped from 0, 12 and 24 feet in drifting boat; filtered with 0.12 mm mesh	Bousfield 1955
June, Aug., Sept. 1979, May-June 1980	47	Inner Bay to Northumberland Strait	ichthyoplankton	Bongo nets 0.505, 0.333 mm mesh, 0.6 m diameter; oblique tows at 2 knots	Messier et al. 1981 (larval herring) Little and Messier 1981 (ichthyoplankton species list)
Apr., July Oct. 1981	3	Dump sites A (between Chatham and Newcastle), B (between Bay du Vin I. and Grand Dune), C (in Outer Bay)	ichthyoplankton	plankton net 0.233 mm mesh 0.74 m diameter oblique tow	McLaren Plansearch 1982 (species list and percent composition in 12 samples)
May-Sept. 1992, June-July 1993, fortnightly	11-24	Red Bank to 6 km outside barrier islands	temperature salinity susp. sediment transparency phytoplankton zooplankton ichthyoplankton	(1)plankton net 0.064 mm mesh 0.5 m diameter vertical tow (2)plankton net 0.505 mm mesh 1.0 m diameter oblique tow at 1-2 knots	Locke and Courtenay 1995 (ichthyoplankton community); this chapter; other documents in preparation (longitudinal distribution, growth rates, gut contents of smelt, tomcod and gaspereau; K. Robichaud M.Sc. thesis on striped bass larvae)

Northwest and Southwest Miramichi Rivers (from the head of tide to their confluence above Beaubears Island), the Miramichi River (from Beaubears Island to Sheldrake Island), the Inner Bay (from Sheldrake Island to the barrier islands) and the Outer Bay (outside the barrier islands, opening into the Gulf of St. Lawrence).

Historical Studies

The "Miramichi Fisheries Expedition", 1918

The first study of plankton in the Miramichi (to our knowledge, the first scientific research of any type in the Miramichi Estuary), was a 1918 "expedition" from the Atlantic Biological Station in St. Andrews, headed by A.G. Huntsman. Huntsman and colleagues collected data on temperature, salinity, phytoplankton, zooplankton and fish (Table 1) in order to "obtain knowledge for increasing food by use of our fishery resources" (Rigby and Huntsman 1958, pg. 156). Invertebrate zooplankton and ichthyoplankton were sampled at monthly (sometimes more frequent) intervals from May to September 1918, from the head of tide at Red Bank on the Northwest Miramichi River to about 50 km seaward of the barrier islands (Fig. 1). The only other methodological information available is that tows of a 0.6 mm mesh, 76 cm diameter plankton net were conducted (Rogers 1940).

Unfortunately, these data were never published, with the exception of an analysis of the distribution of smelt larvae from June to September at 13 stations (Rogers 1940), and a series of four newspaper articles in which major trends in species distributions were mentioned (Huntsman 1945). The species composition of ichthyoplankton in the bay - winter and smooth flounders, sand lance, herring and capelin (for scientific names see Table 2) - was different from that in the river near the upper limit of brackish water - smelt, tomcod and striped bass (Huntsman 1945). Smooth flounder larvae were present in early May "throughout the inner bay and very abundant at its head" (Huntsman 1945), whereas winter flounder larvae were found in early June "chiefly in the middle of the inner bay, but also as far out as a couple of miles outside Portage Island". Larval sand lance, herring and capelin, also present in early June, were most abundant near the mouth of the Inner Bay. These three species occurred mainly at the surface, which according to Huntsman limited the degree to which they were transported into the estuary. Smelt, striped bass and tomcod larvae, present in late June, occurred near the bottom. Huntsman believed that their depth distribution maintained these larvae near the upper limit of brackish water, in the vicinity of Beaubears Island (Huntsman 1945). The role of differential depth distribution in maintaining longitudinal position in the estuary was also noted by Rogers (1940) with respect to smelt larvae.

The only surviving data on ichthyoplankton or invertebrate zooplankton from the 1918 survey are those published by Rogers (1940) on smelt. Larval smelt measured 7 – 10 mm in length when first captured on June 4, 20 – 30 mm by July 1, and up to 50 mm in early August (Rogers 1940). Huntsman (1945) noted that "young smelt" moved down the estuary as they grew larger during the summer, but this observation must be based on juvenile smelt distributions as it is not apparent from the larval abundances (Table II in Rogers 1940). Larval abundance (June-August) was consistently highest between Chatham and Loggieville, and peaked at 20 000 to 40 000 larvae per 30-minute tow near Chatham in early June. Rogers interpreted the observed vertical distribution of smelt larvae as a diel migration pattern - near the surface during the night, near the bottom during the day. Reanalysis of these data by Laprise and Dodson (1989), taking tidal state into account, supports the latter's hypothesis of tidal vertical migrations - near the surface during flood tides, near the bottom during ebbs.

Rogers (1940) noted that the abundance of plankton in general followed that of smelt - higher in the Miramichi River and the upper portion of the Inner Bay (above Oak Point) than in either the Northwest Miramichi River or the lower portion of the Inner Bay.

Neither Huntsman nor Rogers presented any information regarding the invertebrate zooplankton apart from a comment on the occurrence of the barnacle *Balanus improvisus* and the crab *Rhithropanopeus* (Huntsman 1945). Both were reported from the Gulf of St. Lawrence for the first time, and had not previously been found so far

Table 2. List of ichthyoplankton taxa recorded in the Miramichi Estuary (NWR=Northwest Miramichi River, MR=Miramichi River, IB=Inner Bay, OB=Outer Bay) by (1) Huntsman (1945), (2) Little and Messieh (1981), (3) McLaren Plansearch (1982) or (4) Locke and Courtenay (1995).

CLUPEIDAE:
- *Alosa* sp. (*A. pseudoharengus* (Wilson) + *A. aestivalis* (Mitchill)) – gaspereau; NWR, MR, IB; (4)
- *Alosa sapidissima* (Wilson) – American shad; NWR; (4)
- *Clupea harengus* Linnaeus – Atlantic herring; MR, IB, OB; (1,2,3,4)

OSMERIDAE:
- *Mallotus villosus* (Müller) – capelin; IB, OB; (1,2,4)
- *Osmerus mordax* (Mitchill) – rainbow smelt; NWR, MR, IB, OB; (1,2,4)

GADIDAE:
- *Enchelyopus cimbrius* (Linnaeus) – fourbeard rockling; MR, IB, OB; (4)
- *Gadus morhua* Linnaeus – Atlantic cod; MR, IB, OB; (2,4)
- *Merluccius bilinearis* (Mitchill) – silver hake; IB; (2)
- *Microgadus tomcod* (Walbaum) – Atlantic tomcod; NWR, MR, IB; (1,3,4)

ATHERINIDAE:
- *Menidia menidia* (Linnaeus) – Atlantic silverside; IB; (2,4)

GASTEROSTEIDAE:
- *Apeltes quadracus* (Mitchill) – fourspine stickleback; NWR, MR, IB; (4)
- *Gasterosteus aculeatus* Linnaeus – threespine stickleback; NWR, MR, IB; (4)
- *Gasterosteus wheatlandi* Putnam – blackspotted stickleback; MR, IB; (4)

PERCICHTHYIDAE:
- *Morone saxatilis* (Walbaum) – striped bass; NWR, MR; (1,4)
- *Morone americana* (Gmelin) – white perch; NWR; (4)

LABRIDAE:
- *Tautogolabrus adspersus* (Walbaum) – cunner; IB, OB; (2,4)

STICHAEIDAE:
- *Stichaeus punctatus* (Fabricius) – Arctic shanny; OB; (2)
- *Ulvaria subbifurcata* (Storer) – radiated shanny; IB, OB; (2,4)

PHOLIDAE:
- *Pholis gunnelis* (Linnaeus) – rock gunnel; MR, IB, OB; (2,3,4)

AMMODYTIDAE:
- *Ammodytes* sp. – sand lance; MR, IB, OB; (1,2,3,4)

SCOMBRIDAE:
- *Scomber scombrus* Linnaeus – Atlantic mackerel; IB, OB; (4)

COTTIDAE:(2)
- *Myoxocephalus aeneus* (Mitchill) – grubby; IB, OB, (3)
- *Myoxocephalus* sp. – sculpin; IB, OB; (4)

AGONIDAE:
- *Aspidophoroides monopterygius* (Bloch) – alligatorfish; IB, OB, (2)

CYCLOPTERIDAE:
- *Liparis atlanticus* (Jordan and Evermann) – Atlantic snailfish; OB; (3)
- *Liparis* sp. – seasnail; IB, OB; (2)

BOTHIDAE:
- *Scophthalmus aquosus* (Mitchill) – windowpane; MR, IB, OB; (2,4)

PLEURONECTIDAE:
- *Hippoglossoides platessoides* (Fabricius) – American plaice; IB, OB; (2)
- *Limanda ferruginea* (Storer) – yellowtail flounder; OB; (2)
- *Pleuronectes putnami* (Gill) – smooth flounder; IB; (1)
- *Pleuronectes americanus* (Walbaum) – winter flounder; IB, OB; (1,2)
- *Pleuronectes* sp. – flounder; IB; (4)

north. The vertical distribution of the larvae of these two species was presumed to be similar to that of smelt, tomcod and striped bass as their distribution in the upper estuary was similar to the latter species.

Smelt investigations, 1948-1955

The timing and location of spawning, and the abundance of larval smelt were recorded as part of a study of smelt life history in the Miramichi (McKenzie 1964).

Smelt spawned in fresh water, up to 16 km above the head of tide. The spawning run lasted from late April to early June, starting as the water warmed to 4-5 C in the Southwest and Northwest Miramichi, Bartibog River and Mill Stream, and 6-7 C in other tributaries. Eggs deposited in late April usually hatched in about three weeks while those deposited in warmer water later in the season hatched in about ten days. On hatching, larvae were advected downriver into tidal waters.

Collections of larval smelt were made from the Morrissey bridge at Newcastle from 1948 to 1955 (Table 1). Samples were collected daily from early in the season before hatching commenced until larvae could no longer be caught at the end of the spawning season. Sampling was carefully standardized, with all tows being made approximately three hours after the beginning of the ebb tide, and only during daylight hours (from 3 hr after dawn to 3 hr before dark). Two types of tow were conducted daily in water 11 to 12 m deep using a 0.6-cm mesh, 30-cm diameter net - a bottom tow about 1.5 m from the bottom, and an oblique tow from bottom to surface.

The only published data resulting from these larval collections comprise a table of seasonal totals (Table VIII in McKenzie 1964). Total larval counts ranged from 3,930 in 1952 to 119,817 in 1950, spanning two orders of magnitude among nine seasons with similar sampling effort. Total abundance was negatively correlated to total rainfall for the 30 days following the arrival of smelt at the head of tide in the Southwest Miramichi River (R=0.97, N=9). The low abundance of larvae in years of high rainfall was attributed to spawning outside the normal stream bed during periods of freshet, with the eggs drying out and dying after the water receded.

Comparison of the oblique and bottom tows showed that during the day over 90% of the larvae were taken within 1.5 m of the bottom. This is consistent with Laprise and Dodson's (1989) hypothesis of tidal vertical migrations which states that smelt should be near the bottom during ebb tides, but also with the more traditional hypothesis of diel vertical migrations, which states that zooplankton remain near bottom during day and ascend into the water column at night. Nighttime collections also found more larvae near bottom, although some larvae ascended to the surface at night. Since all of McKenzie's samples were collected during ebb tides, it is not possible to use his results to further assess the likelihood of tidal versus diel migrations. McKenzie himself appears to have been unaware of the possibility of tidal vertical migrations and attributed the observed distribution of smelt larvae to diel migrations.

By the time the young smelt reached 3-6 cm in length, they were presumably no longer planktonic as they were being caught in seines along the shore of the river and bay.

McKenzie also tried to relate larval abundance to the numbers of fish from the resultant year-class (based on landings in the fishery), but with no success. He concluded that either the data were not sufficiently detailed, or that year-class abundance in smelt was not determined until after the larval stage.

Barnacles and other invertebrate zooplankton, 1951

The first published study of the invertebrate zooplankton of the Miramichi system was that of Bousfield (1955). In 1951, Bousfield conducted a detailed study of barnacle larvae, with notes on the occurrence of other invertebrate zooplankton (Table 3). Ichthyoplankton were not recorded in this study. Samples were collected weekly from June to August 1951, from Red Bank to approximately 14 km outside the barrier islands (Table 1). A minimum of 2 or 3 stations were sampled in each of 4 areas: the Northwest Miramichi River, the Miramichi River, the Inner Bay and the Outer Bay.

The primary objective of the study was to document the ecological and hydrographical constraints (especially mechanisms permitting retention in the estuary) on distribution of the brackish-water barnacle *Balanus improvisus*. The earliest larval stages were most abundant near the mouth of the river and on the south side of the Inner Bay, corresponding to the areas of highest abundance of the adults. Larvae were advected seawards along the south side of the Inner Bay, crossing to the north shore near the mouth of the bay, then advected landwards along the north shore. Ontogenetic changes in vertical distribution, taking advantage of depth-dependent differences in circulation, were considered to effect this geographic distribution. Nauplius stages I-III (the youngest larvae) were above, nauplius IV-V (intermediate-aged larvae) were at or slightly below, and nauplius VI and cyprid stages (the oldest larvae) were well below the level of no net motion (separating the upper layer of residual seaward drift from the lower layer of landward drift, and estimated to be located at a depth of 3 m in the Inner Bay). Comparable ontogenetic patterns of distribution were observed for the congeneric *Balanus crenatus*, although this species was found further downstream. Seasonal variations in river discharge, in combination with differences in the location of adult populations and timing of reproduction, contributed to interspecific differences in larval distribution.

In total, Bousfield recorded the presence of 24 zooplankton taxa (Table 3) and documented pronounced changes in species composition along the salinity gradient. Freshwater animals such as the cladoceran *Bosmina longirostris* were limited almost entirely to waters of salinity less than 3.

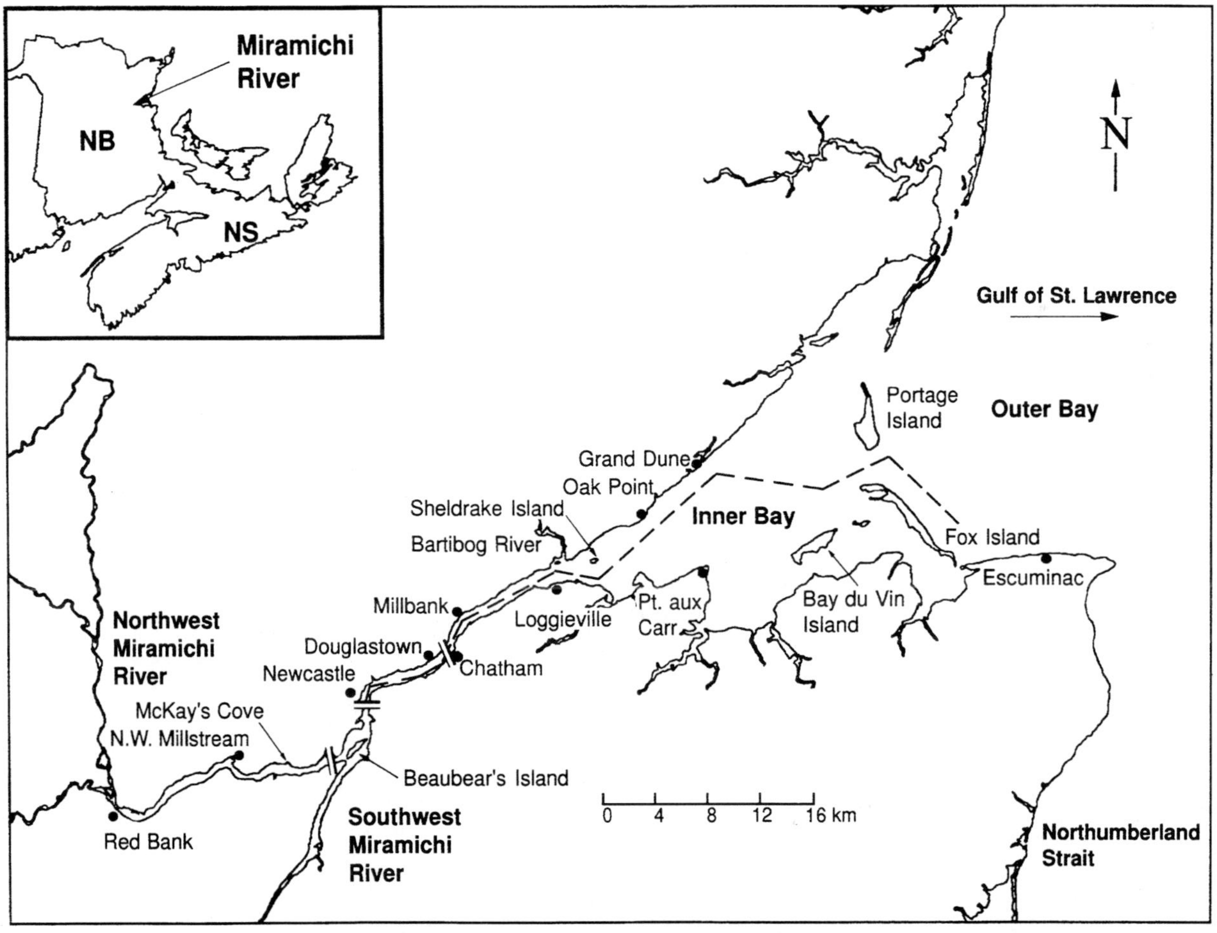

Fig. 1. Miramichi Estuary showing places mentioned in the text. The broken line indicates the dredged channel.

Table 3. List of invertebrate zooplankton collected in the Miramichi Estuary by (1) Bousfield (1955) or (2) Locke and Courtenay (unpublished preliminary data).

CNIDARIA:
- *Aglantha digitale* (Muller); IB; (2)
- *Cyanea capillata* (Linnaeus); MR, IB; (2)
- *Euphysa flammea* (Linko); MR, IB; (2)
- *Hydra* sp.; NWR, MR; (2)
- *Nemopsis bachei* L. Agassiz; NWR, MR, IB; (2)
- *Obelia* sp.; MR, IB, OB; (2)

NEMATODA: NWR, MR, IB; (2)

CHAETOGNATHA:
- *Sagitta elegans* Verrill; MR, IB, OB; (1,2)

MOLLUSCA:
- Lamellibranch veliger; NWR, MR, IB; (2)
- Gastropoda veliger; NWR, MR, IB; (2)

ANNELIDA:
- Polychaeta (Spionidae); NWR, MR, IB, OB; (2)
- Oligochaeta; NWR; (2)
- Hirudinea; NWR, MR; (2)

ARACHNIDA
- Araneae; MR; (2)
- Acari; MWR, MR, IB; (2)

CLADOCERA
- *Bosmina longirostris* (O.F. Müller); NWR, MR; (1)
- *Bosmina* sp.; NWR, MR, IB; (2)
- *Daphnia* sp.; NWR; (2)
- *Evadne nordmanni* Loven; NWR, MR, IB, OB; (1,2)
- *Podon leuckarti* Sars; MR, IB, OB; (1,2)

OSTRACODA; NWR, MR, IB; (2)
- *Cypria* sp.; NWR, MR; (1)

COPEPODA (CALANOIDA)
- *Acartia clausi* Giesbrecht; NWR, MR, IB, OB; (1,2)
- *Acartia hudsonica* (Lilljeborg); NWR, MR, IB; (2)
- *Acartia tonsa* Dana; NWR, MR, IB, OB; (1,2)
- *Acartia* sp.; NWR, MR, IB; (2)
- *Calanus finmarchicus* (Gunnerus); MR, IB, OB; (1,2)
- *Calanus glacialis*; IB; (2)
- *Calanus hyperboreus*; IB; (2)
- *Centropages hamatus* (Lilljeborg); NWR, MR, IB, OB; (1,2)
- *Centropages typicus* Kröyer; MR, IB; (2)
- *Eurytemora affinis* (Poppe); NWR, MR, IB; (2)
- *Eurytemora herdmani* Thompson and Scott; NWR, MR, IB, OB; (1,2)
- *Eurytemora hirundoides* (Nordquist); NWR, MR, IB, OB; (1,2)
- *Labidocera aestiva* Wheeler; NWR, MR, IB, OB; (1,2)
- *Metridia longa* (Lubbock); NWR; (2)
- *Pseudocalanus elongatus* Boeck; MR, IB, OB; (1)
- *Pseudocalanus minutus* Kröyer; MR, IB, OB; (2)
- *Pseudodiaptomus coronatus* Williams; NWR, MR, IB, OB; (1,2)
- *Temora longicornis* (Müller); NWR, MR, IB, OB; (1,2)
- *Tortanus discaudatus* (Thompson and Scott); NWR, MR, IB, OB; (1,2)

COPEPODA (CYCLOPOIDA)
- *Cyclops viridis* (Jurine); NWR, MR; (1,2)
- *Ergasilus chautauquaensis*; NWR, MR, IB; (1)
- *Oithona atlantica*; MR, IB; (2)
- *Oithona plumifera* Baird; IB; (2)
- *Oithona similis* Claus; NWR, MR, IB, OB; (1,2)

Table 3. List of invertebrate zooplankton collected in the Miramichi Estuary by (1)Bousfield (1955b) or (2) Locke and Courtenay (unpublished preliminary data). (*Concluded.*)

COPEPODA (HARPACTICOIDA); NWR, MR, IB; (2)
- *Laophonte huntsmani*; NWR, MR, IB; (1)

BRANCHIURA
- *Argulus* sp.; NWR, MR, IB; (2)

CIRRIPEDIA
- *Balanus crenatus* Bruguière (nauplius); IB, OB; (1)
- *Balanus balanoides* (Linnaeus) (nauplius); OB; (1)
- *Balanus improvisus* Darwin (nauplius); NWR, MR, IB, OB; (1)
- *Balanus* sp. (nauplius, cypris); NWR, MR, IB; (2)

MYSIDACEA; NWR, MR, IB; (2)

CUMACEA; IB; (2)

AMPHIPODA
- Gammaridea; NWR, MR, IB; (2)
- *Hyperia galba* (Montagu); MR, IB; (2)

ISOPODA; IB; (2)

EUPHAUSIACEA
- *Thysanoessa inermis* Kröyer; MR, IB, OB; (2)

DECAPODA
- *Cancer irroratus* Say (zoea); IB, OB; (2)
- *Cancer borealis* Stimpson (zoea, megalopa); IB, OB; (2)
- *Crangon septemspinosa* Say; MR, IB; (2)
- *Neopanope sayi* (Smith) (zoea); IB, OB; (2)
- *Rhithropanopeus harrisii* (Gould) (zoea); NWR, IB, OB; (1,2), (megalopa) (2)

INSECTA
- Coleoptera; NWR; (2)
- Diptera (Sipulidae, Chironomidae, Tendipedidae, Simuliidae); NWR, MR, IB; (2)
- Ephemeroptera (Ephemerella sp., Heptageniidae, *Hexagenia* sp., *Stenonema* sp.) NWR, MR, IB;(2)
- Plecoptera; NWR, MR; (2)
- Trichoptera (Hydropsychidae); NWR; (2)

Brackish water animals, in salinities of 3-26, included the barnacle *Balanus improvisus*, the crab *Rhithropanopeus harrisii*, and the copepods *Eurytemora hirundoides* and *Acartia tonsa*. Essentially marine animals, such as the barnacles *Balanus balanoides* and *B. crenatus*, the arrow worm *Sagitta elegans* and the copepod *Acartia clausi*, were found mainly in the outer portions of Miramichi Bay and the Gulf of St. Lawrence.

Vertical and longitudinal distributions in relation to salinity were discussed for six species (the cladoceran *Bosmina longirostrum*, zoea larvae of the crab *Rhithropanopeus harrisii*, and the copepods *Oithona similis, Tortanus discaudatus, Pseudocalanus elongatus,* and *Centropages hamatus*, listed here in order from freshwater to marine occurrence). *Bosmina* was found mainly near the bottom, above Newcastle. The few taken below this point were mainly near the surface. Those living near the bottom were prevented from moving downriver by the net landward movement of saline bottom water; those transported seaward in the surface water were probably killed upon reaching unsuitable salinities. *Rhithropanopeus* zoeae and adults were found mainly between the head of the bay and the upper limit of salt water. Zoeae occurred mainly near the depth of no net motion, so there was no net horizontal displacement of the population. *Oithona similis* and *Pseudocalanus elongatus* were most abundant in deeper strata in the Gulf. Adults (but not larvae) were also abundant near the bottom as far upriver as Millbank, suggesting that this population was maintained by landward transport from the Gulf. Similarly, adult *Tortanus discaudatus* were collected at and below the level of no net motion almost to Newcastle, although adults and larvae were most abundant in the Gulf. By contrast, adults and copepodites of

Centropages hamatus were located almost entirely in the Gulf and near the surface. Bousfield proposed that the two-layer circulation typical of the Miramichi resulted in net landward movement of deep-swimming species, and exclusion from the estuary of surface-swimming species.

Finally, Bousfield compared the relative roles played by tidal and non-tidal circulation. Non-tidal residual drift served to retain larvae of the estuarine barnacle *Balanus improvisus* in the estuary, to bring in from the Gulf the deep-swimming larvae of *B. crenatus*, and to exclude from the estuary the surface-swimming larvae of *B. balanoides*. According to Bousfield, tidal currents affected barnacle larvae mainly through physical control of vertical distribution since all stages were closer to the surface at mid-flood than at mid-ebb. Thus, Bousfield was clearly aware of the principles underlying tidal vertical migration.

Pre-dredging ichthyoplankton, 1979-1980

The need for environmental assessment of the effects of channel dredging in the early 1980's prompted further ichthyoplankton research, especially to determine the potential effects on larval herring. To establish a pre-dredging baseline for larval herring abundance, Messieh et al. (1981) sampled 47 stations, about half of which were located in the Inner or Outer Bay and the remain-

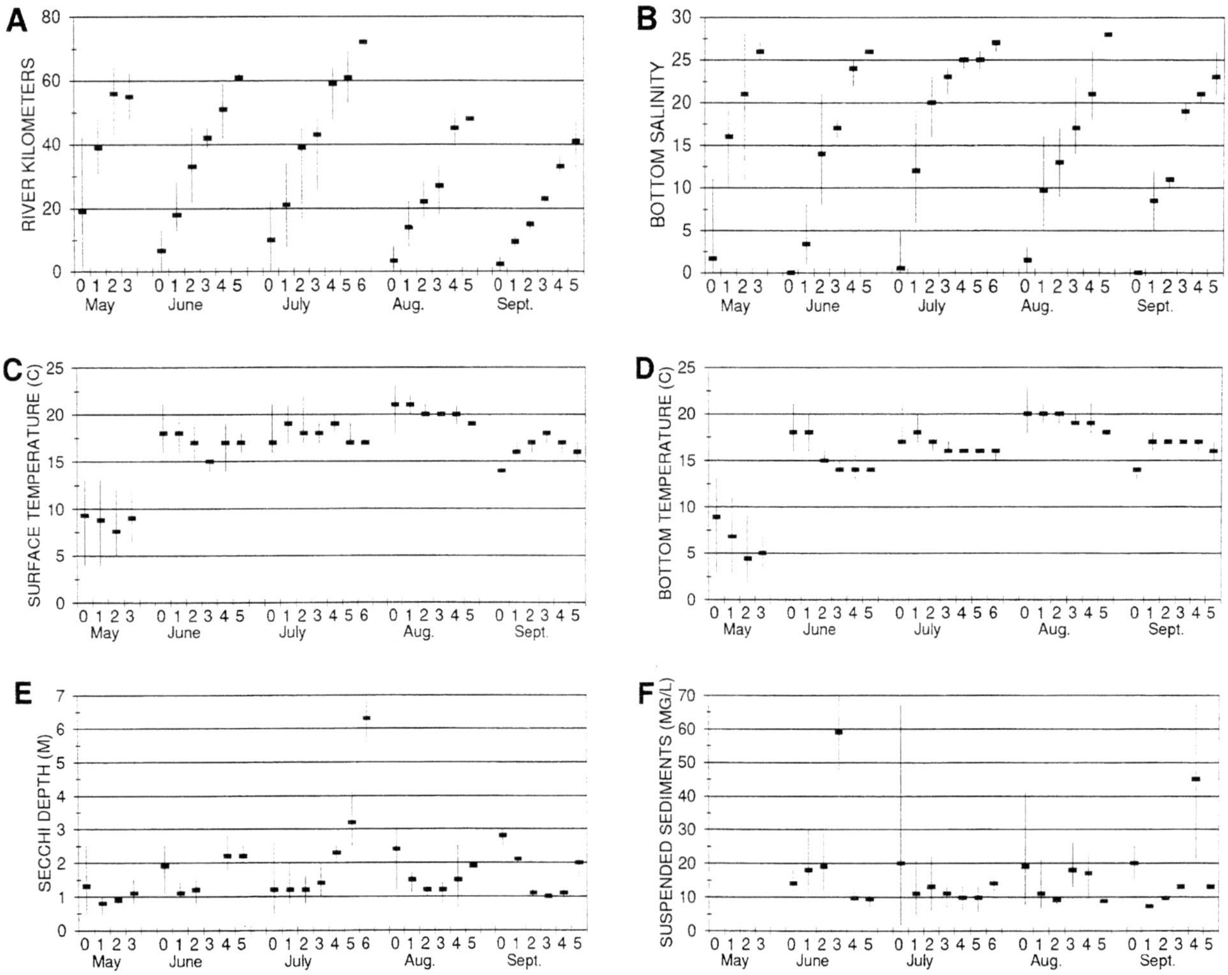

Fig. 2. Means and ranges of environmental conditions (bottom salinity, surface and bottom temperatures, Secchi depth, suspended sediments) in the Miramichi Estuary, summarized by month of sampling (1992 data only) and salinity stratification interval (0 = surface salinity 0, 1 = surface salinity 1-5, 2 = surface salinity 6-10, 3 = surface salinity 11-15, 4 = surface salinity 16-20, 5 = surface salinity 21-25, 6 = surface salinity>25). Relationship of surface salinity intervals with distance from Red Bank (river km) over time is also shown.

der in the Northumberland Strait as far south as Richibucto. No samples were collected upstream of a line drawn between Point aux Carr and Oak Point. During surveys in May-June, August and September 1979 and May-June 1980 (Table 1), 16 species of ichthyoplankton were collected from the Inner and Outer Bay (Little and Messieh 1981; Table 1). (The 25 species listed in Messieh et al. (1981) include those found only in the Northumberland Strait.) Once again the published data are very limited - in addition to the species lists, there are distribution maps of four species, for May-June 1979 only.

In May-June 1979, Messieh et al. (1981) collected larvae of five species of commercial value in the Miramichi area - smelt (total count 34,818 individuals), winter flounder (8,051), herring (2,630), Atlantic cod (1,276) and mackerel (766). Three of these – herring, smelt and winter flounder – along with sand lance, were more abundant in the Inner or Outer Bay than in the adjacent Gulf, and distribution maps were presented. Herring and winter flounder larvae were most abundant off the southeast shore of Fox Island. Sand lance larvae were also abundant in this area as well as further seawards. Smelt were most abundant along the north shore of the Inner Bay near Grand Dune. Cod and mackerel were more abundant in the Gulf than the Bay, and their distributions were not mapped.

Herring eggs and larvae were found to be extremely abundant near dump site C outside the barrier islands. Experiments (Johnson and Wildish 1982) showed them to be sensitive to the elevated sediment levels typical of dredging. Messieh et al. (1981) recommended that dredging effects be minimized by stopping work in May-June, especially in the Fox Island-Escuminac area adjacent to the spawning beds.

Dredging effects, 1981

A further assessment of dredging effects on larval fish was conducted by McLaren Plansearch (1982). Ichthyoplankton were sampled in three areas corresponding to the locations of the proposed dump sites (site A - Miramichi River near Douglastown, site B - Miramichi Bay between Grand Dune and Bay du Vin Island, site C - outside the barrier islands), during the first year of dredging (Table 1). Ichthyoplankton were sampled using an oblique tow of a 0.74 m diameter net with 233 µm mesh net. This survey included only 12 samples (5 at site A, 3 at site B, 4 at site C), collected mainly in April and October 1981. Most sites and dates were sampled during both day and night. No baseline data had been collected for comparison and ichthyoplankton sampling was not continued after the first year of dredging.

In total, 7 fish taxa were recorded. In April, sand lance, Atlantic cod, tomcod and rock gunnel were found at dump site A, sand lance, tomcod, rock gunnel and sculpin at dump site B, and sand lance, herring, snailfish and sculpin at dump site C. Herring were found at dump site A in July, and dump site C in October. No ichthyoplankton were found at dump sites A or B in October.

It is likely that the larval cod were misidentified in this study. Recent surveys have collected Greenland cod (*Gadus ogac*) inside the Miramichi system in fall and winter (Hanson and Courtenay 1995). Ripe and running fish were caught in late February. No Atlantic cod, *G. morhua*, were caught in the Miramichi Estuary at any time, and in any case this species spawns in late May and June. The larvae were almost certainly Greenland cod.

Conclusions and limitations of these studies

1. *Taxonomic diversity*: The Miramichi contains a diverse assemblage of invertebrate zooplankton (24 taxa) and ichthyoplankton (23 taxa). The species composition is similar to that of other Atlantic estuaries.

The accuracy and completeness of the species lists in Tables 2 and 3 could be improved by (1) expanding geographic and seasonal coverage of sampling (see point 5, below) and (2) by resolving differences in the taxonomy used by different investigators (e.g., changes in species nomenclature in genera such as *Gadus, Bosmina*, and *Acartia*, and variations between investigators in the taxonomic level to which organisms were identified). Since samples collected by previous investigators have evidently been lost or destroyed, only option (1) is feasible at present.

2. *Larval stages of economically important fish species*:

(a) *Distribution:* At least six commercially fished species (smelt, tomcod, two species of gaspereau, winter flounder, herring) and a variety of non-commercial species (including the recreationally important striped bass) spawn in Miramichi Bay or connecting rivers, and spend at least part of the larval stage as ichthyoplankton within the Miramichi system.

(b) *Environmental impacts:* The larvae about which most is known are smelt and herring. Huntsman's (1945) newspaper articles and McKenzie's (1964) larval smelt collections place early stages of smelt and other anadromous species (tomcod, striped bass) in or above the Newcastle area, which receives effluents from several industrial and municipal sources (Miramichi River Environmental Assessment Committee, 1992). Without more detailed data on the distribution of the eggs and larvae of anadromous fishes with respect to these point sources of pollutants, the potential for environmental effects of these effluents cannot be evaluated. Messieh et al. (1981) predicted that dredging activities near the barrier islands could negatively affect herring larvae. No followup study has been conducted to look at the seasonal and geographic distribution of dredging, or its impact on eggs and larvae of herring and other species.

3. *Effects of salinity gradient on species distribution*: Like most other estuaries, the Miramichi contains planktonic organisms of both freshwater and marine origin and those with brackish salinity preferences. The species assemblage varies along the salinity gradient. Salinity tolerances and longitudinal distribution of certain species within the estuary (for example, smelt and barnacle larvae) change with the ontogenetic stage of development.

4. *Tidal influences on longitudinal and vertical distribution*: The two-layer circulation system (in combination with ontogenetic changes

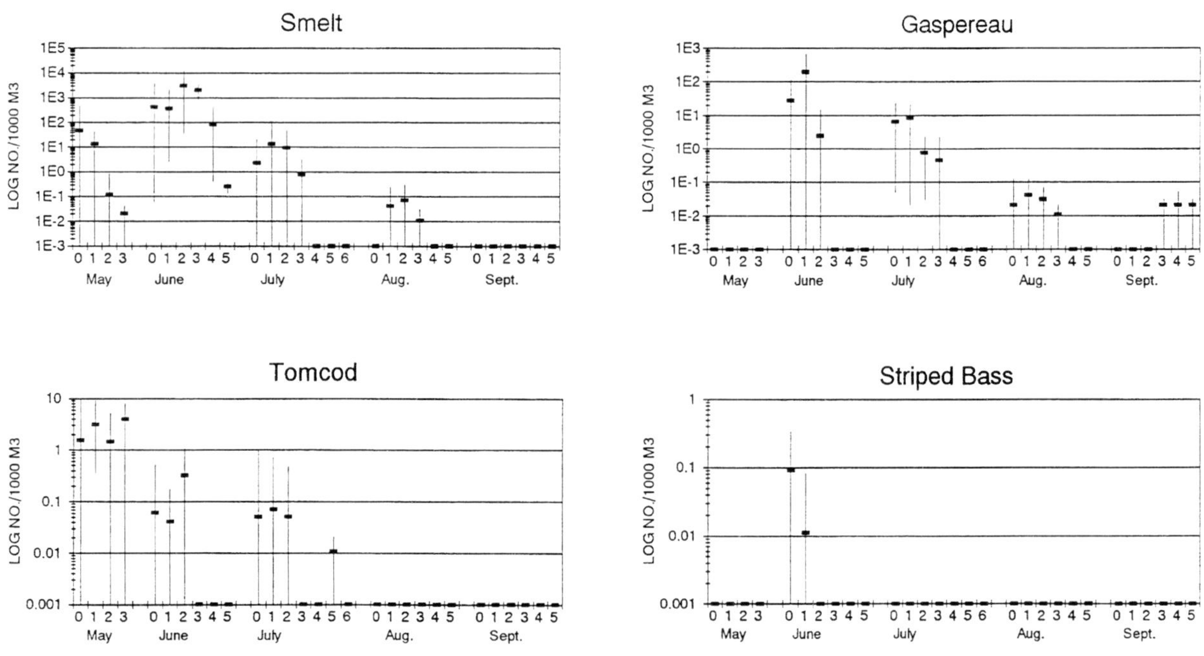

Fig. 3. Spatial and temporal trends in log-transformed abundance of commonly collected anadromous ichthyoplankton taxa - smelt, gaspereau, tomcod and striped bass. Surface salinity intervals are defined as in Fig. 2.

in vertical distribution) is instrumental in determining the position of organisms in the estuary. The findings of Bousfield (1955) and the reanalysis of Rogers' (1940) data by Laprise and Dodson (1989) suggest that many zooplankters in the Miramichi system utilize tidal, rather than diel, vertical migration in order to maintain position or reach an appropriate habitat.

5. *Geographic and seasonal coverage of sampling*:

(a) *Invertebrate plankton*: Bousfield's (1955) study of the invertebrate zooplankton was comprehensive in its geographic and taxonomic coverage of the estuary during the summer of 1951, but covered only the period of June through August. The invertebrate plankton of other seasons has not been documented. In particular, there has been no documentation of winter zooplankton.

(b) *Ichthyoplankton*: The 1918 study headed by Huntsman appears to have been comprehensive in its coverage of the system during the ice-free season and in the range of variables sampled, but sadly almost none of these data have ever been published. Likewise, both McKenzie and Messieh and co-workers appear to have extensively sampled the composition, abundance and distribution of Miramichi ichthyoplankton, but their respective publications summarize only a fraction of the data collected. As far as we have been able to determine, none of these data or samples still exist.

The existing data on ichthyoplankton include only a species list and limited distributional data for the Inner and Outer Bays (Messier), fragmented data on smelt larval distributions and abundance (Rogers, McKenzie), a few estimates of larval abundance near the dredging dump sites (McLaren Plansearch), and some anecdotal accounts of larval distributions (Huntsman).

Ichthyoplankton in the river portion of the estuary have not been documented. Relationship between larval abundance data and environmental conditions have not been determined. There has been no sampling for ichthyoplankton during the ice-covered season or during ice formation or breakup.

Neither ichthyoplankton nor invertebrate zooplankton have been documented in the Southwest Miramichi River, or in the Northwest Miramichi River above Red Bank.

1992-93 STUDIES

Introduction

In 1992, when we initiated a study of Miramichi zooplankton, no plankton work had been done in the Miramichi system for over 10 years. Our intention was to complete an ecological survey of the Miramichi plankton, assessing its current status and answering some of the questions which were not addressed by previous work. The scope and objectives of the study were:

(1) comprehensively survey the ichthyoplankton and invertebrate zooplankton of the entire estuary during the ice-free season
(2) examine the relationship of dominant fish and invertebrate species with their physical environment; in particular, identify important habitats for larval fishes (nursery areas)
(3) examine feeding interactions of major ichthyoplankton species with their zooplankton prey
(4) identify geographic areas where ichthyoplankton may be negatively affected by human activities.

The results of these investigations are currently being analysed, and some are to be published in other venues (Locke and Courtenay 1995). In this chapter, we will describe the methodology of the study, and summarize the results obtained to date.

Methods

In May-September 1992, plankton and water samples were collected from Red Bank to Newcastle two or three times a week until July and fortnightly thereafter, from Newcastle to Oak Point at fortnightly intervals, and from Oak Point to just outside the barrier islands at approximately monthly intervals (Table 1). Additional samples were collected between Newcastle and the Gulf of St. Lawrence outside the barrier islands during two surveys in June and July, 1993. The results summarized in this document are those of the 1992 season only, as the 1993 data have not been analysed. Sampling locations of each survey varied, as they were selected according to surface salinity rather than by geographic locality. Fixed stations are of limited use in estuarine plankton studies where environmental conditions vary substantially with tidal displacement of water masses. Selection of sampling locations was stratified by 5-unit surface salinity intervals. Eleven to 24 samples were collected during each sampling period, with a minimum of two samples collected from each salinity interval, when possible.

Ichthyoplankton were collected during daylight hours with a 10-minute stepped oblique tow of a 500-µm-mesh metre net from approximately 1 m above the bottom to the surface. Flowmeter readings were used to determine the volume of water filtered. Tow speed (approximately 1-2 knots) was adjusted to compensate for changes in current velocity with depth, so as to maintain a constant tow wire angle. Smaller zooplankton were collected using a 64-µm-mesh net, 0.5 m in diameter, towed vertically from bottom to surface from a drifting vessel. All plankton samples were immediately preserved in 10% formaldehyde in seawater, buffered with calcium carbonate. Samples were returned to the laboratory where plankton were counted and identified to species where possible.

Concurrently with the zooplankton samples, we collected data on turbidity, salinity and temperature. An SBE-19 Seacat Profiler (Seabird Electronics Inc., Bellevue, WA) was used to record

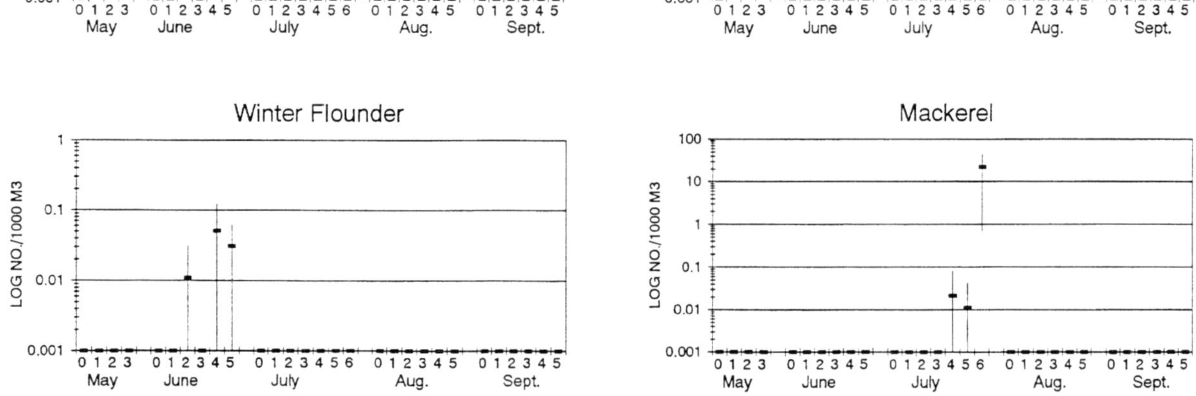

Fig. 4. Spatial and temporal trends in log-transformed abundance of representative marine ichthyoplankton species - herring, smooth and winter flounders, and mackerel. Surface salinity intervals are defined as in Fig. 2.

temperature and salinity profiles of the water column. A water sample (0.5 – 1 L) for total suspended sediment (TSS) determination was collected using a weighted Tygon tube to sample the water column from surface to near-bottom (to a maximum depth of 9 m; few stations exceeded 10 m in depth). Upon return to the laboratory, this sample was frozen, then later thawed and filtered onto a pre-weighed 0.45-µm cellulose nitrate filter, dried at 60°C for 24 h, and weighed. Water transparency (Secchi depth) was also measured at each station using a standard 20 cm black-and-white Secchi disk. At most stations, phytoplankton samples were collected using the weighted Tygon tube and preserved with FAA (formalin-acetic acid-alcohol solution).

Laboratory analyses presently underway or already completed include determination of the species composition and environmental relationships of ichthyoplankton and invertebrate zooplankton, and size structure, diet and distribution of three dominant anadromous taxa (smelt, gaspereau and tomcod). Dietary composition of these taxa will be compared to the availability of zooplankton prey using selectivity indices.

Results and Discussion

Environmental Conditions

(1) Salinity

Surface salinities ranged from 0 to 26, and sampling was stratified into seven salinity categories (surface salinities of 0, 1-5, 6-10, 11-15, 16-20, 21-25, >25). Salinity increased from upstream to downstream, and from surface to bottom (Fig. 2), as is typical of estuaries. The upriver sites near Red Bank were essentially freshwater throughout the ice-free season. Surface salinity at Red Bank was 0 throughout the study, as was bottom salinity on all dates except August 13-14, when salinity was 2. The downstream extent of fresh water varied seasonally with discharge. In early May, surface fresh water extended into the main Miramichi River almost to Sheldrake Island. At this time, bottom waters were also fresh as far downstream as Millbank. Two

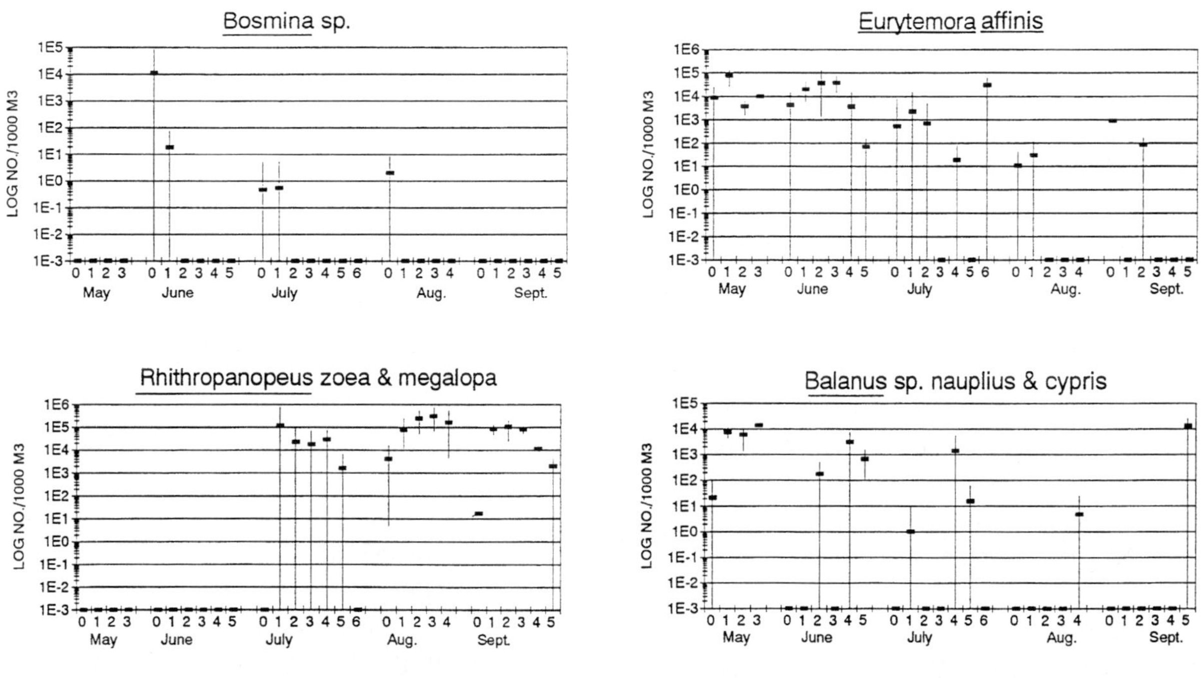

Fig. 5. Spatial and temporal trends in log-transformed abundance of representative zooplankton taxa - *Bosmina* sp., *Eurytemora* sp., *Rhithropanopeus harrisi* larvae, *Cirripedia* larvae. Surface salinity intervals are defined as in Fig. 2.

weeks later, freshwater bottom salinities extended only to the mouth of the Northwest Miramichi River. Bettignies and St-Hilaire (1993), sampling stations below Newcastle in mid-May of 1991, found no fresh water near the bottom. By August and September, surface fresh water was also restricted to the Northwest Miramichi.

(2) Temperature

Surface temperatures ranged from 4°C in early May to a maximum of 23°C in August (Fig. 2), similar to the maximum of 22°C observed by Bousfield (1955) in July and August, 1951. Likewise, St-Hilaire et al. (1992) found that surface temperatures reached the maximum temperature capacity of their thermographs (23.5°C) on numerous occasions in July and August, 1991. In general, temperatures were higher in the Northwest and main Miramichi rivers than in the Inner Bay. However, in late August and September the Northwest Miramichi River was colder than the Miramichi River.

Bottom temperatures ranged from 2°C in May to 22.5°C in August (Fig. 2). Temporal and geographic trends in bottom temperature were similar to those described above for surface temperature.

(3) Secchi depth

In all months sampled, Secchi depths were higher in the Northwest Miramichi River and in Miramichi Bay than in the main Miramichi River (Fig. 2). Seasonal variation in Secchi depth was very limited. Typical values for the Northwest Miramichi and Inner Bay were 3 m, and for the main Miramichi River, 1 m or less. Outside the barrier islands, Secchi depths increased to 7 m. These results are consistent with those reported by Bousfield (1955) for Miramichi Bay (usually 1.8 m in the upper end and south side of the Inner Bay, 2.5 m in the Outer Bay and along the north side of the Inner Bay) and adjacent Gulf of St. Lawrence (typically 6.1 m, maximum value 7.7 m).

(4) Suspended solids

Typical suspended solid concentrations in June-September 1992 were in the range of 5 to 30 mg/L (Fig. 2). Occasionally samples were collected with suspended solids of 70 mg/L. There was no strong trend with location or seasonal variation. The range of suspended sediment concentrations we recorded is consistent with the observations of Winters (1981). Suspended sediment concentrations in May 1976 ranged from 0.5 to 90 mg/L, and in February 1977 from 0.1 to 80 mg/L. Maxima of suspended particulate material occurred at regular intervals during tidal cycles, generally near bottom, but occasionally at the surface near the end of ebb tides (Winters 1981).

(5) Relationships between physical variables

Physical variables were not independent of one another. In general, waters of the lower estuary were more saline, colder and more transparent than those of the upper estuary. Correlations of surface with bottom temperatures, surface with bottom salinities, and surface and bottom salinities with distance from Red Bank were consistently positive and significant at $P<0.05$ (Table 4). Significant negative correlations of Secchi depth with bottom temperature, surface temperature with surface and bottom salinities, bottom temperature with bottom salinity, and surface and bottom temperatures with distance from Red Bank were seen in at least half of the sampling periods (Table 4).

Ichthyoplankton

Twenty-two taxa (19 identified to species, 3 to genus) representing 14 families were collected from May to September 1992 (Table 2). With a few exceptions, all taxa were present as larvae. Tomcod were present in both larval and juvenile stages. The stickleback *Gasterosteus wheatlandi* was collected as juveniles and adults, and *Gasterosteus aculeatus* and *Apeltes quadracus* were present as larvae, juveniles and adults.

Following the pattern commonly observed in estuaries (deLafontaine 1990), a few species - in this case, smelt and the two species of gaspereau - strongly dominated the ichthyoplankton. In the Miramichi, the dominant species are anadromous. Maximum mean abundances of 2000-3000 larvae/m^3 in June are the result of peak smelt abundances at that time. Diversity (H) is very low

Table 4. (a) Pearson correlation coefficients of physico-chemical variables in Miramichi Estuary, for all 1992 dates combined. Values in bold type are significant at $P<0.05$.

	Temperature		Salinity		Susp.	Distance from
	Surface	Bottom	Surface	Bottom	Sed.	Red Bank
Secchi depth	**0.21**	**0.18**	**0.41**	**0.19**	**-0.22**	**0.19**
Surface temperature	–	**0.96**	0.12	0.001	**-0.31**	**-0.27**
Bottom temperature	–	–	-0.0003	**-0.18**	-0.11	**-0.43**
Surface salinity	–	–	–	**0.87**	-0.09	**0.80**
Bottom salinity	–	–	–	–	-0.18	**0.88**
Suspended sediment	–	–	–	–	–	-0.08

(b) Occurrence of significant ($P<0.05$) positive and negative correlations between physico-chemical variables, by date of sampling. Dates are designated by numbers (0=May 6-8, 1=May 19-20, 2=June 3-4, 3=June 16-18, 4=July 2-3, 5=July 15-17, 6=July 21-23 and 27, 7=Aug. 13-14, 8=July 25 and 27, 9=Sept. 24-25), and the sign of the correlation coefficient is indicated by + (positive) or - (negative).

	Temperature		Salinity		Susp.	Distance
	Surface	Bottom	Surface	Bottom	Sediment	from Red Bank
Secchi depth	+:8 -:4679	+:8 -:34579	+:36 -:28	+:3 -:28	+:no correlation -:3	+:6 -:128
Surface temperature	–	+:123456789 -:0	+:0 -:123689	+:0 -:12368	+:3 -:6	+:04 -:12368
Bottom temperature	–	–	+:no correlation -:1267	+:no correlation -:01268	+:no correlation -:6	+:no correlation -:012678
Surface salinity	–	–	–	+:0123456789 -:no correlation	+:68 -:3	+:0123456789 -:no correlation
Bottom salinity	–	–	–	–	+:6 -:3	+:0123456789 -:no correlation
Suspended sediment	–	–	–	–	–	+:no correlation -:36

Table 5. Number of species (*S*), Shannon-Weiner diversity index (*H*) and total abundance (mean±standard deviation) of ichthyoplankton collected from the Miramichi Estuary in 1992. Data are summarized by month and surface salinity (0=0, 1=1-5, 2=6-10, 3=11-15, 4=16-20, 5=21-25, 6=>25). *N* is number of stations sampled.

MONTH	SAL	*N*	*S*	*H*	Abundance·1000m^{-3} (Mean±SD)
May	0	12	2	0.14	48.5 ± 135.4
	1	4	6	0.52	16.7 ± 23.6
	2	7	12	1.62	3.7 ± 1.9
	3	2	9	1.26	7.5 ± 5.9
Jun	0	9	6	0.23	445.4 ± 1161.4
	1	8	5	0.64	568.7 ± 637.3
	2	6	6	0.01	3095.9 ± 4888.8
	3	3	1	0.00	2007.4 ± 1262.7
	4	5	6	0.11	86.1 ± 184.0
	5	3	4	0.96	0.7 ± 0.3
Jul	0	19	5	0.65	8.9 ± 9.7
	1	10	5	0.68	22.2 ± 37.7
	2	9	6	0.35	10.3 ± 16.7
	3	5	4	0.71	1.2 ± 2.3
	4	4	6	1.01	0.4 ± 0.4
	5	4	6	0.41	1.4 ± 2.5
	6	2	5	0.40	368.7 ± 393.8
Aug	0	5	2	0.69	<0.01 ± 0.1
	1	6	2	0.69	0.1 ± 0.1
	2	4	2	0.57	0.1 ± 0.2
	3	4	4	1.38	<0.01 ± <0.01
	4	5	3	0.82	0.1 ± 0.1
	5	1	0	0.00	0.0
Sep	0	2	0	0.00	0.0
	1	2	0	0.00	0.0
	2	2	0	0.00	0.0
	3	2	0	0.00	0.0
	4	2	2	0.64	0.0 ± 0.1
	5	2	2	0.41	0.1 ± 0.2

during these peaks (Table 5). Diversity is consistently highest in marine waters in spring. By August and September, larval abundance and diversity were very low. In September, ichthyoplankton were no longer found in the upper part of the estuary.

Larvae may be separated into species which spawn in fresh water or marine environments, and further subdivided by their season of appearance. Details of larval distribution and abundance, and relationships with environmental variables are included in Locke and Courtenay (1995).The following summary will emphasize the data collected on species of commercial or recreational importance.

(1) Freshwater – spring

Larvae of anadromous species, collected in spring, included smelt, gaspereau and tomcod.

Smelt (Fig. 3) were by far the most abundant larva and were present in the ichthyoplankton longer than any other species, from May 6 to August 25. In May this species was most abundant in waters with surface salinities of 0 and bottom salinities of 0 to 8 but peak abundances in June occurred in waters with surface salinities of 1-10 and bottom salinities of 1-21. This supports Huntsman's (1945) observation that smelt move downriver into more saline waters as they develop. The preference of smelt for bottom waters, stated by both Huntsman and MacKenzie, could not be assessed as we did not collect vertically stratified samples.

Gaspereau were the second most abundant larva. They first appeared on June 16, 2 weeks after the peak abundances of smelt (Fig. 3). In almost every sampling period, the peak abundances of gaspereau occurred in waters with surface salinities of 1-5 (bottom salinities ranged from 1 to 15), suggesting that there was little or no tendency for the larvae to move downstream.

Tomcod were present from May 6 to July 21, in abundances two to four orders of magnitude lower than those of smelt or gaspereau (Fig. 3). In May, they were more uniformly distributed across surface salinities of 0-15 than were smelt, and on average were found further downstream. In June and July, tomcod were concentrated in waters of surface salinity 0-10. July collections were of juveniles only.

(2) Freshwater – summer

Striped bass (Fig. 3) and white perch were associated with tidal fresh water. Striped bass eggs and larvae were collected for a short period in late May-early June. Larvae were concentrated in waters with surface salinities of 0-5, and were most common near the fresh-brackish water interface in the Northwest Miramichi River. White perch larvae were captured only sporadically, at the uppermost stations sampled, in the Northwest Miramichi River near Red Bank.

(3) Marine – spring

'Spring larvae' were first collected in May, and a few remained in the estuary until mid-June. These included herring (Fig. 4), sand lance, winter and smooth flounder, rock gunnel, sculpins and radiated shanny. Most specimens were collected in the Inner Bay.

Herring were present through May and June. Some specimens collected in May were in late post-larval stages, close to metamorphosis to the juvenile form. These are probably fall-spawned individuals advected to the Miramichi, perhaps from northern waters (H. Dupuis, personal communication).

Smooth flounder were collected in waters of surface salinity 1-15 in the Inner Bay, only in early May (Fig. 4). Winter flounder, collected in approximately the same geographic area, but in waters of higher salinity in June, were an order of magnitude less abundant (Fig. 4).

(4) Marine – summer

'Summer larvae' did not appear in the estuary until June or July. Species found in the bay included cunner, windowpane, capelin, silversides, four-bearded rockling and mackerel. Cunner, rockling and mackerel occurred primarily in the waters adjacent to or just outside the barrier islands, unlike the other species which were more commonly collected in the Inner Bay. Mackerel occurred in the Inner Bay in waters of surface salinity 16-25, but were three orders of magnitude more abundant in the higher-salinity waters of the Outer Bay and adjacent Gulf of St. Lawrence (Fig. 4).

Nursery areas:

The upper Miramichi Estuary (in general, the region between McKay's Cove and Sheldrake Island) meets three criteria as a nursery area for smelt, gaspereau and tomcod: higher abundances relative to surrounding areas (Pearcy and Myers 1974), and presence of larvae and/or juveniles over an extended time period and through a large range of sizes (Faber 1976). All three are anadromous species whose adults enter tidal fresh waters to spawn, so the upper estuary is a logical location for their nursery grounds. Data collected by Bradford et al. (1995) show that this area also serves as a nursery area for striped bass larvae and juveniles.

The lower Miramichi Estuary and adjacent Gulf waters probably provide nursery habitat for typically marine species such as winter and smooth flounders, sand lance, sculpin, mackerel and fourbearded rockling. The latter two species were more abundant outside the barrier islands and therefore habitat in the Inner Bay is probably less important for their early development than that in offshore waters.

The identification of the upper Miramichi Estuary as a major nursery ground for anadromous species is particularly important because of its potential for environmental degradation, especially by effluents from the wood processing plants in Newcastle and Nelson-Miramichi. To date the direct effects of toxins potentially present in effluents, or indirect effects mediated through the food chain, on ichthyoplankton in this area are unknown. This is a question of special interest, since the three species which we have identified as using the area immediately adjacent to the wood processing plants as a nursery ground for very early life history stages are important components of the commercial fisheries of the Miramichi Estuary. Striped bass, also utilizing this nursery area, is a recreational species of increasing public interest.

Invertebrate Zooplankton in 1992

Approximately 73 different invertebrate taxa were collected in plankton tows throughout the Miramichi system in 1992 (Table 2), over three times as many taxa as previously recorded by Bousfield (1955). The most commonly collected planktonic invertebrates were copepods (the calanoids *Eurytemora affinis, Acartia hudsonica,* and *Temora longicornis*, the cyclopoids *Cyclops* sp. and *Oithona similis*, an unidentified harpacticoid, and copepod nauplius larvae), the cladoceran *Bosmina* sp., barnacle nauplius and cypris larvae, polychaetes, ostracods, nematodes and a variety of insects with aquatic stages (Plecoptera, Ephemeroptera, Trichoptera and Diptera). By far the most abundant of these were *Eurytemora affinis* and copepod nauplii, collected consistently throughout the season in the entire system with abundance up to 30 ,000·m^{-3}, and barnacle nauplii, sporadically collected in large numbers (maximum 50 ,000·m^{-3}) and especially prevalent in areas of higher salinity (>16 at surface).

Distributions of the dominant taxa with respect to salinity were similar to those described by Bousfield. These will be described in more detail in a later publication, with major trends summarized here for a few typical species.

In general, *Bosmina* (Fig. 5), *Cyclops*, ostracods and insects were typical of freshwater or slightly brackish (generally waters with surface salinities less than 10) portions of the system. The recorded salinity preference of these organisms is fresh water. *Bosmina* (Fig. 5) first appeared in samples in June (probably the time of hatching from overwintering eggs in fresh water) and was found until August in waters of surface salinities of 0-5. *Eurytemora* occurred throughout the full salinity range sampled but in most sampling periods was most abundant in waters with surface salinities of 1 to 15 (Fig. 5). *Eurytemora* is typically the dominant copepod genus of temperate estuarine habitats, and this was also the case in the Miramichi Estuary.

Zoea and megalopa larvae of the estuarine crab *Rhithropanopeus harrisi* (Fig. 5) first appeared in the plankton in July and were present in virtually all waters of surface salinities of 0-20 through August and September. In general, the species was present in lower abundance in waters of surface salinity 0. The estuarine habitat is typical for this species, but New Brunswick is the northern limit of its range (Williams 1974).

Acartia, Temora, Oithona, and polychaetes were distributed in a manner more typical of marine forms and were primarily present in waters with surface salinities of 11 and higher. Nauplius and cypris larvae of *Balanus* sp. were found in waters with surface salinities as low as 0 in May, when a surface freshwater layer extended into the Inner Bay (Fig. 5). Larvae moved into waters of higher salinity as the summer progressed. By August barnacle larvae were found only in waters of surface salinity >15, and by September, surface salinity >20. Since we did not identify barnacle larvae to species or ontogenetic stage, it is not possible to reproduce Bousfield's distributional observations for barnacles.

Feeding of smelt, tomcod, and gaspereau

The most important prey item in the gut contents of tomcod and smelt is the calanoid copepod *Eurytemora affinis*, which was found in approximately 70% of the guts analysed (Table 6). Immature smelt eat *Eurytemora* almost exclusively, with other prey taxa present in comparatively few (<10%) guts. Tomcod diets were more diverse – other calanoids, especially *Acartia hudsonica*, harpacticoid copepods, mysids, and unidentified invertebrate pieces were found in approximately 1/3 of the guts examined. *Eurytemora* copepodites are less important in the diet of gaspereau (18%) than those of smelt or tomcod. Other calanoids occurred with similar frequency (26%), along with smaller prey items, such as copepod nauplii (21%) and the freshwater cladoceran *Bosmina* (21%), which were not heavily utilized by smelt or tomcod.

These preliminary gut content analyses show that *Eurytemora affinis*, which was probably the most abundant zooplankter in the estuary, was an important component of the diet of all three fish species. This was especially evident in the diet of smelt, where *Eurytemora* was practically the only common food item. The diets of tomcod and

Table 6. Gut contents of smelt (TL 4.6-33.0 mm, N=571, empty guts in 53.4% of individuals), tomcod (TL=3.1-59.5 mm, N=161, empty guts in 55.3%) and gaspereau (TL=3.5-20.0 mm, N=402, empty guts in 57.1%) caught in plankton samples in the Miramichi system. Prey items present in fewer than 1% of guts are marked as "+".

	Proportion of guts containing prey item		
Prey taxon	Smelt	Tomcod	Gaspereau
Eurytemora affinis	0.71	0.70	0.18
Acartia hudsonica	0.09	0.38	0.03
unidentified	0.09	0.06	0.01
Bosmina sp.	0.08	0.01	0.21
Temora longicornis	0.06	0.07	0
Calanoida	0.04	0.38	0.04
Copepoda	0.03	0.17	0.26
Harpacticoida	0.03	0.29	0.21
Polychaeta - adult	0.01	0.03	0
Centric diatom	0.01	0.01	0.05
unidentified Insecta	0.01	0.01	0.02
Invertebrata pieces	+	0.29	0
Ostracoda	+	0.03	0.06
Oithona sp.	+	0	0.01
Polychaeta - larva	+	0	0
unidentified trochophore	+	0	0
plant seed	+	0	0
Cyclopoida	+	0.10	0
Decapoda - larva	+	0	+
Neomysis americana	0	0.35	0
Labidocera aestiva	0	0.06	0
Osmerus mordax – larva	0	0.06	0
Cirripedia	0	0.04	0
Mallotus villosus – larva	0	0.01	0
Amphipoda	0	0.01	0
unidentified fish - larva	0	0.01	0
Copepoda - nauplius	0	0	0.21

gaspereau were more diversified. The rapidly growing immature tomcod tended to prey upon larger species such as the mysid *Neomysis americana*. Gaspereau, in contrast, utilized smaller prey items including copepod nauplii, which were not eaten by smelt or tomcod. Differences in prey utilization probably resulted from differences in the mouth size of larvae and in their position within the estuary at any given time.

Conclusions and Limitations

1. *Species distributions/Geographic and seasonal coverage*:

(a) *Ichthyoplankton*: Our study confirms the general trends noted by previous distributional studies but a great deal of supplemental information is now available, especially on the distribution and ecology of ichthyoplankton in the river portion of the estuary. There is clear separation of the marine and freshwater-brackish habitats in terms of their usage by ichthyoplankton. Our coverage of the Northwest Miramichi River was centred in the vicinity of the freshwater-brackish interface and most of the species represented there are anadromous. Numerically, these ichthyoplankters dominate the Miramichi system. Fishes carrying out the entire life cycle in fresh water were generally not collected by us. It should also be noted that we did not conduct any sampling of the Southwest Miramichi River, which has also been neglected by previous studies. Our coverage of the marine habitat was much less complete than that of the main and Northwest Miramichi rivers, and we almost certainly missed collecting species which may have occurred there for short periods of time in between our summer surveys. However, our survey of Miramichi Bay was sufficient to identify it as a nursery area for a variety of marine species.

The distribution of larvae of anadromous species is of particular interest. These species are numerically very important and their early development is centred around a part of the river which receives a major portion of the industrial and municipal wastewater entering the system. In particular, the effect of wood processing plants upriver of Newcastle on early development of anadromous fishes should be further investigated. This is especially important considering that in recent years, landings of several of these species have declined (Chaput 1995).

(b) *Invertebrate zooplankton*: The main contribution of our preliminary analysis of Miramichi zooplankton is the expansion of the existing species list from 24 to approximately 70 taxa. Our data confirm the distributions described by Bousfield. We will utilize the data collected on invertebrate zooplankton mainly in the context of assessing the food supply of larval fishes in the Miramichi.

The same restrictions noted for ichthyoplankton with respect to geographic and temporal coverage of the estuary also apply to our invertebrate zooplankton collections. We have also not addressed the lack of sampling during winter - with the exception of a few (as yet unanalysed) vertical hauls taken under the ice in February 1992, all our sampling has been conducted during the ice-free season.

2. *Larval fish feeding*: *Eurytemora* is clearly an important component of the diet of larval smelt, tomcod and gaspereau. Smelt, in particular, appear to specialize on *Eurytemora*. The diets of tomcod and gaspereau are more diversified. Further analysis of these data will focus on selectivity of prey items relative to their abundance in the environment.

3. *The Miramichi as larval fish habitat/ Possible conflicts with human activities*: We cannot provide final conclusions regarding the quality of larval fish habitat in the Miramichi Estuary until our analyses are complete. However, the preliminary results discussed in this document have identified nursery areas in two discrete habitats: the freshwater-brackish interface (anadromous species), and Miramichi Bay (marine species). The environmental concerns of the two habitats are also

distinct - pulp mill effluents and other industrial wastes in the upper estuary, and dredging-related sediments in the lower estuary. Chemical analysis of water samples, ichthyoplankton and perhaps of their prey will probably be required in order to determine whether pulp mill effluents on the nursery grounds are problematic for ichthyoplankton of anadromous species. Analysis of our 1993 data collected adjacent to dredging operations will assist in determining the impact of sediments on larvae in the Inner Bay (dredging in 1992 was confined to the Outer Bay). Much work remains to be done on the plankton of the Miramichi Estuary; we hope that the information presented in this document will provide a good starting point for future researchers to address relevant and interesting questions.

Acknowledgements

Special thanks are due to K. Robichaud for assistance and participation in all stages of this project. C. Daigle, G. Klassen, C. Vardy, A. Smith, R. Laprise, P. LeBlanc, A. St-Hilaire, and R. Eisner assisted in the field. L. Chisholm, G. Pohle, D. McCullough and staff of the Atlantic Reference Centre processed plankton and gut content samples. D. McCullough, J. Matthews and M. Freeborn entered data to the computer. J. Leger helped produce final versions of graphics. We appreciate L. Van Guelpen of the Atlantic Reference Centre's taxonomic advice on difficult ichthyoplankton specimens. Sampling would not have been possible without K. Davidson's loan of the *Queen Larry*, D. Moore and C. MacDougall's assistance with boat maintenance and launching, and the participation of the captains and crews of the *Navicula* and *Osborne*. M. Powers and D. Swain kindly loaned plankton sampling gear and a CTD, respectively. During the field component of this project, A. Locke was funded through the Visiting Fellowship program of Fisheries and Oceans. Laboratory work was funded by the Dept. of Fisheries and Oceans and the Green Plan Sustainable Fisheries Program. This study was part of the Miramichi Estuary component of the Gulf Community Project of Fisheries and Oceans, Gulf Region.

References

Bettignies, C., and St-Hilaire, A. 1993. Variations mensuelles de la salinité et de la stratification dans l'estuaire de la Miramichi - influence des marées, des debits et du vent. Proc. Can. Soc. Civil Eng. Ann. Conf., Fredericton, N.B., 1993: 495-504.

Bousfield, E.L. 1955. Ecological control of the occurrence of barnacles in the Miramichi Estuary. Nat. Mus. Can. Bull. 137: 69p.

Bradford, R.G., Robichaud, K.A., and Courtenay, S.C. 1995. By-catch in commercial fisheries as an indicator and regulator of striped bass (*Morone saxatilis*) abundance in the Miramichi River Estuary, p.249–262. *In* E.M.P. Chadwick [*editor*]. Water, science, and the public: the Miramichi ecosystem. Can. Spec. Publ. Fish. Aquat. Sci. 123.

Chaput, G.J. 1995. Temporal distribution, spatial distribution, and abundance of diadromous fishes in the Miramichi River watershed, p. 121–140. *In* E.M.P. Chadwick [*editor*]. Water, science and the public: the Miramichi ecosystem. Can. Spec. Publ. Fish. Aquat. Sci. 123.

Clark, J. 1967. Fish and man, conflict in the Atlantic estuaries. Amer. Littoral Soc. Spec. Publ. 5: 78p.

de Lafontaine, Y. 1990. Ichthyoplankton communities in the St. Lawrence estuary: composition and dynamics. *In* M.I. El-Sabh and N. Silverberg [*editors*]. Oceanography of a large-scale estuarine system: the St. Lawrence. Coastal Estuarine Stud. 39: 321-343.

de Lafontaine, Y., Demers, S., and Runge, J. 1991. Pelagic food web interactions and productivity in the Gulf of St. Lawrence: a perspective. p. 99-123. *In* J.-C. Therriault [*editor*]. The Gulf of St. Lawrence: small ocean or big estuary? Can. Spec. Publ. Fish. Aquat. Sci. 113.

de Lafontaine, Y., El-Sabh, M.I., Sinclair, M., Messieh, S.N., and Lambert, J.-D. 1984. Structure océanographique et distribution spatio-temporelle d'oeufs et de larves de poissons dans l'estuaire maritime et la partie ouest du Golfe Sainte-Laurent. Sci. Tech. Eau 17 : 43-50.

Faber, D.J. 1976. Hyponeustonic fish larvae in the Northumberland Strait during summer 1962. J. Fish. Res. Board Can. 33: 1167-1174.

Hanson, J.M., and Courtenay, S.C. 1995. Seasonal abundance and distribution of fishes in the Miramichi Estuary, p. 141–160. *In* E.M.P. Chadwick [*editor*]. Water, science, and the public: the Miramichi ecosystem. Can. Spec. Publ. Fish. Aquat. Sci. 123.

Huntsman, A.J. 1945. Miramichi fisheries investigation twenty-five years ago. The Commercial and The World, Chatham. Jan. 18, p. 2; Jan. 25, p. 2; Feb. 1, p. 7; Feb. 8, p. 6.

Johnson, D.D., and Wildish, D.J. 1982. Effect of suspended sediment on feeding by larval herring (*Clupea harengus harengus* L.). Bull. Environ. Contam. Toxicol. 29: 261-267.

Kohler, A.C., Faber, D.J., and McFarlane, N.J. 1975. Eggs, larvae and juveniles of fishes from plankton collections in the Gulf of St. Lawrence during 1969. Fish. Mar. Serv. Tech. Rep. 490: 105p.

Kohler, A.C., Faber, D.J., and McFarlane, N.J. 1976. Eggs, larvae and juveniles of fishes from plankton collections in the Gulf of St. Lawrence during 1970-1971. Fish. Mar. Serv. Tech. Rep. 645: 139p.

Kohler, A.C., Faber, D.J., and McFarlane, N.J. 1977. Eggs, larvae and juveniles of fishes from plankton collections in the Gulf of St. Lawrence during 1972 to 1975. Fish. Mar. Serv. Tech. Rep. 747: 180p.

Laprise, R., and Dodson, J.J. 1989. Ontogeny and importance of tidal vertical migrations in the retention of larval smelt *Osmerus mordax* in a well-mixed estuary. Mar. Ecol. Prog. Ser. 55: 101-111.

Little, D.B., and Messieh, S.N. 1981. Manual for identifying larval fish of the Miramichi Bay, Gulf of St. Lawrence. Laboratory Reference No. 81/1. Dept. of Fisheries and Oceans, Bedford Inst. of Oceanography, Dartmouth N.S. and Biological Station, St. Andrews, N.B. 53p.

Locke, A., and Courtenay, S.C. 1995. Effects of environmental factors on ichthyoplankton communities in the Miramichi Estuary, Gulf of St. Lawrence. J. Plankton Res. 17:333–349.

McHugh, J.L. 1966. Management of estuarine fisheries. Amer. Fish. Soc. Spec. Publ. 3: 133–154.

McHugh, J.L. 1967. Estuarine nekton. *In* G.H. Lauff [*editor*]. Estuaries, p. 581-620. Amer. Assoc. Adv. Sci. Publ. 83.

McKenzie, R.A. 1964. Smelt life history and fishery in the Miramichi River, New Brunswick. Bull. Fish. Res. Board Can. 144: 77p.

McLaren Plansearch. 1982. Environmental monitoring of the dredging of the Miramichi River. Summary of 1981 Operations. Draft report for Public Works Canada, Feb. 1982.

Messieh, S.N., Wildish, D.J., and Peterson, R.H. 1981. Possible impact from dredging and spoil disposal on the Miramichi Bay herring fishery. Can. Tech. Rep. Fish. Aquat. Sci. 1008: 33p.

Miramichi River Environmental Assessment Committee. 1992. Miramichi River Environmental Assessment Committee - Summary, Final Report 1989-1992. 157p.

Pearcy, W.G., and Myers, S.S. 1974. Larval fishes of Yaquina Bay, Oregon: a nursery ground for marine fishes? Fish. Bull. U.S. 72: 201-213.

Rigby, M.S., and Huntsman, A.G. 1958. Materials relating to the history of the Fisheries Research Board of Canada (formerly the Biological Board of Canada) for the period 1898-1924. Fish. Res. Board Can. Manuscr. Rep. Ser. (Biol.) 660: 272p.

Rogers, H.M. 1940. Occurrence and retention of plankton within the estuary. J. Fish. Res. Board Can. 5: 164-171.

St-Hilaire, A., Bettignies, C., Booth, D., and Chadwick, E.M.P. 1992. Tidal influence on the stratification of the Miramichi Estuary. *In* M. Jennings and N.G. Bhowmik [*editors*]. Proc. Amer. Soc. Chem. Eng., Water Forum '92: Hydraulic Engineering. pp. 953-958.

Williams, A.B. 1974. Marine flora and fauna of the northeastern United States. Crustacea: Decapoda. U.S. Dep. Commer., Nat. Oceanic Atmos. Admin., April 1974. 47p.

Winters, G.V. 1981. Environmental geology of the Miramichi Estuary: suspended sediment transport. Geol. Surv. Can. Pap. 81-16: 12p.

CHAPTER 8

Temporal Distribution, Spatial Distribution, and Abundance of Diadromous Fish in the Miramichi River Watershed

Gérald J. Chaput

Department of Fisheries and Oceans, Science Branch, P.O. Box 5030, Moncton, New Brunswick, Canada E1C 9B6

Chaput, G.J. 1995. Temporal distribution, spatial distribution, and abundance of diadromous fish in the Miramichi River watershed, p. 121-139. *In* E.M.P. Chadwick [*editor*]. Water, science, and the public: the Miramichi ecosystem. Can. Spec. Publ. Fish. Aquat. Sci. 123.

Abstract

The Miramichi River drains about 14 000 km^2 and is the most important producer of diadromous fish in Atlantic Canada. A total of 11 diadromous fish utilize the river, of which seven are exploited by fishers: Atlantic salmon (*Salmo salar*), rainbow smelt (*Osmerus mordax*), two species of gaspereau, blueback herring and alewife (Alosa aestivalis and A. pseudoharengus), American shad (*Alosa sapidissima*), Atlantic tomcod (*Microgadus tomcod*), and the American eel (*Anguilla rostrata*). The spawning populations of several of these species are at the northern or southern limits of their distribution in the western Atlantic. The proportion of time spent in fresh, estuarine and sea water varies among and within species. The diadromous species which spend the greatest proportion of their life cycle in fresh water (salmon, eel and sea lamprey) also penetrate farthest upriver. There is also large variation in the sea distribution among species. The most diverse fauna is present in the estuary during the months of April to June. Activity in the winter period is restricted to smelt and tomcod which can tolerate sub-zero water temperatures. In terms of number of fish, smelt are 10 times more abundant than gaspereau, 20 to 30 times more abundant than tomcod, 100 times more abundant than eels, 1 000 times more abundant than salmon and shad, and 10 000 times more abundant than striped bass. The total exploitable biomass of diadromous fish passing through the estuary has exceeded 16 000 t in recent years. Of this biomass, smelt are the dominant species at over 50% of the total, followed closely by gaspereau at 42%. Smelt dominated the commercial catches prior to 1945 but gaspereau became more important in 1950 and now make up the largest proportion of the total landings.The commercial fisheries can be categorized into three groups: (1) smelt and tomcod fisheries are declining, (2) gaspereau and eel fisheries are increasing, and (3) shad, striped bass and salmon fisheries are in a precarious state or terminated. Management measures have improved the status of Atlantic salmon in the Miramichi River. It would appear that the nature of the estuary has changed and under current levels of industrial development, the abundance of diadromous fish will remain below that observed in the early 1900's.

Résumé

La Miramichi draine une zone d'environ 14 000 km^2 et vient au premier rang de la production de poissons diadromes du Canada atlantique. Au total, onze espèces diadromes utilisent le fleuve, et sept d'entre elles sont exploitées par les pêcheurs : le saumon atlantique (*Salmo salar*), l'éperlan arc-en-ciel (*Osmerus mordax*), le gaspareau (Alosa pseudoharengus, pêché avec l'alose d'été, (*Alosa aestivalis*), l'alose savoureuse (*Alosa sapidissima*), le poulamon (*Microgadus tomcod*) et l'anguille (*Anguilla rostrata*). Les populations reproductrices de plusieurs de ces espèces se trouvent ici à la limite septentrionale ou méridionale de leur aire de répartition dans l'Atlantique ouest. La proportion du temps passé en eau douce, en estuaire et en eau de mer varie entre les espèces et même à l'intérieur d'une même espèce. Les espèces diadromes qui passent la plus grande partie de leur cycle vital en eau douce (saumon, anguille et lamproie) sont également celles qui remontent le plus loin en amont. On observe aussi une grande variation dans la distribution marine parmi les espèces. C'est d'avril à juin que la faune est la plus diverse dans l'estuaire. Pendant l'hiver, l'activité est limitée à celle de l'éperlan et du poulamon, qui peuvent tolérer des températures de l'eau inférieures à zéro. Pour ce qui est des effectifs, les éperlans sont 10 fois plus abondants que les gaspareaux, 20 à 30 fois plus abondants que les poulamons, 100 fois plus abondants que les anguilles, 1 000 fois plus abondants que les saumons et les aloses savoureuses, et 10 000 fois plus abondants que les bars rayés. La biomasse exploitable totale des poissons diadromes qui passent dans l'estuaire dépassait 16 000 t ces dernières années. Sur cette biomasse, l'éperlan est l'espèce dominante avec plus de 50 % du total, suivi de près par le gaspareau à 42 %. L'éperlan dominait les prises commerciales avant 1945, mais le gaspareau l'a supplanté en 1950, et constitue maintenant la plus grande proportion du total des débarquements. On peut classer les pêches commerciales en trois catégories : 1) les pêches de l'éperlan et du poulamon, en déclin; 2) les pêches du gaspareau et de l'anguille, en hausse; 3) les pêches de l'alose savoureuse, du bar rayé et du saumon, qui sont dans un état précaire ou ont disparu. Des mesures de gestion ont amélioré la situation du saumon atlantique dans la Miramichi. Il semble que la nature de l'estuaire ait changé et, au niveau actuel de développement industriel, l'abondance des poissons diadromes va rester inférieure à celle du début du siècle.

Introduction

The Miramichi River is a north-temperate river which empties into the Gulf of St. Lawrence at 47°N latitude and whose freshwater and estuarine areas freeze over during the winter months (December through early April) (Philpott 1978). On a global scale, the Miramichi River is a small system. At a maximum axial length of about 250 km, it is 4.4% as long as the Amazon River (6 276 km) and 20% as long as the Fraser River in western Canada (Northcote and Larkin 1989). The Miramichi River drains an area of 13 799 km2 (Randall et al. 1989) and is the second largest river system in the Maritimes, after the St. John River in New Brunswick. Its freshwater input represents about half of all the freshwater inputs of the southern Gulf of St. Lawrence (that area including Chaleur Bay, the Gulf shores of New Brunswick, all of Prince Edward Island, and the Gulf shores of Nova Scotia up to the northern tip of Cape Breton Island) (Fig. 1). Because of its size and its accessibility to migratory fish along its entire length, it is the largest producer of diadromous fishes in Atlantic Canada.

Diadromous fish are those which migrate between sea and fresh water (McDowall 1987) of which three types can be described from the Miramichi:

Anadromous fish: those which spend a significant portion of their lives in the sea and migrate to freshwater to spawn. The best example is the Atlantic salmon.

Catadromous fish: those which spend most of their lives in freshwater and migrate to the sea to spawn. The best example of this is the American eel.

Amphidromous fish: those whose migrations between sea and freshwater are not for the sole purpose of breeding but can occur at different times in the life cycle. An example of such a fish would be the striped bass.

Table 1 lists diadromous species that utilize the Miramichi River (classification from Scott and Scott 1988).

Brook trout (family Salmonidae, *Salvelinus fontinalis* Mitchill) has diadromous forms but the extent of movements between fresh water and marine areas is unknown. Brook trout are exploited intensively in recreational fisheries but no distinction is made in the recreational fisheries statistics between diadromous and freshwater forms. The Atlantic sturgeon (family Acipenseridae, Acipenser oxyrhynchus Mitchill) has been recorded from the Miramichi River on occasion (McKenzie 1959; Randall et al. 1989).

Several diadromous fishes are prized for angling while others are prized for commercial uses. Several species are of special ecological interest because their spawning populations in the Miramichi represent the northern or southern limits of their distributions in the western Atlantic. The two gaspereau species, shad, and striped bass are at the northern limits of their distribution. In terms of abundance, tomcod is at the southern limit of its distribution. Atlantic salmon is also located towards the southern limit but within the most productive area of its distribution. Finally, smelt appear in the centre of their distribution (Chaput and LeBlanc 1991; Scott and Scott 1988).

Fish production in the freshwater fluvial habitat is low but the fisheries yield is high because of the high biomass produced in the marine environment (Randall et al. 1989). The population size of most diadromous species is unknown. Information is available for gaspereau and Atlantic salmon stocks where stock assessments describing the characteristics of the spawning population, the fishery exploitation rates and the levels of escapement have been prepared annually (Mowbray et al. 1993; Chaput et al. 1994). The status of the other species is poorly known or based on studies from several decades previously (Chaput and Randall 1990; McKenzie 1964).

In their migration between fresh water and the sea, diadromous fish invariably pass through the estuary. An estuary, by simple definition, is a place where the freshwater discharge from a river meets the saltwater of the ocean. Estuaries can be further described on the basis of the geomorphology or the stratification (St. Hilaire 1993) but for our purposes, the estuary defined as that area where mixing of freshwater and saltwater occurs will suffice. The estuary can have very diverse salinity and temperature profiles depending upon the season and the tidal cycle in contrast to freshwater areas which have a more stable structure, at least in terms of salinity and direction of water currents.

The purpose of this paper is to provide an overview of the spatial and temporal utilization of the Miramichi River and Bay by diadromous fishes. The first part describes life history characteristics as

Table 1. Diadromous species that utilize the Miramichi River.

Family	Scientific Name	Common Name
Petromyzontidae	*Petromyzon marinus* Linneaus	Sea lamprey
Anguillidae	*Anguilla rostrata* Lesueur	American eel
Clupeidae	*Alosa pseudoharengus* Wilson	Alewife
	Alosa aestivalis Mitchill	Blueback herring
	Alosa sapidissima Wilson	American shad
Salmonidae	*Salmo salar* Linneaus	Atlantic salmon
Osmeridae	*Osmerus mordax* Mitchill	Rainbow smelt
Gadidae	*Microgadus tomcod* Walbaum	Atlantic tomcod
Percichthyidae	*Morone saxatilis* Walbaum	Striped bass

well as seasonal and spatial differences in habitat utilization. The second part looks at the relative biomass of fish moving into and out of Miramichi River. I also examine fisheries for these resources and how catches have changed over time. I hope to provide a picture of the importance of the estuary in the life history of these species.

Sources of Data and Analysis

Spatial and temporal occurrence data are available from research programs, commercial fisheries, and recreational fisheries. Fish species have been identified, enumerated and sampled for biological characteristics at the index trapnet at Millbank fished annually from mid-May until the end of October, between 1954 and 1992. Since 1985, the trapnet has been constructed of 5 cm stretched mesh whereas the leaders are constructed of 12.7 cm stretched mesh. In the earlier years, the Millbank trapnet was constructed of 9 cm stretched mesh (Kerswill 1971).The catches of the smaller individuals of all species and of the smaller species, mainly smelt, at Millbank may not reflect the true abundance or size characteristics of these in the estuary.

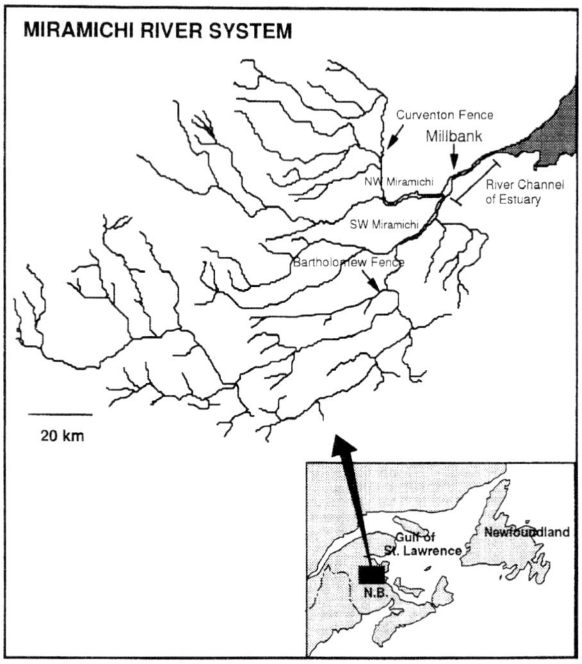

Fig. 1. Miramichi River and inner Miramichi Bay indicating location of Millbank and Curventon sampling facilities.

Data were available from electrofishing surveys conducted annually since 1971 at fixed sites in tributaries and in the main stems of both branches of the Miramichi River (Chadwick et al. 1985; Locke et al. 1993) as well as from counting fences operated in the Miramichi River system: Curventon fence in the Northwest Miramichi for 1951 to 1974 and Bartholomew River fence for 1976 to 1990 (Fig. 1).

Length, weight and age data were obtained at the Millbank trapnet, from sampling of the commercial fisheries and at counting fences. Sampling programs for Atlantic salmon are described in the annual stock status reports (Chaput et al. 1994). Gaspereau sampling was conducted both at Millbank and from the commercial fisheries (Chaput and LeBlanc 1990). Opportunistic sampling for striped bass and shad occurred at the Millbank trapnet. Samples from the smelt commercial fishery were obtained opportunistically during 1989 and 1990 (DFO unpubl. data). When biological characteristics were not available directly from the Miramichi River, values were obtained from the literature.

Estimates of biomass and population size of the species passing through the estuary annually were calculated using one of three methods: 1 - from published estimates of population size based on Sequential Population Analysis (SPA), 2 - catches and estimated or assumed efficiencies at the Millbank trapnet, or, 3 - using commercial landings and assumed exploitation rates in the fisheries. The estimates are not intended as exact figures but rather as relative indicators of the population size and biomass of diadromous species passing through the estuary.

The size of the spawning populations of gaspereau for the years 1982 to 1990, alewife and blueback herring combined, were obtained from the prefishery population estimates generated by SPA (LeBlanc et al. 1991). The fishery for gaspereau closes on June 15 in the Miramichi River but the spawning migration of gaspereau continues into July. At the Millbank trapnet, which sampled the entire gaspereau migration, on average 71% of the annual migration of gaspereau occurred prior to

June 16 (range 53–86% for 1982-90). Consequently, the population numbers and biomass were adjusted upward by the average proportion of the run which was available to the fishery.

The population sizes of small (fork length < 63 cm) and large (fork length >= 63 cm) Atlantic salmon were obtained from the annual stock status report (Moore et al. 1991). Between 1982 and 1984, the efficiency of the Millbank trapnet for Atlantic salmon was assumed to be 2.2% for large salmon and 3.4% for small salmon. Mark and recapture experiments, conducted between 1985 and 1987, indicated that the average efficiency of the Millbank trapnet was 1.5% for both small and large salmon (Randall et al. 1989) Between 1982 and 1990, the efficiency of the trapnet has averaged 1.9% ranging between 1.5 and 3.0% (Moore et al. 1991). The biomass of salmon was calculated using the estimated population size and an average weight of 1.55 kg for small salmon and 4.46 kg for large salmon (Randall and Chadwick 1983).

Population sizes of shad and striped bass were estimated from the catches of each species at the Millbank trapnet and an assumed catch efficiency. Shad and bass migrating into the estuary are midway in size between gaspereau and salmon (see Results). The efficiency of the Millbank trapnet for gaspereau averaged 0.7% (range 0.2–2.0%) between 1982 and 1990 (calculated as the ratio of the catch of gaspereau at Millbank prior to June 16 relative to the prefishery population estimates from SPA). The efficiency for salmon ranged between 1.5 and 3.0%. In the absence of better estimates, the efficiency of the Millbank trapnet for both shad and bass was assumed to be between that for gaspereau and salmon. The efficiencies for gaspereau and salmon for 1982-90 averaged 1.3% (range 0.8–2.0%) and this point estimate was assumed to represent the efficiency for shad and bass. The biomass of shad was calculated from the estimated population size and an average weight of 1.87 kg (DFO unpublished data). Striped bass biomass was calculated assuming an average weight of 1.25 kg per bass (Chaput and Robichaud 1995).

For smelt, tomcod, and eels, population size and biomass was estimated from the catches in the commercial fishery divided by the exploitation rate. (McKenzie 1964) indicated that between 3 and 5% of the smelt population was exploited by the commercial fishery in the Miramichi River. Because the fishery closes at the end of February, only part of the stock would have moved into the inner bay and been exploited (McKenzie 1964). The estimated biomass of the smelt population for 1982-90 was calculated from the landings in the commercial fishery and a fishing mortality of 5%. Population numbers of smelt were estimated assuming an average weight in the commercial fishery of 50 g per fish (DFO unpubl. data). Exploitation rates in the tomcod and eel fisheries are unknown but assumed to be at least as high as for smelt since these fisheries use similar gear and fishing practices. For tomcod and eels, an annual fishing mortality rate of 10% was assumed. Numbers of fish were obtained from average weights for eels of 0.2 kg per fish (Jessop 1987) and for tomcod, 0.15 kg per fish.

Landings from Northumberland County (statistical districts 70 to 73) were obtained from LeBlanc and Chaput (1991). Recent data for 1989 to 1992 were obtained from Statistics Branch, Dept. of Fisheries and Oceans, Moncton. Salmon landings were augmented with additional data for the years 1949 to 1965 from Allen and Lindsey (1967) and for 1971 to 1984 from the Redbook series of commercial catches (O'Neil et al. 1984).

Life Histories of Diadromous Fish

The early life history characteristics of most diadromous fish in the Miramichi River are poorly known. With exception of Atlantic salmon, whose juvenile stages have been extensively studied over the last several decades (Elson 1974; Locke et al. 1993), spawning areas, growth rates, sizes at migration, feeding habits and seasonal distributions can only be described in general terms and often based on studies from other areas. We have substantially more information about the adults.

There are large differences in the life histories and extent of diadromy (Table 2). The proportion of time spent in freshwater, estuarine and sea water varies among and even within a species (Fig. 2). Gaspereau spend comparatively

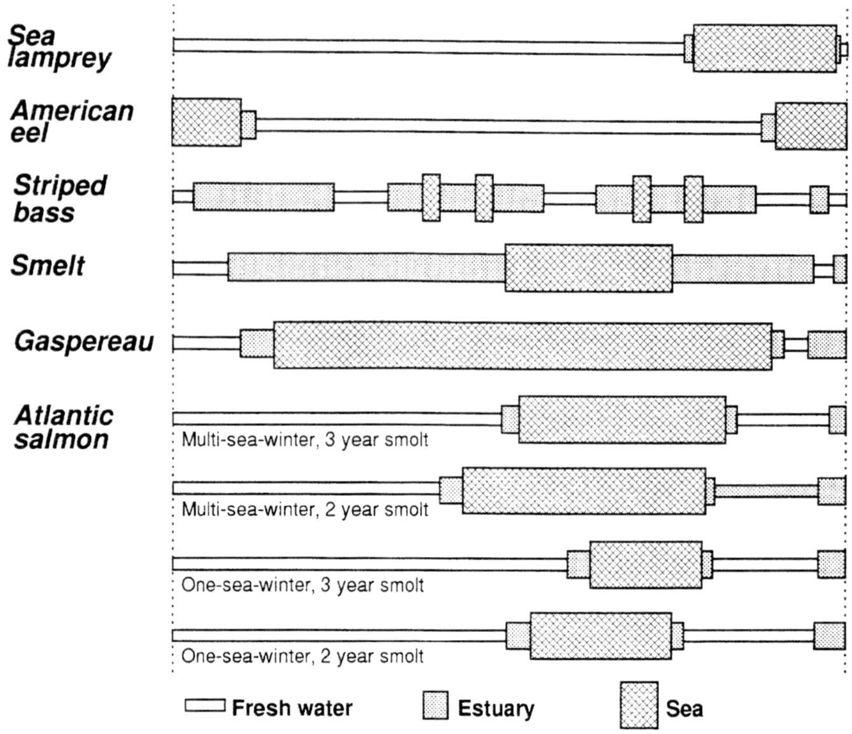

Fig. 2. Comparative life histories of diadromous fish from the Miramichi River in terms of timing and proportion of the life cycle spent in fresh water, estuary and the sea.

Table 2. Life history characteristics of diadromous fish of the Miramichi River.

Species	Spawning habitat	Nursery habitat	Average age at first maturity	Fecundity (eggs per female)	Spawning potential	Spawning season	Stock fidelity
Sea lamprey	fresh water	fresh water	8–10[a]	210 000[b]	semelparous[b]	June & July[b]	unknown
American eel	marine	marine	6–22 or more years[e]	unknown	semelparous[c]	Feb.–Apr.[j]	panmictic population[c]
Blueback herring	fresh water and brackish	fresh water	3–5 years[h]	185 000[f]	iteroparous	June & July	very high >95% (DFO unpublished data)
Alewife	fresh water	fresh water and brackish	3–4 years[h]	150 000[f]	iteroparous	June & July	very high >95% (DFO unpublished data)
American shad	fresh water	fresh water and brackish	4–6 years[i]	129 000[i]	iteroparous[i]	June	very high >95%[l]
Smelt	fresh water	brackish[k]	2–3 years[k]	8 000–60 000[k]	probaby iteroparous	Apr.– June[k]	high[k]
Striped bass	fresh water and brackish	brackish	3–6 years[d]	500 000–3 000 000[n]	iteroparous	May–June	very high >95%[m]
Atlantic salmon	fresh water	fresh water	3–5 years	3 000–7 000[o]	iteroparous	Oct. & Nov.	very high >95%

References: [a]Beamish (1980);[b]Beamish and Potter (1975); [c]Helfman et al. (1987); [d]Hogans and Melvin (1984); [e]Jessop (1987); [f]Jessop (1993); [g]Johnston and Cheverie (1988); [h]LeBlanc et al. (1991); [i]Leggett and Carscadden (1978); [j]McCleave et al. (1987); [k]McKenzie (1964); [l]Melvin et al. (1986); [m]Nichols and Miller (1967); [n]Rago and Goodyear (1987); [o]Randall (1989).

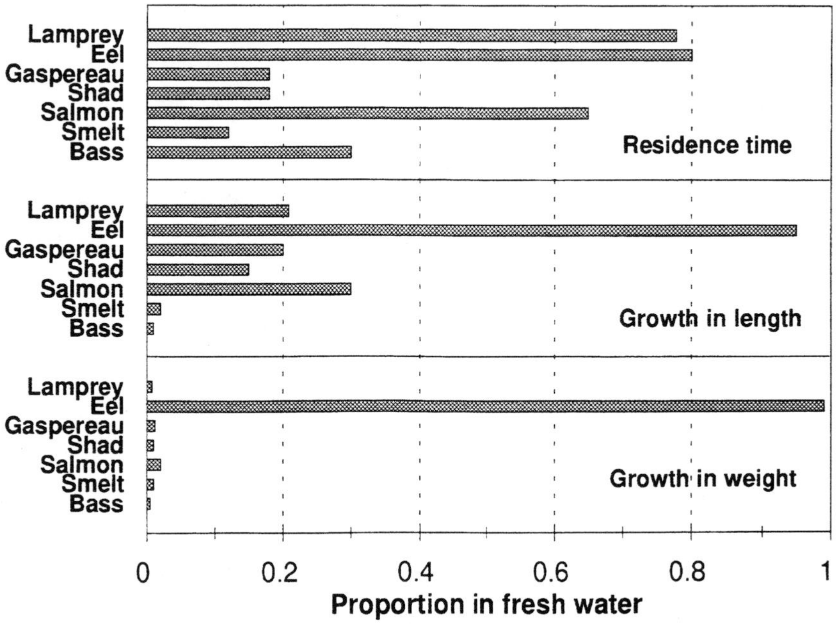

Fig. 3. Proportion of the life cycle, growth in length and growth in weight that occurs in fresh water for selected diadromous fish from the Miramichi River.

little time in fresh or brackish water and complete 80-90% of their growth in the sea (Fig. 3). By contrast, Atlantic salmon may spend from 40 to 60% of their life cycle in fresh water (Fig. 2, 3). Marine growth in salmon is high: 75% of the growth in length and 98% of the growth in weight of small salmon, 99% of the growth for large salmon. Small salmon are mostly one-sea-winter maiden salmon spawners whereas large salmon are a mixture of multi-sea-winter maiden and previous spawners. Smelt and striped bass complete most of their growth in the estuary (Fig. 2, 3); even for these species, the freshwater environment is used mainly for spawning and hatching. The eel, except for the first year or two during the elver stage, completes most of its growth in freshwater, returning to the sea to spawn (Fig. 2, 3).

The diadromous species which spend the greatest proportion of their life cycle in fresh water are also the ones which penetrate farthest into the freshwater environment. Atlantic salmon and lamprey are found throughout the Miramichi River system; eggs are laid and juveniles grow in both main branches and all the tributaries of the river (Fig. 4). Eels are less abundant than the salmon juveniles, but are distributed as extensively (Fig. 4). The distribution of salmon, eels and lamprey is not limited by elevation above sea level although lamprey were seen at a fewer number of sites (50%) above 200 m elevation than were eels (75%) and salmon (100%). The upper elevation sites are characterized by steeper gradients with proportionally less of the sand and silt substrate favoured by lamprey for spawning and early growth. By contrast, striped bass, gaspereau and smelt do not penetrate far into the river and use the lower reaches for spawning and overwintering. The freshwater penetration of these species is about 30 m above sea level (Fig. 4).

Species with extensive distribution in fresh water have equally extensive distributions at sea (Fig. 5). For example, Atlantic salmon has the deepest freshwater penetration and also ranges widely at sea, feeding and growing in waters from the Grand Banks to the coasts of Newfoundland and Labrador and Greenland (Fig. 5) (Kerswill 1971; Saunders 1969; Turner 1975). The extensive marine migrations undertaken by Atlantic salmon are nearly paralleled by the spawning migration of eels to the Sargasso Sea, southwest Atlantic (Kleckner et al. 1983). Lamprey are also widely distributed in the western Atlantic, from north of

Newfoundland, on the Grand Banks and on the Scotian Shelf (Dempson and Porter 1993; Halliday 1991). In contrast to eel and salmon, lamprey are parasitic on teleost fish and their distribution at sea is likely determined by the movements of the host species, which include Atlantic salmon as well as anadromous and marine clupeids (Beamish 1980).

Shad and gaspereau are less extensive migrants but leave the Gulf of St. Lawrence during the winter months, overwintering in the western Atlantic off Nova Scotia and perhaps further south (Crawford and Tully 1989; Dadswell et al. 1987). Gaspereau return in the spring to the Gulf of St. Lawrence rivers arriving progressively later in an east to west direction (Chadwick and Claytor 1989; Chaput and LeBlanc 1991). Striped bass are highly migratory undertaking coastal feeding migrations from spring to fall and overwintering in fresh water (Hogans and Melvin 1984; Melvin 1991). Smelt and tomcod are likely the least migratory diadromous fish. Studies from other areas in the Gulf of St. Lawrence indicate that there is minimal exchange of smelt (Fréchet et al. 1983; McKenzie 1964) and tomcod (Vladykov 1955) populations between neighbouring large embayments.

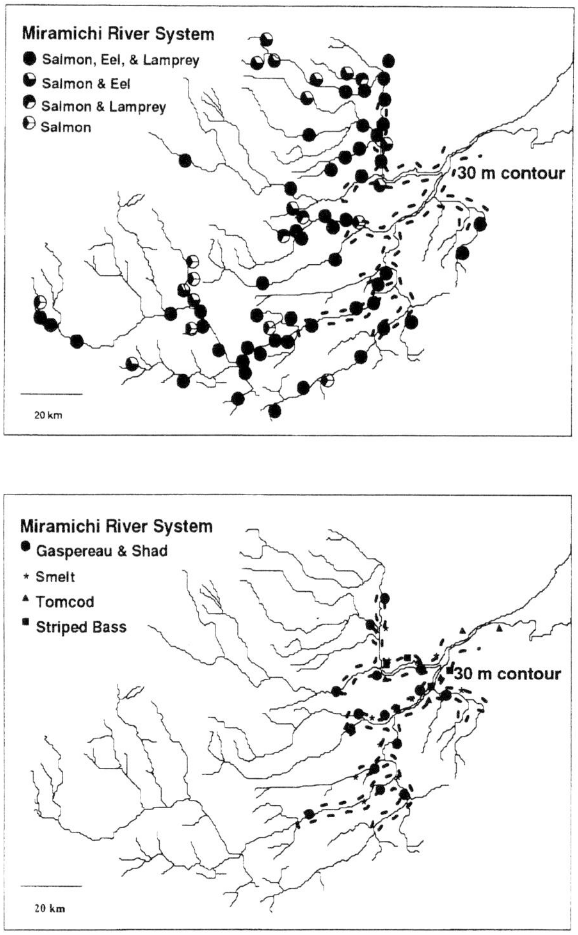

Fig. 4. Freshwater penetration of diadromous fish in the Miramichi River. Distribution of salmon, eels and lamprey (*upper*) based on electrofishing data from 1971 to 1992. Distribution of gaspereau, shad, tomcod, smelt, and bass (*lower*) based on observations at counting fences and anecdotal information.

Seasonal Migrations

Spring (April to June) is the most intense period in terms of diversity and abundance of both upstream and downstream migrating fish (Fig. 6). Gaspereau and shad have intense runs of relatively short duration with peak daily catches exceeding 4% of the total catch for the year (Fig. 6). Sea lamprey was infrequently reported from the Millbank trapnet but many were enumerated at the Curventon fence. Spawners migrated upstream in June and descended shortly after spawning before dying (Fig. 6). Eels also migrated downstream through the Curventon fence with peak movements occurring from the end of May to the end of June. Atlantic salmon were generally enumerated at Millbank from the end of May onwards in two modes: the first in June and July followed by a second run in September. Large salmon tend to return earlier in the year than small salmon. The maximum daily run represents about 1% of the annual run for the large salmon component whereas small salmon daily catches exceeded more than 2% of the total annual run in the summer period (Fig. 6). Striped bass, although captured at the trapnet throughout the fishing season (early May to end of October), had a distinct migration component in the months of May and June.

Although only the large fish were caught at Millbank, other species and life history stages pass through the estuary during April through June. Spent smelt descend from the spawning tributaries in May and June (McKenzie 1964; Chadwick et al. 1985; DFO Unpubl. Data from Bartholomew River counting fence) in large quantities at the same time as Atlantic salmon smolts are migrating downstream to the sea (Saunders 1967). Atlantic salmon kelts or black salmon, survivors of the previous year's spawning migration also descend to the sea and are actively feeding in tidal waters, probably on smelt.

The summer period is characterized by a less diverse large fish fauna but a species rich juvenile fish component. Bright salmon continue their upstream migrations from the sea, striped bass are feeding and migrating between estuaries and eels move out to sea. Salmon would be passing through or staging in the upper tidal areas (Randall et al. 1991). Young-of-the-year of smelt, gaspereau and shad are abundant in the tidal waters (Locke and Courtenay 1995).

Diversity and abundance increases in the fall as salmon complete their spawning migration, eels continue their seaward migration, sea lamprey ammocoetes move downstream into the estuary (Beamish and Potter 1975) while smelt and tomcod return to the estuary from the marine areas (McKenzie 1964; LeBlanc and Chaput 1991). Striped bass juveniles and older fish are using the estuary at this time, actively feeding and returning to fresh water for overwintering.

Activity in the winter period is dominated by smelt and tomcod. Tomcod spawn under the ice towards head of tide in late winter and smelt school in preparation for the upstream spawning migrations in early April. These two species are able to resist sub-zero water temperatures because they manufacture blood anti-freeze proteins (Ewart and Fletcher 1990; Reisman et al. 1984).

Biomass and Fisheries Exploitation

Generally, smaller fish are more abundant. The modal fork lengths of spawning populations from the smallest to the largest species are (Fig. 7): smelt (18 cm), tomcod (about 25 cm), gaspereau (28 cm), striped bass (38 cm), shad (40 cm), salmon (56 cm), eels (about 60 cm), and sea lamprey (70 cm). Smelt are the most abundant species in the estuary; they are 10 times more abundant than gaspereau, 20 to 30 times more abundant than tomcod, 100 times more abundant than eels, 1 000 times more abundant than salmon and shad, and 10 000 times more abundant than striped bass (Fig 8).

The total biomass of commercially exploitable diadromous fish passing through the

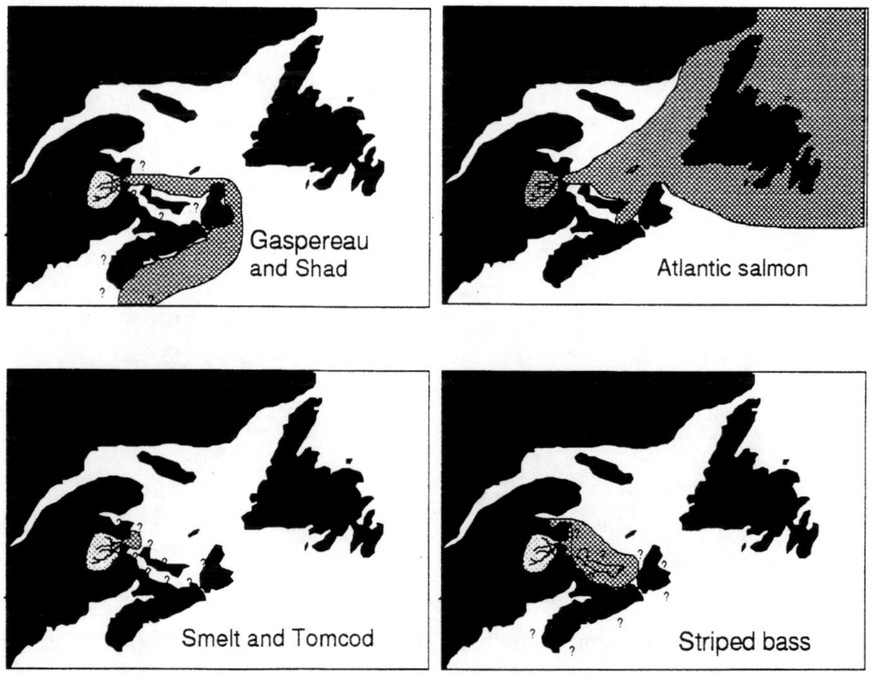

Fig. 5. Marine distributions of diadromous fish from Miramichi River.

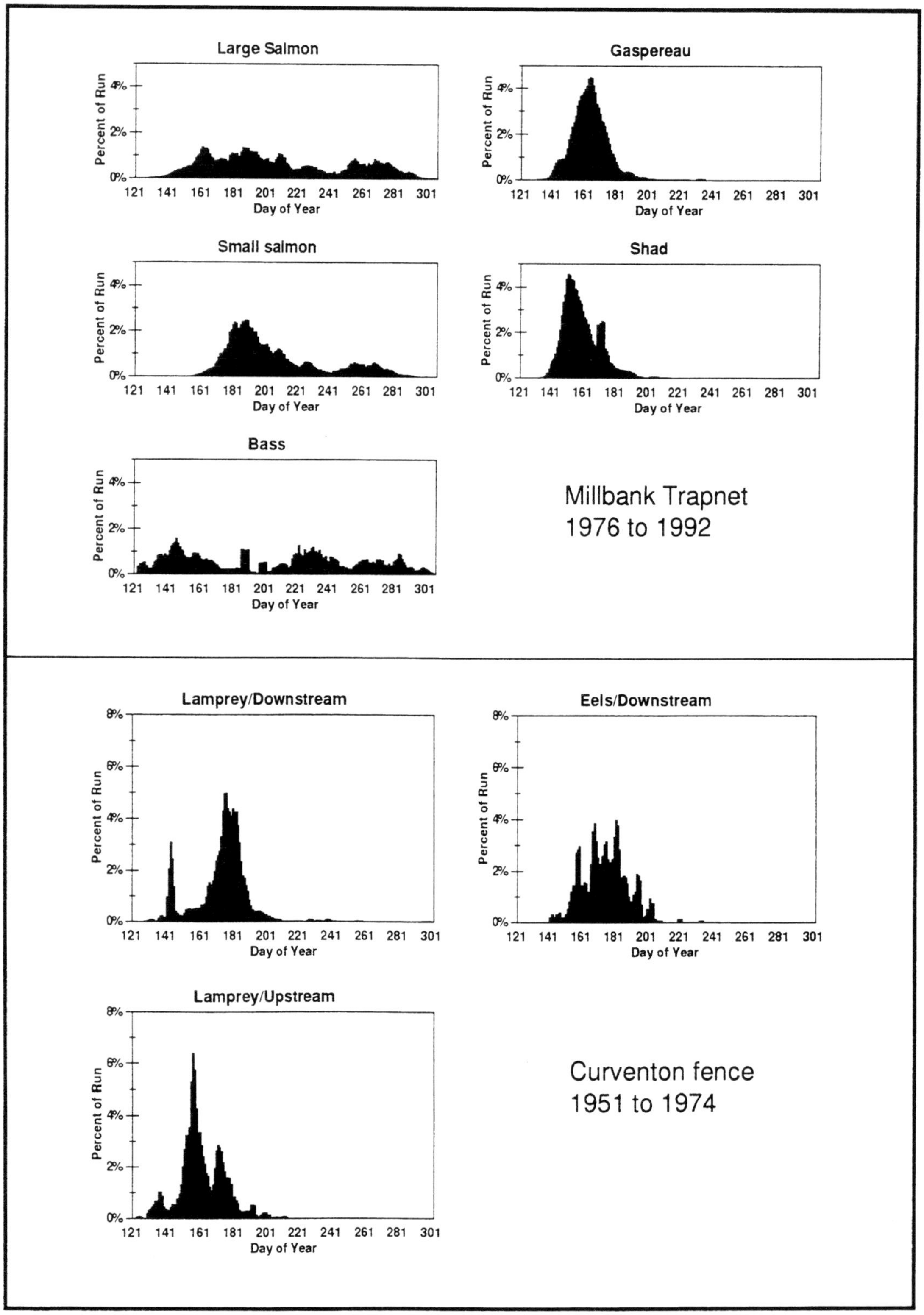

Fig. 6. Timing of upstream catches of diadromous fish at the Millbank trapnet (*upper*) between 1976 and 1992 and timing of upstream and downstream catches at Curventon fence (*lower*) during 1951 to 1974. Day of year corresponds to the following: 121 = May 1, 201 = July 20, 261 = Sept. 18.

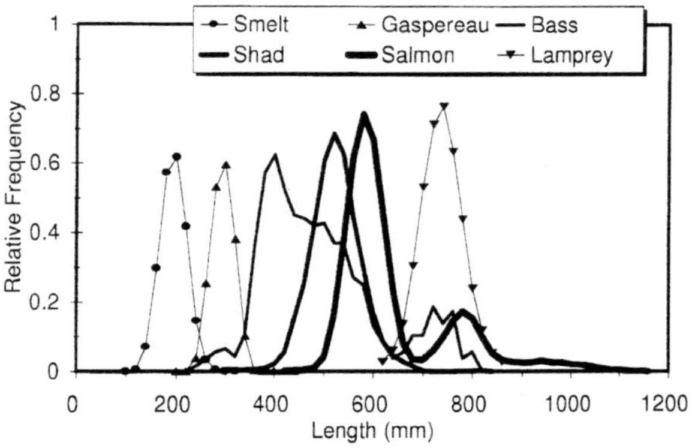

Fig. 7. Relative length frequencies of the spawning populations of diadromous fish migrating through the Miramichi River Estuary. Lamprey data from Beamish (1975).

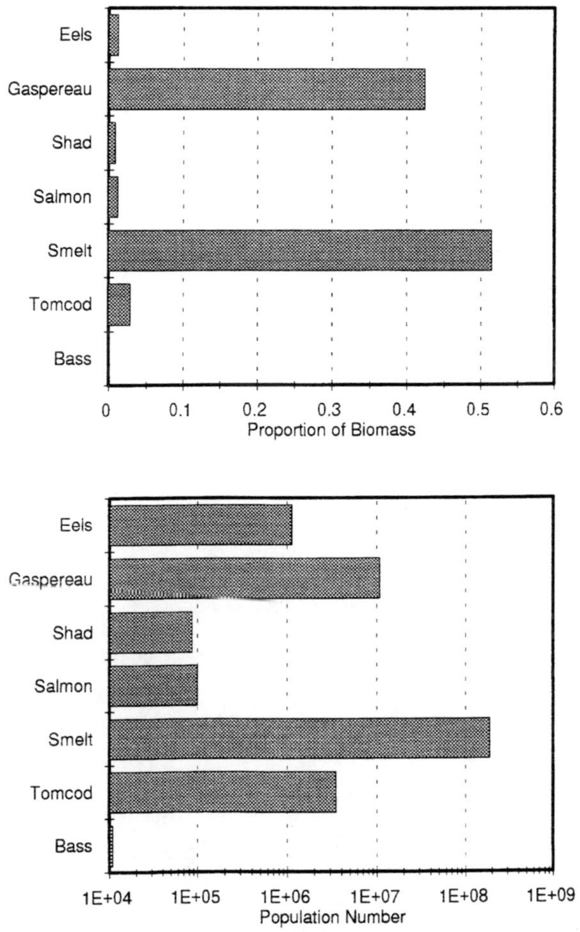

Fig. 8. Relative biomass and approximate population size of diadromous species passing annually through the Miramichi River Estuary.

estuary has probably exceeded 16 000 t in recent years. Again smelt dominate at over 50% of the total, followed closely by the two species of gaspereau at 42% (Fig. 8). Of the gaspereau species, blueback herring are about three times more abundant than alewife (LeBlanc et al. 1991). The remaining species represent a small proportion of total biomass. For example, mature Atlantic salmon represent on average about 1% of the spawning anadromous biomass passing through the estuary annually.

The biomass of diadromous fish passing through the estuary has varied over the years. In 1952, the commercial catch of all the diadromous species from the Miramichi area totalled 14 000 t, of which 90% was gaspereau. Except for the very high catches of gaspereau recorded between 1951 and 1953, commercial catches since 1980 have generally been higher than those recorded in the 1960's and 1970's and similar to those in the first half of the 20th century (Fig. 9). Whereas the Miramichi area contributed to about 50% of the total landings from the southern Gulf of St. Lawrence prior to 1940, the contribution today is just over 30% (Fig. 9).

The relative importance of the different diadromous species to the total landings has also changed. Prior to 1945, smelt dominated the catches and were 50% of the total landings. Gaspereau became more important in 1950 and now make up, at 80%, the largest proportion of the landings (Fig. 9). Catches of shad, eels, tomcod, salmon and bass have always been small relative to smelt and gaspereau, but particularly since 1980.

Trends in the landings of the fisheries can be categorized into three groups: (1) declining fisheries, (2) increasing fisheries, and (3) fisheries in a precarious state or that have essentially disappeared. But regardless of their state, large annual fluctuations are evident in all the fisheries.

Declining fisheries are marked by species that reside mainly in the estuary, smelt and tomcod. The catches of smelt for example, have declined steadily since 1917 to remain at low levels between 1960 and 1975 (Fig. 10). Sharp drops in catches were seen after 1940 and again after 1955. The rise after 1975 was short-lived and catches in the last three years have been about 400 t annually. Historically, the catches from the Miramichi River area contributed 40% of the smelt landings of the southern Gulf; this proportion has since declined to 30% (Fig. 10). Declining catches between 1931 and 1963 corresponded to a sharp decrease in licensed effort, dropping from a high of 3 000 licensed nets to 464 by 1962 (McKenzie 1964). There were 250 licenses in districts 70 to 73 in 1987 although 10 or more boxnets and bagnets may be fished with each license (Cairns 1989).

Tomcod and smelt catches are synchronous. The catches of tomcod have also declined since the early 1900's from about 500 t annually to current low levels of less than 50 t (Fig. 10). Tomcod are taken mainly as a bycatch in the smelt fishery between September and March (LeBlanc and Chaput 1991). Landings are generally determined by markets and very little of the product is consumed locally. The Miramichi River fishery constitutes the largest share of tomcod catches from the southern Gulf but has declined from 75% to less than 50% in the 1970's and 1990's (Fig. 10).

Gaspereau and eels support increasing fisheries (Fig. 11). Gaspereau now represent the largest landed catch of diadromous fish in the Miramichi River at just under 2,000 t annually. Except for very high reported catches from 1950 to 1955, the fishery in the last decade has produced the highest landings since 1917. As with the other

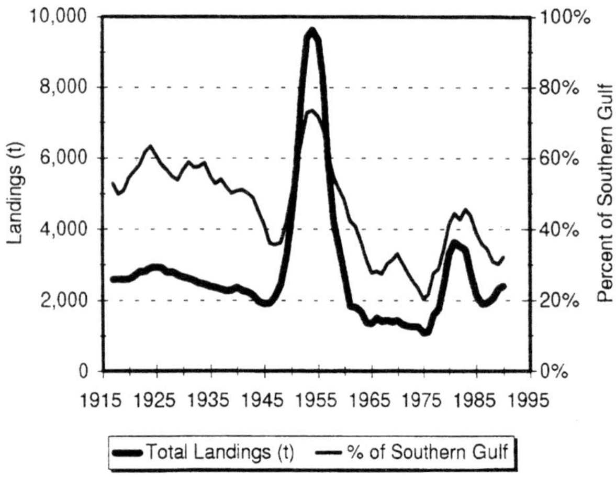

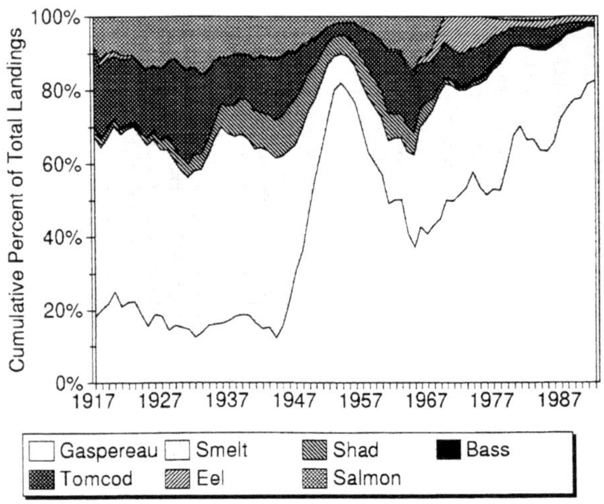

Fig. 9. Total commercial landings (*upper*) and relative change in species composition of the landings (*lower*) from Northumberland County, 1917 to 1992. All data were smoothed using a 5-year moving average.

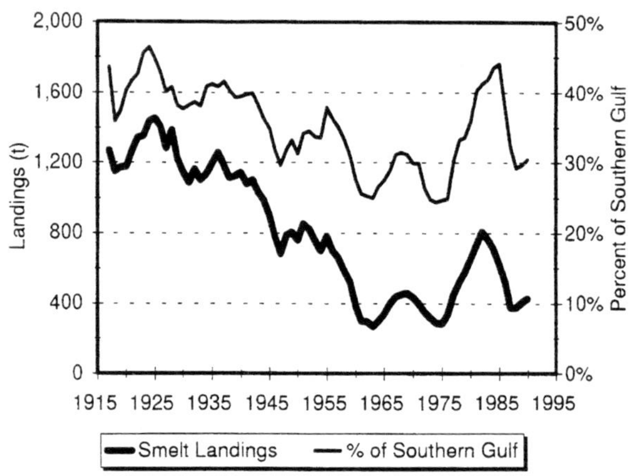

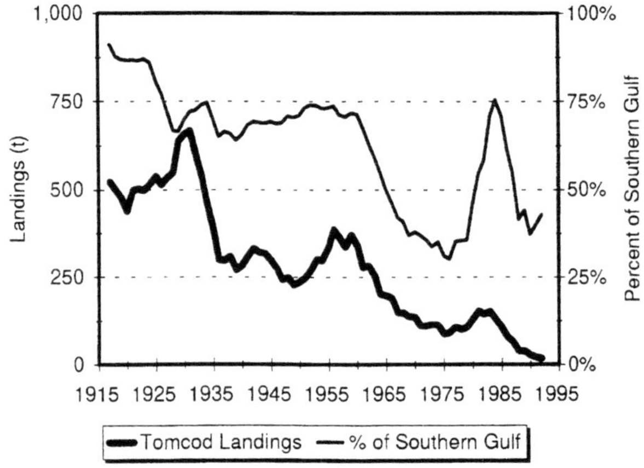

Fig. 10. Total commercial landings (t) of smelt (*upper*) and tomcod (*lower*) from Northumberland County, 1917 to 1992. All data were smoothed using a 5-year moving average.

species, the Miramichi River fishery now contributes less to the total landings from the southern Gulf than it did historically; current landings are between 20 and 45% of the total landings compared to 60% in the early part of the century. Eel catches in the last 20 years have been higher than those reported since 1917. Catches peaked between 1970 and 1972 and have since declined to about 40 t annually. Eel fisheries in other parts of the southern Gulf have been increasing; where the Miramichi once contributed 30% of the total Gulf catch, it now contributes between 10% and 20% of the total.

Precarious fisheries can be defined for shad, striped bass, and Atlantic salmon. The shad fishery has undergone the most dramatic rise and collapse of all the fisheries in the Miramichi (Fig. 12). Landings peaked in 1955 at 450 t but dropped to 100 t within ten years and then to 25 t annually since 1970. In recent years, shad are landed as bycatch from gaspereau trapnets. The catches of shad in the southern Gulf in the early portion of the time series were dominated by the Miramichi River fishery but since 1965, the Miramichi River contributes between 40% and 70% of the total catch. The striped bass fishery was never large in terms of landings and there were no reported landings between 1939 and 1967 (Fig. 12). The Miramichi River fishery has contributed as much as 75% but generally less than 50% of the total landings from the southern Gulf. Bass were historically harvested between May to December, as bycatch in the gaspereau fishery and from directed fisheries during the winter months (LeBlanc and Chaput 1991; McKenzie 1959). Atlantic salmon was exploited commercially in

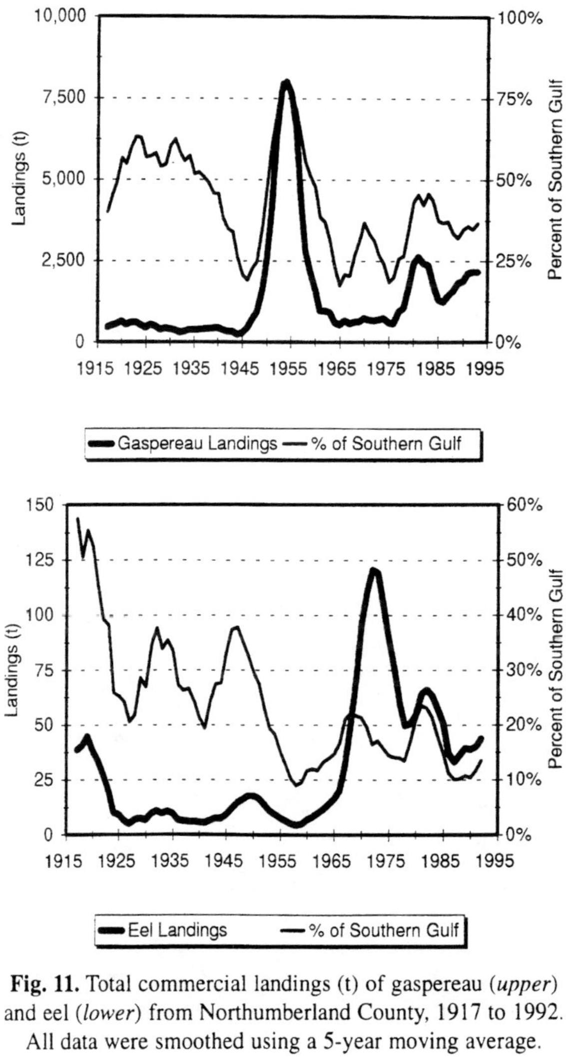

Fig. 11. Total commercial landings (t) of gaspereau (*upper*) and eel (*lower*) from Northumberland County, 1917 to 1992. All data were smoothed using a 5-year moving average.

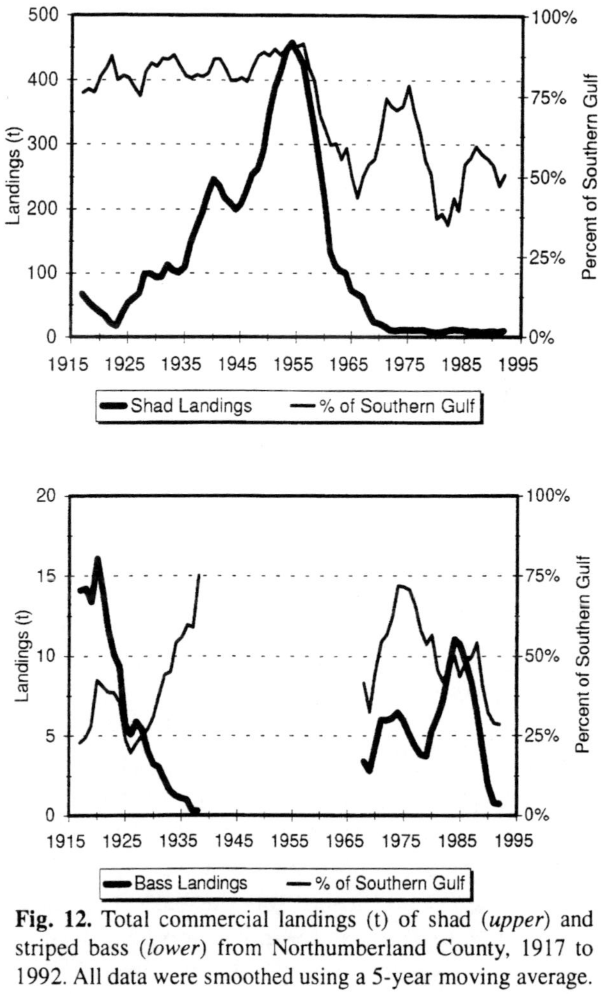

Fig. 12. Total commercial landings (t) of shad (*upper*) and striped bass (*lower*) from Northumberland County, 1917 to 1992. All data were smoothed using a 5-year moving average.

Miramichi Bay until 1970 and again for a brief period in 1981 to 1983. The annual commercial catch of salmon from Northumberland County peaked at 571 t in 1924 (Fig. 13). Annual catches prior to 1935 were generally above 300 t but decreased to less than 150 t between 1955 and 1965. Catches increased substantially in the late 60's only to decrease again to less than 100 t by 1971. Reported landings of salmon between 1971 and 1980 were legal bycatch from other fisheries. The commercial fishery has been closed since 1984 and returns to the estuary have increased to levels of about 300 t with 1992 returns at 400 t. Returns to the river in the absence of commercial fisheries are increasing but have yet to reach levels equal to the commercial catches during 1920 to 1935 (Fig. 13).

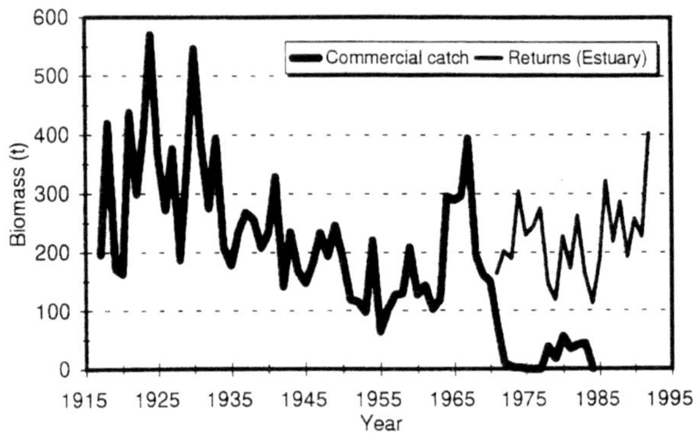

Fig. 13. Commercial catch (t) from Northumberland County and returns (t) of Atlantic salmon to the Miramichi River Estuary, 1917 to 1992.

Current Status and Conclusions

The status of Atlantic salmon in the Miramichi River is evaluated annually. Our understanding of Atlantic salmon life history and population dynamics, resulting from decades of study, accessibility of life stages which are easily sampled, combined with monitoring programs, have provided a reliable indication of the status of the Atlantic salmon resource. Atlantic salmon fisheries have changed dramatically. Commercial fisheries have been closed for the last 10 years. Recreational fisheries are limited to seasonal quotas of small salmon and mandatory release of all large salmon (defined earlier). As a result of these management measures, returns and spawning escapements of small salmon have doubled from the levels noted in the 1970's and large salmon returns have increased every year since 1986 (Fig. 14; Chaput et al. 1994). These management changes have also resulted in a four-fold increase in the survival rate to a second spawning migration of two-sea-winter maiden spawners since 1989 (Moore et al. 1995).

Control of fishing is not always sufficient to ensure the continued vitality of the population. Several man-made perturbations between 1954 and 1960 impacted on the freshwater stages of Atlantic salmon and on the migration and survival of spawners. There were spraying with DDT for the control of spruce budworm (Zitko 1995) piling pulpwood in the river bed and subsequent blockage to fish passage, and pollution from a base-

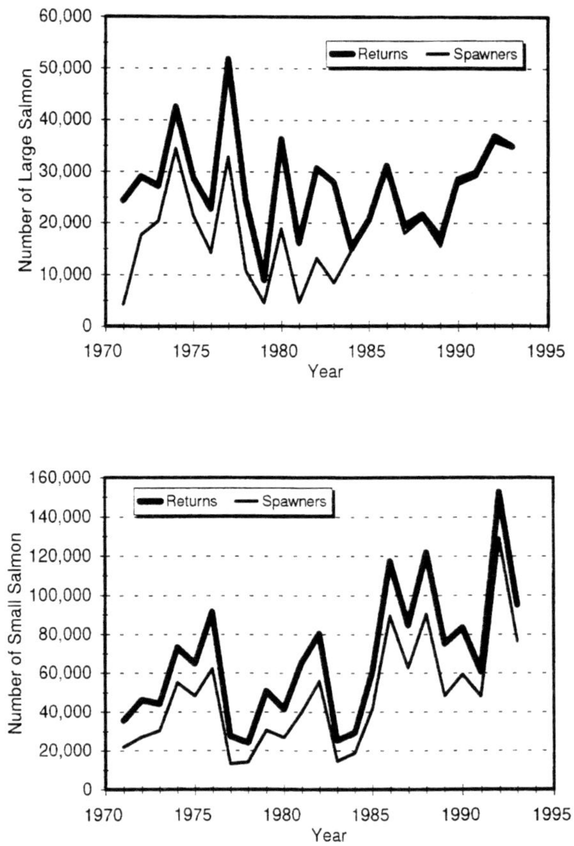

Fig. 14. Total returns to the Miramichi River Estuary and number of spawners of large (*upper*) and small (*lower*) salmon, 1971 to 1992.

metal mine (Elson 1967; Saunders 1969). Today, the quality of freshwater habitat has potential to be affected by accidental spills of toxic wastes and to alterations of stream profiles by forestry practices (Anon. 1992).

Gaspereau assessments have provided indications of the abundance of spawning runs, the age structure and estimates of the exploitation rate in the commercial fishery (LeBlanc et al. 1991). Exploitation occurs at a higher rate on alewife than on blueback herring; this is attributable to the difference in migration timing of the two species. Alewife enter the river almost two weeks before blueback herring and usually comprise more than half of the landings The alewife run finishes in early June while the blueback herring run continues to several weeks after the seasonal closure of the fishery (Chaput and LeBlanc 1991). The fishery also exploits a high proportion of new recruitment, varying from 37 to 85% for alewife and 17 to 66% for blueback herring. Spawning grounds, larval abundance and distribution, growth rates and other early life history characteristics of gaspereau from the Miramichi remain to be described.

The status of the striped bass resource of the Miramichi River, and of the Gulf of St. Lawrence is unknown but the population size is thought to be small (Chaput and Randall 1990). The catches at the Millbank trapnet fluctuated annually and the catches of bass during May and June have declined from the peak observed in 1980 (Chaput and Randall 1990). There is a spawning population in the Miramichi River (Bradford et al. 1995) and these bass are genetically distinct from those of the Bay of Fundy and the eastern seaboard of the United States of America (Wirgin et al. 1993) Striped bass are very susceptible to anthropogenic inputs from industrial developments. High levels of mercury were measured in the musculature of striped bass from the Miramichi (Anon. 1992). Declines of striped bass populations in other maritime locations have been attributed to excessive mortality of fertilized eggs and larvae caused by the interactive effects of low pH, aluminum, water hardness and salinity (Jessop 1990). Several causes for the extinction of the striped bass population from the St. Lawrence River by the mid-1960's (Robitaille et al. 1991) have been proposed: habitat degradation primarily from river channel dredging in the vicinity of spawning areas, water pollution and intensive fishery exploitation (Beaulieu 1985).

Both reduced fishing effort and habitat degradation could explain the declines in smelt landings. Declines between 1931 and 1963 correspond to reductions in licensed fishing effort rather than declining abundance because the catch rates remained unchanged and even improved (McKenzie 1964). Catches of smelt today are about one-third the level of the 1920's and a large part of the reduced catch is probably the result of reduced effort. A total of 221 license holders in the Miramichi River area potentially fish from 7 to 10 box nets and 4 to 5 bag nets each (Cairns 1989). About two-thirds of the fishers are classified as occasional, spending less than five months per year fishing (Philpott 1978). The decline that occurred in the 1950's correspond to the period of rapid development of the commercial smelt fishery in the Great Lakes. In 1966, the freshwater smelt fishery of Ontario landed 7,200 t, the highest of all the Canadian commercial landings from the Great Lakes (Scott and Crossman 1973), dwarfing by a factor of 15 times the landings from Northumberland County of the same year.The precipitous decline in the Miramichi catches between 1955 and 1960 also coincide with the development of a zinc, lead and copper mine in the upper reaches of the Northwest Miramichi which combined with pulp mill effluent caused severe pollution problems in the Northwest Miramichi and the estuary (Philpott 1978).

Because it is primarily a bycatch fishery, reported tomcod landings do not correspond to the actual abundance of the stock. Tomcod and smelt are closely linked in terms of the method by which they are exploited and the extensive use which they make of the estuarial waters of the Miramichi River. Compared to the other diadromous species, tomcod and smelt spend the greatest proportion of the time in the estuary, especially in the fall and winter. Spawning areas of tomcod are believed to be at the head of tide and smelt larvae are quickly transported downstream into the tidal areas soon after hatching (McKenzie 1964).

Size distribution, age distribution and indicators of abundance are absent for eel and shad. Although the commercial landings of eels fluctuate annually and have been higher in the last decade than at the turn of the century, continued high catches in the Miramichi may not be sustained. Recruitment of juvenile eels to Lake Ontario has undergone an 81-fold decline between 1985 and 1992 (Castonguay et al. 1994). Many possible causes for the recruitment decline have been considered; pollution, overfishing, and oceanographic changes. Because of panmixia (one common spawning population for the entire species), the recruitment decline could also have occurred in the Miramichi. Assuming that the eel fishery in the Miramichi exploits eels of about the same age as those from Nova Scotia, between 8 and 17 years old (Jessop 1987), then reduced abundance and probably reduced landings would be expected after 1993.

Shad abundance in the Miramichi is also unknown but based on counts of shad at the Millbank trapnet between 1976 and 1992, shad were most abundant during 1986 to 1990. The extremely high catches recorded in the Miramichi in the early 1950's were also observed in other areas of the Gulf of St. Lawrence, including the fisheries of the St. Lawrence River (Provost et al. 1984). The St. Lawrence River fishery is now essentially non-existent (LeBlanc and Chaput 1991), the loss attributed to a combination of impediments to migration, pollution and overfishing (Provost et al. 1984).

Most diadromous species utilizing the Miramichi River and Estuary have sustained populations in spite of the numerous stresses imposed by the environment and human activity. Atlantic salmon have a low post-spawner survival, in part because of human exploitation (Moore et al. 1995). Sea lamprey die after spawning. These two species make the deepest excursions into fresh water for spawning, the eggs are deposited in gravel redds protected from predators and ensured of relatively constant environmental conditions during incubation. They make limited use of the estuary. There have been major pollution stresses in the Miramichi (Zitko 1995) but when these were removed, the populations were able to recover and Atlantic salmon are now increasing in abundance.

In contrast, striped bass and shad are two species at the northern limit of their distribution which are sustained at low and widely fluctuating levels. Gaspereau are also at the northern limit and the present level of abundance would not sustain the exploitation rates which generated the high landings recorded in the 1950's. The larvae and juvenile stages of these species make extensive use of the estuary (Locke and Courtenay 1995). The adults are spring spawners, the eggs hatch quickly and the larvae become pelagic and are washed downstream soon after hatching. These species are also iteroparous; a life strategy to counteract the greater environmental variation expected in northern locations thereby producing greater population stability (Leggett and Carscadden 1978).

Smelt is very abundant in the Miramichi as compared to bass, gaspereau and shad. It is not at the northern limit of its distribution and yet appears to have declined in abundance. Smelt larvae are washed down to the estuary soon after hatching and the majority of smelt tend to die after spawning (McKenzie 1964). This species does not have the benefit of several cohorts spawning in a given year as do bass, gaspereau and shad and the larvae are exposed to large environmental variation in the estuary in contrast to salmon and lamprey.

Landings alone may be insufficient for determining the exact status of the resource but when the landings of all the diadromous species which make extensive use of the estuary have declined over time, changes in the nature of the estuary could be a possible cause. Industrial development is concentrated in the estuary (Philpott 1978) and discharges from their activities have detectable effects on movements of fish into and through the estuary and on spawning success and survival (Elson 1974; Stasko 1975). Under current levels of human activity, the abundance of these diadromous fish will remain below the levels observed in the early 1900's. Efforts should be made to reduce and eliminate habitat perturbations, many of which are known to have or are suspected of having detrimental consequences on the sustainability of the populations.

References

Allen, K.R., and Lindsey, J.K. 1967. Commercial catches of Atlantic salmon in the Maritimes area 1949-1965. Fish. Res. Board Can. Tech. Rep. No. 29.

Anonymous. 1992. Summary - Final Report. Miramichi River Environmental Assessment Committee 1989-1992. 157p.

Beamish, F.W.H. 1980. Biology of the North American anadromous sea lamprey, *Petromyzon marinus*. Can. J. Fish. Aquat. Sci. 37: 1924-1943.

Beamish, F.W.H., and Potter, I.C. 1975. The biology of the anadromous Sea lamprey (*Petromyzon marinus*) in New Brunswick. J. Zool., Lond. 177: 57-72.

Beaulieu, H. 1985. Rapport sur la situation du bar rayé (*Morone saxatilis*). Faune et flore à protéger au Québec. Association des biologistes du Québec, Publication No. 7. 53p.

Bradford, R.G., Robichaud, K.A., and Courtenay, S.C. 1995. By-catch in commercial fisheries as an indicator and regulator of striped bass (*Morone saxatilis*) abundance in the Miramichi River Estuary, p. 249–259. *In* E.M.P. Chadwick [*editor*]. Water, science, and the public: the Miramichi ecosystem. Can. Spec. Publ. Fish. Aquat. Sci. 123.

Cairns, D.K. 1989. Gear types, seasonal distribution of effort, and bycatch of the smelt fishery in the southern Gulf of St. Lawrence. Can. Tech. Rep. Fish. Aquat. Sci. No. 1668: v + 20p.

Castonguay, M., Hodson, P.V., Couillard, C.M., Eckersley, M.J., Dutil, J.-D., and Verreault, G. 1994. Why is recruitment of the American eel, *Anguilla rostrata*, declining in the St. Lawrence River and Gulf? Can. J. Fish. Aquat. Sci. 51: 479-488.

Chadwick, E.M.P., Alexander, D.R., Gray, R.W., Lutzac, T.G., Peppar, J.L., and Randall, R.G. 1985. 1983 Research on anadromous fishes, Gulf Region. Can. Tech. Rep. Fish. Aquat. Sci. No. 1420: xi + 69p.

Chadwick, E.M.P., and Claytor, R.R. 1989. Run timing of pelagic fishes in Gulf of St Lawrence: area and species effects. J. Fish. Biol. 35 (Suppl. A): 215-223.

Chaput, G.J., and LeBlanc, C.H. 1990. Evaluation of the 1989 gaspereau fishery (*Alosa aestivalis* and *A. pseudoharengus*) from the Miramichi River, New Brunswick. CAFSAC Res. Doc. 90/32: 43p.

Chaput, G.J., and LeBlanc, C.H. 1991. Les pêches commerciales de poissons dans les baies, estuaires et rivières du sud-ouest du golfe du Saint-Laurent, p. 293-301. *Dans* J.-C. Therriault [*éditeur*] Le golfe du Saint-Laurent : petit océan ou grand estuaire? Publ. spéc. can. sci. halieut. aquat. 113.

Chaput, G.J., Moore, D., Biron, M., and Claytor, R. 1994. Stock status of Atlantic salmon (*Salmo salar*) in the Miramichi River, 1993. DFO Atl. Fish. Res. Doc. 94/20: 80p.

Chaput, G.J., and Randall, R.G. 1990. Striped bass (*Morone saxatilis*) from the Gulf of St. Lawrence. Can. Atl. Fish. Sci. Adv. Comm. Res. Doc. 90/71: 29p.

Chaput, G.J., and Robichaud, K.A. 1995. Size and growth of striped bass, *Morone saxatilis*, from the Miramichi River, p. 161–176. *In* E.M.P. Chadwick [*editor*]. Water, science, and the public: the Miramichi ecosystem. Can. Spec. Publ. Fish. Aquat. Sci. 123.

Crawford, R., and Tully, D. 1989. The biology of gaspereau from Pictou Harbour, Nova Scotia. NS Dep. Fish. Manuscr. Tech. Rep. Ser. Proj. No. 89-01: 41p.

Dadswell, M.J., Melvin, G.D., Williams, P.J., and Themelis, D.E. 1987. Influences of origin, life history, and chance on the Atlantic coast migration of American shad. Amer. Fish. Soc. Symp. 1: 313-330.

Dempson, J.B., and Porter, T.R. 1993. Occurrence of sea lamprey, *Petromyzon marinus*, in a Newfoundland river, with additional records from the northwest Atlantic. Can. J. Fish. Aquat. Sci. 50: 1265-1269.

Dodson, J.J., Leggett, W.C., and Jones, R.A. 1972. The behavior of adult American shad (*Alosa sapidissima*) during migration from salt to fresh water as observed by ultrasonic tracking techniques. J. Fish. Res. Board Can. 29: 1445-1449.

Elson, P.F. 1967. Effects on wild young salmon of spraying DDT over new Brunswick forests. J. Fish. Res. Board Can. 24: 731-767.

Elson, P.F. 1974. Impact of recent economic growth and industrial development on the ecology of Northwest Miramichi Atlantic salmon (*Salmo salar*). J. Fish. Res. Board Can. 31: 521-544.

Ewart, K.V., and Fletcher, G.L. 1990. Isolation and characterization of antifreeze proteins from smelt (*Osmerus mordax*) and Atlantic herring (*Clupea harengus harengus*). Can. J. Zool. 68: 1652-1658.

Fréchet, A., Dodson, J.J., and Powles, H. 1983. Use of variation in biological characteristics for the classification of anadromous rainbow smelt (*Osmerus mordax*) groups. Can. J. Fish. Aquat. Sci. 40: 718-727.

Halliday, R.G. 1991. Marine distribution of the sea lamprey (*Petromyzon marinus*) in the northwest Atlantic. Can. J. Fish. Aquat. Sci. 48: 832-842.

Helfman, G.S., Facey, D.E., Stanton Hales, L., Jr., and Bozeman, E.L., Jr. 1987. Reproductive ecology of the American eel. Amer. Fish. Soc. Symp. 1: 42-56.

Hogans, W., and Melvin, G. 1984. Kouchibouguac National Park striped bass (*Morone saxatilis* Walbaum) fishery survey. Aquatic Industries Limited, St. Andrews, NB. 91p.

Jessop, B.M. 1987. Migrating american eels in Nova Scotia. Trans. Amer. Fish. Soc. 116:161-170.

Jessop, B.M. 1990. The status of striped bass in Scotia-Fundy region. Can. Atl. Fish. Sci. Adv. Comm. Res. Doc. 90/36: 22p.

Jessop, B.M. 1993. Fecundity of anadromous alewives and blueback herring in New Brunswick and Nova Scotia. Trans. Amer. Fish. Soc. 122: 85-98.

Johnston, C.E., and Cheverie, J.C. 1988. Observations on the diel and seasonal drift of eggs and larvae of anadromous rainbow smelt, *Osmerus mordax*, and blueback herring, *Alosa aestivalis*, in a coastal stream in Prince Edward Island. Can. Field-Nat. 102(3): 508-514.

Kerswill, C.J. 1971. Relative rates of utilization by commercial and sport fisheries of Atlantic salmon (*Salmo salar*) from the Miramichi River, New Brunswick. J. Fish. Res. Board Can. 28: 351-363.

Kleckner, R.C., McCleave, J.D., and Wippelhauser, G.S. 1983. Spawning of American eel, *Anguilla rostrata*, relative to thermal fronts in the Sargasso Sea. Environ. Biol. Fishes 9: 289-293.

LeBlanc, C.H., and Chaput, G.J. 1991. Landings of estuarine fishes in the Gulf of St. Lawrence 1917-1988 / Débarquements de poissons estuariens dans le golfe du Saint-Laurent 1917-1988. Can. Data Rep. Fish. Aquat. Sci. No. 842: 101p.

LeBlanc, C., Chaput, G., and Nielsen, G. 1991. Evaluation of the 1990 gaspereau fishery *(Alosa pseudoharengus)* and *(A. aestivalis)* from the Miramichi River, New Brunswick. CAFSAC Res. Doc. 91/4: 40p.

Leggett, W.C., and Carscadden, J.E. 1978. Latitudinal variation in reproductive characteristics of American shad (*Alosa sapidissima)*: evidence for population specific life history strategies in fish. J. Fish. Res. Board Can. 35: 1469-1478.

Locke, A., and Courtenay, S.C. 1995. Ichthyoplankton and invertebrate zooplankton of the Miramichi Estuary: 1918-1993, p. 97–120. *In* E.M.P. Chadwick [*editor*]. Water, science, and the public: the Miramichi ecosystem. Can. Spec. Publ. Fish. Aquat. Sci. 123.

Locke, A., Courtenay, S., and Chaput, G. 1993. Juvenile Atlantic salmon (*Salmo salar*) densities and egg deposition in the Restigouche and Miramichi rivers, New Brunswick. DFO Atl. Fish. Res. Doc. 93/26: 30p.

McCleave, J.D., Kleckner, R.C., and Castonguay, M. 1987. Reproductive sympatry of American and European eels and implications for migration and taxonomy. Amer. Fish. Soc. Symp. 1: 286-297.

McKenzie, R.A. 1959. Marine and freshwater fishes of the Miramichi River and estuary, New Brunswick. J. Fish. Res. Board Can. 16: 807-833.

McKenzie, R.A. 1964. Smelt life history and fishery in the Miramichi River, New Brunswick. Fish. Res. Board Can. Bull. No. 144: 77p.

McDowall, R.M. 1987. The occurrence and distribution of diadromy among fishes. Amer. Fish. Soc. Symp. 1: 1-13.

Melvin, G.D. 1991. A review of striped bass, (*Morone saxatilis*), population biology in eastern Canada, p.1-11. *In* R.H. Peterson [*editor*]. Proceedings of a workshop on biology and culture of striped bass (*Morone saxatilis*). Can. Tech. Rep. Fish. Aquat. Sci. No. 1832.

Melvin, G.D., Dadswell, M.J., and Martin, J.D. 1986. Fidelity of American shad, *Alosa sapidissima* (Clupeidae), to its river of previous spawning. Can. J. Fish. Aquat. Sci. 43: 640-646.

Mowbray, F., Chaput, G., and Courtenay, S. 1993. Assessment of the Miramichi River gaspereau fishery, 1991 and 1992. DFO Atl. Fish. Res. Doc. 93/51: 27p.

Moore, D.S., Courtenay, S., and Pickard, P.R. 1991. Status of Atlantic salmon in the Miramichi River during 1990. Can. Atl. Fish. Sci. Adv. Comm. Res. Doc. 91/8: 33p.

Moore, D.S., Chaput, G.J., and Pickard, P.R. 1995. The effects of fisheries on the biological characteristics and survival of mature Atlantic salmon (*Salmo salar*) from the Miramichi River, p. 229–247. *In* E.M.P. Chadwick [*editor*]. Water, science, and the public: the Miramichi ecosystem. Can. Spec. Publ. Fish. Aquat. Sci. 123.

Nichols, P.R., and Miller, R.V. 1967. Seasonal movements of the striped bass, *Roccus saxatilis* (Walbaum), tagged and released in the Potomac River, Maryland, 1959-61. Chesapeake Sci. 8: 102-124.

Northcote, T.G., and Larkin, P.A. 1989. The Fraser River: a major salmonine production system, p. 172-204. *In* D.P. Dodge [*editor*]. Proceedings of the International Large River Symposium. Can. Spec. Publ. Fish. Aquat. Sci. 106.

O'Neil, S.F., Bernard, M., and Gallop, P.A. 1984. 1984 Atlantic salmon commercial catch statistics Maritime provinces. Fisheries and Oceans, Freshwater and Anadromous Fish Division, Fisheries Research Branch, Scotia-Fundy Region. 57p.

Philpott, K.L. 1978. Miramichi Channel Study. Public Works Canada. 284p.

Provost, J., Verret, L. et Dumont, P. 1984. L'alose savoureuse au Québec : synthèse des connaissances biologiques et perspectives d'aménagement d'habitats. Rapp. manuscr. can. halieut. aquat. 1793 : xi + 114p.

Rago, P.J., and Goodyear, C.P. 1987. Recruitment mechanisms of striped bass and Atlantic salmon: comparative liabilities of alternative life histories. Amer. Fish. Soc. Symp. 1: 402-416.

Randall, R.G. 1989. Effect of sea-age on the reproductive potential of Atlantic salmon (*Salmo salar*) in eastern Canada. Can. J. Fish. Aquat. Sci. 46: 2210-2218.

Randall, R.G., and Chadwick, E.M.P. 1983. Assessment of the Miramichi River salmon stock in 1982. CAFSAC Res. Doc. 83/21: 24p.

Randall, R.G., O'Connell, M.F., and Chadwick, E.M.P. 1989. Fish production in two large Atlantic coast rivers: Miramichi and Exploits, p. 292-308. *In* D.P. Dodge [*editor*]. Proceedings of the International Large River Symposium. Can. Spec. Publ. Fish. Aquat. Sci. 106.

Randall, R.G., Pickard, P.R., and Moore, D. 1989. Biological assessment of Atlantic salmon in the Miramichi River, 1988. Can. Atl. Fish. Sci. Adv. Comm. Res. Doc. 89/73: 36p.

Randall, R.G., Wright, J.A., Pickard, P.R., and Warren, W.G.

1991. Effect of run timing on the exploitation by anglers of Atlantic salmon in the Miramichi River. Can. Tech. Rep. Fish. Aquat. Sci. No. 1790: viii + 46p.

Reisman, H.M., Kao, M.H., and Fletcher, G.L. 1984. Antifreeze glycoprotein in a "Southern" population of Atlantic tomcod. Comp. Biochem. Physiol. A, 78: 445-447.

Robitaille, J.A., Choinière, L. et Vigneault, Y. 1991. Identification des populations de poissons d'intérêt économique en situation précaire dans le réseau du Saint-Laurent et sélection des espèces pour des interventions immédiates. Rapp. tech. can. sci. halieut. aquat. 1810 : ix + 24p.

Saunders, R.L. 1967. Seasonal pattern of return of Atlantic salmon in the Northwest Miramichi River, New Brunswick. J. Fish. Res. Board Can. 24: 21-32.

Saunders, R.L. 1969. Contributions of salmon from the Northwest Miramichi River, New Brunswick, to various fisheries. J. Fish. Res. Board Can. 26: 269-278.

Scott, W.B., and Crossman, E.J. 1973. Freshwater fishes of Canada. Bull. Fish. Res. Board Can. 184: 966p.

Scott, B.V., and Scott, M.G. 1988. Atlantic Fishes of Canada. Can. Bull. Fish. Aquat. Sci. No. 219: 731p.

St-Hilaire, A. 1993. Étude hydrodynamique de l'estuaire de la rivière Miramichi durant la saison sans glace - 1991. M.Sc., Univ. de Moncton, Moncton, N.-B. 166p.

Stasko, A.B. 1975. Progress of migrating Atlantic salmon (*Salmo salar*) along an estuary, observed by ultrasonic tracking. J. Fish Biol. 7: 329-338.

Turner, G.E. 1975. Migration route and timing of Miramichi River salmon (*Salmo salar*) as indicated from recaptures of tagged smolts and adults. Environment Canada, Fish. Mar. Serv. Tech. Rep. Ser. No. MAR/T-75-7: 11p.

Vladykov, V.D. 1955. Poissons du Québec : les morues. Dépt. des Pêches, Prov. du Québec Album 4 : 12p.

Wirgin, I.I., Ong, T.-L., Maceda, L., Waldman, J.R., Moore, D., and Courtenay, S. 1993. Mitochondrial DNA variation in striped bass (*Morone saxatilis*) from Canadian rivers. Can. J. Fish. Aquat. Sci. 50: 80-87.

Zitko, V. 1995. Fifty years of research on the Miramichi River, p. 29–41. *In* E.M.P. Chadwick [*editor*]. Water, science, and the public: the Miramichi ecosystem. Can. Spec. Publ. Fish. Aquat. Sci. 123.

Beach seines are used to collect fish in shallow aquatic habitats. This photograph shows the gently-sloping shoreline habitat in the lower reaches of the Miramichi Estuary.

CHAPTER 9

Seasonal Abundance and Distribution of Fishes in the Miramichi Estuary

J. Mark Hanson and Simon C. Courtenay

Marine and Anadromous Fish Division, Science Branch, Gulf Fisheries Centre, P.O. Box 5030, Moncton, New Brunswick, Canada E1C 9B6

Hanson, J.M., and Courtenay, S.C. 1995. Seasonal abundance and distribution of fishes in the Miramichi Estuary., p. 141-160. *In* E.M.P. Chadwick [*editor*]. Water, science, and the public: the Miramichi Ecosystem. Can. Spec. Publ. Fish. Aquat. Sci. 123.

Abstract

The fish community of the Miramichi Estuary was sampled by beach seine, beam trawl, and bycatch from commercial trapnet fisheries between January 1991 and December 1993. Forty-seven species from 27 taxonomic families were captured, of which 12 species were abundant. Of the large-bodied species, smooth flounder (*Pleuronectes putnami*) was abundant all year and completed its life cycle in the Estuary. Adult winter flounder (*Pleuronectes americanus*), Atlantic tomcod (*Microgadus tomcod*), and rainbow smelt (*Osmerus mordax*) entered the Estuary during autumn, remained overwinter, and left the Estuary during early summer. Unlike Atlantic tomcod and rainbow smelt, winter flounder did not spawn in the Miramichi Estuary. Adult Atlantic salmon (*Salmo salar*) were common in the Estuary during spring and summer but were only captured by gaspereau trapnets during spring. Atlantic salmon smolts and parr were seldom caught in the Miramichi Estuary. Adult blueback herring (*Alosa aestivalis*), alewife (*Alosa pseudoharengus*), American shad (*Alosa sapidissima*), and striped bass (*Morone saxatilis*) moved through the Estuary to spawn in fresh water or near-fresh water during spring. Large white sucker (*Catostomus commersoni*) were very common in much of the Estuary during spring but the purpose for this movement from fresh to brackish water is unknown. During summer, adults of most large-bodied species left the Estuary; it functioned primarily as a nursery for blueback herring, alewife, rainbow smelt, Atlantic tomcod, and striped bass. During early autumn, the young blueback herring and alewife left the Estuary. In contrast, juvenile Atlantic tomcod and rainbow smelt remained in the estuary during autumn and winter. Of the small-bodied species, all sizes of Atlantic silverside (*Menidia menidia*) were present during spring, summer, and autumn. Fourspine stickleback (*Apeltes quadracus*) and blackspotted stickleback (*Gasterosteus wheatlandi*) were abundant during spring, when they spawned, and all sizes appeared to leave the Estuary during summer. Large numbers of juvenile Atlantic herring (*Clupea harengus*) and white hake (*Urophycis tenuis*) were present in the Estuary during autumn but it is unknown whether they stayed to overwinter. The Miramichi Estuary is a critical nursery area and overwinter habitat for many valuable fish species and any commercial or recreational development of the Miramichi ecosystem that might affect fish habitat must consider this fact during the planning stages.

Résumé

Nous avons échantillonné la communauté ichthyenne de l'estuaire de la Miramichi entre janvier 1991 et décembre 1993 à la senne de plage et au chalut à perche, et à partir des prises accessoires des pêches commerciales à la trappe. Quarante-sept espèces, appartenant à 27 familles taxinomiques, ont été capturées, parmi lesquelles 12 étaient abondantes. Parmi les espèces de bonne taille, la plie lisse (*Pleuronectes putnami*) était abondante toute l'année et passait tout son cycle vital dans l'estuaire. Les adultes de plie rouge (*Pleuronectes americanus*), de poulamon (*Microgadus tomcod*) et d'éperlan arc-en-ciel (*Osmerus mordax*) entraient dans l'estuaire en automne, y passaient l'hiver et le quittaient pendant l'été. À la différence du poulamon et de l'éperlan arc-en-ciel, la plie rouge ne frayait pas dans l'estuaire de la Miramichi. Les saumons atlantiques (*Salmo salar*) adultes étaient communs dans l'estuaire au printemps et en été, mais n'étaient capturés qu'au printemps dans les trappes à gaspareau. Les smolts et les tacons de saumon atlantique n'ont été que rarement capturés dans l'estuaire. Les adultes d'alose d'été (*Alosa aestivalis*), de gaspareau (*Alosa pseudoharengus*), d'alose savoureuse (*Alosa sapidissima*) et de bar rayé (*Morone saxatilis*) traversaient l'estuaire pour pondre en eau douce ou presque douce au printemps. Les gros meuniers noirs (*Catostomus commersoni*) étaient très communs dans une bonne partie de l'estuaire au printemps, mais on ne connaît pas la raison de ce déplacement de l'eau douce à l'eau saumâtre. Pendant l'été, les adultes de la plupart des espèces de bonne taille quittaient l'estuaire, qui fonctionnait principalement comme une nourricerie pour l'alose d'été, le gaspareau, l'éperlan arc-en-ciel, le poulamon et le bar rayé. Au début de l'automne, les juvéniles d'alose d'été et de gaspareau quittaient l'estuaire. Par contre, les juvéniles de poulamon et d'éperlan arc-en-ciel y restaient pendant l'automne et l'hiver. Parmi les petites espèces, on observait toutes les tailles de capucette (*Menidia menidia*) dans l'estuaire pendant le printemps, l'été et l'automne. L'épinoche à quatre épines (*Apeltes quadracus*) et l'épinoche tachetée (*Gasterosteus wheatlandi*) étaient abondantes au printemps, période de la fraye, et toutes les tailles semblaient quitter l'estuaire pendant l'été. De grands nombres de juvéniles de hareng (*Clupea harengus*) et de merluche blanche (*Urophycis tenuis*) étaient présents dans l'estuaire pendant l'automne, mais on ne sait pas s'ils y passaient l'hiver. L'estuaire de la Miramichi est une nourricerie et un habitat d'hivernage d'un intérêt critique pour de nombreuses espèces de poissons de valeur, et tout projet de développement commercial ou récréatif de l'écosystème de la Miramichi qui pourrait avoir des effets sur l'habitat du poisson doit tenir compte de cette réalité dès la phase de planification.

Introduction

Apart from sampling required for stock assessments of Atlantic salmon (*Salmo salar*) and gaspereau (alewife, *Alosa pseudoharengus*, and blueback herring, *Alosa aestivalis*) and information on migrating fish (Chadwick and Claytor 1989; Mowbray et al. 1993; Chaput et al. 1994; Chaput 1995; Moore et al. 1995), little work has been done to document the use of the Miramichi Estuary by any fish species. Unlike most estuaries studied elsewhere, the Miramichi freezes over during winter and the fish community is dominated by diadromous species. McKenzie (1959) provided a check list of fish species found in the Miramichi River, Estuary, and Outer Bay. Since then, little work has been done on the biology of fish inhabiting the Estuary with the exception of McKenzie's (1964) work on rainbow smelt (*Osmerus mordax*).

As a first step in describing the importance of the Estuary as a nursery area, feeding area, and overwinter refuge for a large number of fish species, this study provides an overview of the seasonal distributions and qualitative estimates of abundance of the fish community. Detailed analyses of the biology of some of the more abundant species, e.g., Atlantic tomcod (*Microgadus tomcod*), striped bass (*Morone saxatilis*), smooth flounder (*Pleuronectes putnami*), and winter flounder (*Pleuronectes americanus*), are ongoing.

Material and Methods

The fish community was sampled in an area from just outside the Barrier Islands to near the upstream boundary of salt water at Red Bank

(Fig. 1). We divided the Estuary into six locations that represent a gradient in salinity from near-fresh water at Red Bank to nearly full salt water at the Barrier Islands. This is a crude simplification because the salt wedge moves upstream and downstream with the tides and the seasonal location of the upstream edge is greatly affected by the amount of freshwater runoff (see Lafleur et al. 1995; St.-Hilaire et al. 1995). The near-freshwater location was defined as from Red Bank to the highway bridge on the Northwest Miramichi River. The Newcastle location was defined from the highway bridge to the Waferboard Plant near the Town of Chatham. The Chatham location was defined as from the Waferboard plant to Gordon Point. The Sheldrake Island location went from Gordon Point to Point aux Carr. The Inner Bay location was defined as from Point aux Carr to the inside of Fox Island. The Outer Bay location was defined as the shipping channel from between Portage and Fox Islands to about 5.5 km outside of the Barrier Islands. The seasons were defined as: spring (21 March to 20 June), summer (21 June to 20 September), autumn (21 September to 20 December), and winter (21 December to 20 March).

Fish samples were collected during all four seasons between January 1991 and February 1993. Four sources of samples were used in this study: seining; trawling; and bycatch from the smelt and gaspereau trapnet fisheries. These methods primarily sample demersal species. To sample more

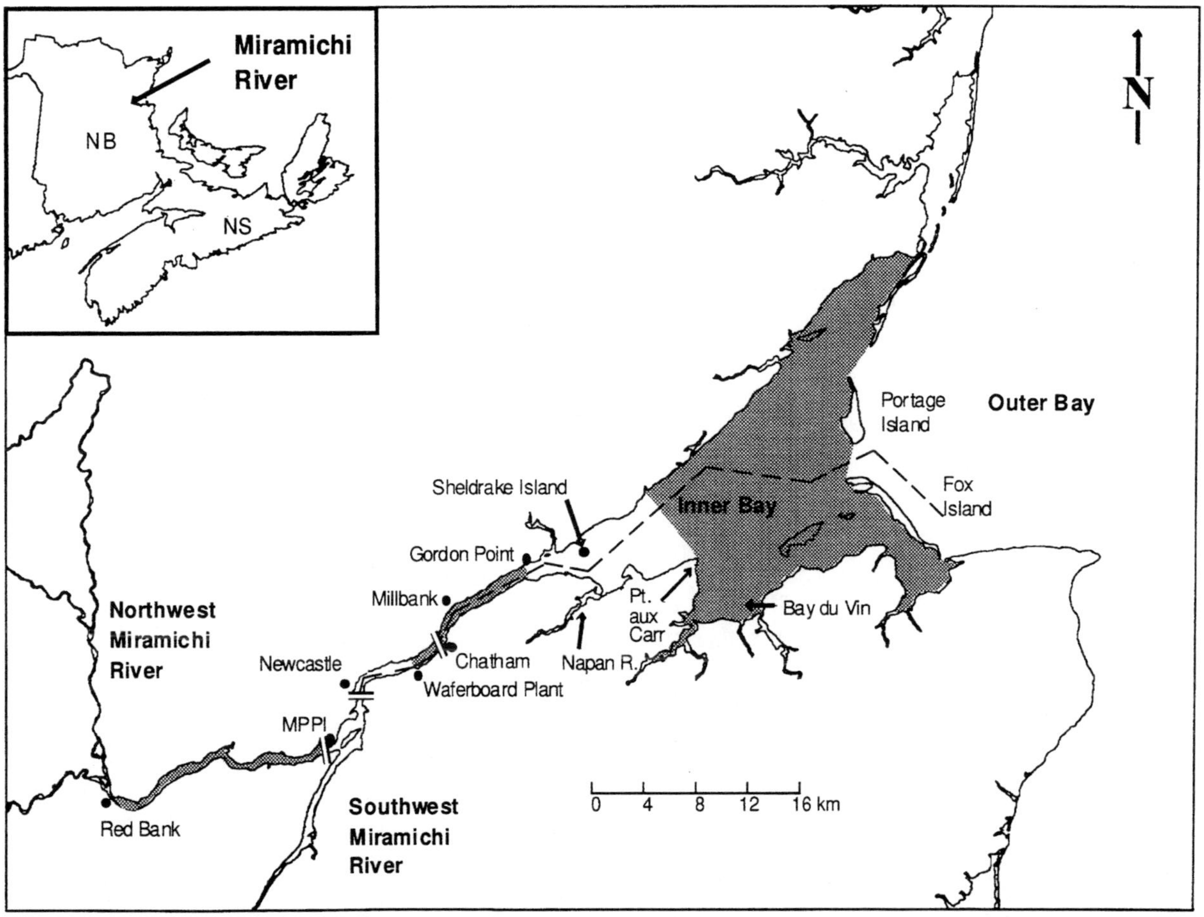

Fig. 1. Map of the Miramichi Estuary showing the place names and sampling locations mentioned in the text. The dashed line shows the main shipping channel. Moving from upstream to the Outer Bay, the first shaded section is the near freshwater location, the first clear section is the Newcastle location, the second shaded section is the Chatham location, the second clear section is the Sheldrake Island location, and the last two locations (Inner and Outer Bay) are marked on the map. The location of the Miramichi Pulp and Paper Incorporated mill is indicated by MPPI.

pelagic species (e.g., Atlantic herring *Clupea harengus*, Atlantic mackerel *Scomber scombrus*, American shad *Alosa sapidissima*) during all seasons, we would have had to deploy large numbers of gillnets. This would have resulted in unacceptable mortalities of Atlantic salmon and striped bass.

Seining was done with a 30 × 1.8 m beach seine (6-mm mesh) set from a boat to cover an area of about 500 m^2 (Hanson and Leggett 1985; Boisclair and Leggett 1985; Pierce et al 1990). As used, the seine was effective to a depth of 2 m and for a distance of about 40 m from shore. Twenty-four sites were chosen based on accessibility by boat between Red Bank and Point aux Carr. Because of the distance from the laboratory at Millbank, no sites were seined in the Inner Bay or Outer Bay locations. Seining was conducted from May to October 1991 from the junction of the two branches of the Miramichi River to Point aux Carr. Red Bank to Point aux Carr was sampled from June to October 1992. Seining was attempted at least monthly at each site. We successfully completed 368 seine hauls (Table 1) but coverage of the various locations was least complete during spring because severe weather conditions (especially wind) frequently prevented access to the sites downstream of Chatham. If the catch was small (< 5 kg), all fish were returned to the laboratory for detailed examination. In a few cases, very large numbers (> 1,000 individuals) of Atlantic silverside (*Menidia menidia*) were caught and a subsample of about 10% of the volume was returned to the laboratory. The fish samples were immediately placed on ice and transported to the laboratory where they were frozen for future analysis. All fish were identified to species, measured (fork length, if possible) to the nearest mm, and at least two fish per cm were weighed to the nearest 0.1 g.

Eight research surveys were conducted in the Miramichi Estuary: July 1991, 1992 and 1993; end of June 1992; May 1992 and 1993; and October 1991 and 1992. Trawling was done by means of a 2.5 × 0.5 m beam trawl (25-mm diamond mesh) fitted with a 6-mm mesh codend liner. Surveys were conducted from Newcastle to outside of the Barrier Islands on the m/v *Osborne*, CSS *Navicula,*

Table 1. Summary of number of samples collected seasonally during trawl surveys, seine surveys, and bycatch collections in the Miramichi Estuary, 1991 to 1993. The sampling locations were near freshwater (NFW), Newcastle (NEW), Chatham (CHA), Sheldrake Island (SHE), Inner Bay (INB), and Outer Bay (OUT).

	Location						
Season	NFW	NEW	CHA	SHE	INB	OUT	Sum
			Seine surveys (number of sets)				
Spring	18	42	6	28	0	0	94
Summer	48	46	20	28	0	0	142
Autumn	32	34	22	44	0	0	132
			Trawl surveys (number of sets)				
Spring	0	9	5	11	7	5	37
Summer	0	8	11	24	15	10	68
Autumn	0	12	17	21	13	8	71
			Smelt bycatch (number of sites)				
Autumn	0	0	7	23	4	0	34
Winter	0	0	0	28	14	0	42
			Gaspereau bycatch (number of sites)				
Spring	6	0	6	12	0	0	24

or CSS *Opilio*. The near-freshwater location was not accessible to the trawlers. Each tow was done at a speed of 6.5 km h^{-1} and lasted 10 minutes from the time the net reached bottom to the start of haul back. Trawling was restricted to depths ≥ 5 m, which resulted in most sets being done in, or adjacent to, the shipping channel. There is a large area with water > 5 m deep in the Inner Bay and this area was fully covered by the surveys. At all locations, sampling stations were selected arbitrarily to cover as wide an area as possible during each survey. In practice, the sampling design was that of a fixed station survey because only a very limited number of trawlable sites were possible in each sampling location due to the distance needed to deploy, fish, and recover the trawl. We successfully completed 176 sets (Table 1). The lower number of sets done during spring was due to one survey scheduled for early May 1992 being cancelled because severe ice conditions in the southern Gulf of St. Lawrence prevented the research vessel from reaching the Estuary. Extra sets were taken in the Sheldrake Island location during each survey to permit collection of additional live specimens for ongoing laboratory studies. For each set, the fish captured were sorted to species and either frozen for later detailed sampling (most occasions) or sampled on board. When sampled on board, all fish were measured (fork length, if possible) to the nearest mm, individual weights were measured using calibrated spring balances (to the nearest g) for at least two fish per cm length group, sex and maturity were determined, and stomach contents were preserved (either frozen or in 4% formalin solution). Weights of fish returned to the laboratory were measured to the nearest 0.1 g.

The bycatch from the smelt trapnet fishery (about 9 mm mesh in the nets) was sampled during the open-water period (15 October to early December 1991 and 1992) from upstream of Chatham, near Millbank, near Sheldrake Island, near Napan Bay, and in the Inner Bay (Fig. 1). The under ice fishery (January to early March 1991 to 1993) was sampled from near Sheldrake Island to the Inner Bay. No nets were set upstream of the Sheldrake Island site during the period of ice cover. Smelt nets were set in water 2.5 m deep to the edge of the shipping channel (about 6 m deep). In some instances, we were present when the nets were fished but usually we only sampled the bycatch brought back to shore. The bycatch collected on a specific date for an individual site was defined as a sample. The number of fish counted for each sample was divided by the number of nets fished that day, which permitted comparisons based on a common unit of effort. The samples collected on three dates during autumn in the Inner Bay were not included in this study because it was clear that the trap workers were only landing a very small fraction of the bycatch. We conducted detailed examination on 76 samples (Table 1) with the highest level of coverage in the Sheldrake Island location because this area had high fishing effort and was easily accessible for sampling for the entire duration of the fishery. The fish samples were returned to the laboratory and sampling was done as described for trawl samples.

The bycatch from the gaspereau trapnet fishery (50 mm mesh in the nets) was sampled from 20 May to 14 June 1991 at sites near: Redbank and the highway bridge on the Northwest Miramichi River; Chatham; Millbank; Sheldrake Island; and the Napan River (Fig. 1). Each site was visited weekly. Gaspereau traps were usually set with the box in water about 4 m deep and a long leader (30 to 70 m) extending shoreward to a depth of about 1.5 m. In most case, we were present when the nets were fished, otherwise, we met the boats while unloading the catch at the wharf. A sample consisted of the bycatch from a single site on each date. The sample numbers were divided by the number of nets fished at that site that day, which permitted comparisons based on a common unit of effort. We collected 24 fish samples between the near-freshwater and Sheldake Island locations (Table 1). We were unable to obtain samples in the Newcastle location and there were no nets set downstream the Napan River in the Sheldrake Island location (Mobray et al. 1993). The fish that we collected were sampled as described for onboard sampling of trawl samples except that Atlantic salmon and brook char (*Salvelinus fontinalis*) were measured (fork length cm), weighed on spring scales (to the nearest g), and returned alive to the water. We were permitted to sample fork lengths, wet weights, scales, and stomach contents of American shad and large striped bass before they were sold by the fishermen.

Relative abundances were used throughout the paper. While trawling and seining are quantitative methods in that the area sampled is known, the catchability of the various species to the gear was unknown (see, for example, DeAlteris et al. 1989; Pierce et al. 1990; Allen et al. 1992; Buijse et al. 1992), thus unbiased estimates of density were not possible. Nevertheless, the numbers of each species per tow from our seine and trawl surveys represent indices of abundance and can be used to detect changes in relative abundance between the present and future studies provided a similar sampling scheme is used. Indeed, these indices of relative abundances are being used in our ongoing studies on the biology of the important fish species in the Estuary. Bycatch sampling, however, provides less precise estimates of abundance because fish capture is passive, the area sampled by a net is unknown, and unknown proportions were discarded before we sampled. Furthermore, unknown numbers of the larger individuals of some species were marketed, thus, we often had samples that consisted only of small individuals of American shad, striped bass, winter flounder, and Atlantic tomcod. In contrast, smooth flounder are not marketed because of small size and poor flesh quality during winter (McKenzie 1959), therefore, we had reasonably unbiased samples of all size-classes of this species. In order to combine data collected by different methods, we arbitrarily defined the following levels of relative abundance: abundant (numerical score = 4) as when a species exceeded 75 individuals per unit of capture; common (numerical score = 3) as 20 to 74 individuals per unit of capture; scarce (numerical score = 2) as 5 to 19 individuals per unit of capture; and rare (numerical score = 1) was 1 to 4 individuals per unit of capture. Total fish abundance was calculated as the sum of relative abundances across all species for a specific sampling location and season.

Results and Discussion

Community composition

We collected 47 species of fish from 27 taxonomic families in the Miramichi Estuary (Table 2). In judging the relative overall abundance of a species, no attempt was made to distinguish between adults and young, e.g., adult striped bass were sometimes caught in moderate numbers in gaspereau traps during spring but the young were abundant in seine catches during summer and in the smelt bycatch during early autumn, thus, we ranked striped bass as abundant. We considered 12 species abundant, of these, rainbow smelt, blueback herring, alewife, and Atlantic tomcod are targets of commercial fisheries in the Estuary. Although winter flounder were abundant in the Estuary, commercial fisheries directed at this species were only conducted outside of the Estuary. Similar to American shad and large striped bass, however, large winter flounder were sold commercially when caught as bycatch in one of the fisheries within the Estuary. American eel (*Anguilla rostrata*) are common in the Estuary and fished commercially (LeBlanc and Chaput 1991; Chaput 1995) but were not susceptible to most of our collection gear.

A number of rare species, such as brown bullhead (*Ameiurus nebulosus*) and yellow perch (*Perca flavescens*), represent species normally only found in fresh water (Scott and Crossman 1973) but probably were washed downstream during spring runoff. Two white perch (*Morone americana*) were caught during the study. They likely represented fish that strayed from nearby rivers (e.g., Kouchibouguac and Richibucto Rivers) where white perch are common (Scott and Scott 1988; S. C. Courtenay, pers. obs.).

Atlantic cod (*Gadus morhua*) and American plaice (*Hippoglossoides platessoides*) were not caught in the Estuary but were caught on one occasion in the shipping channel within six kilometres of the Barrier Islands. Both species are abundant in the Gulf of St. Lawrence waters adjacent to the Estuary (Powles 1965; Swain and Wade 1993). We did not capture American plaice or Atlantic halibut (*Hippoglossus hippoglossus*) in the Estuary during this study. Earlier reports of capture of moderately large numbers of American plaice and juvenile Atlantic halibut in the Estuary (Addison et al. 1991) were in error (R. F. Addison, Bedford Institute of Oceanography, Dartmouth, N.S., pers. com.). Winter skate (*Raja ocellata*), fourbeard rockling (*Enchelyopus cimbrius*), daubed shanny (*Lumpenus maculatus*), ocean pout (*Macrozoarces americanus*), sea raven (*Hemtripterus americanus*), longhorn sculpin (*Myoxocephalus octodecemspinosus*), and yellowtail flounder (*Pleuronectes ferrugineus*) represent species that are common just outside of the Estuary and occasionally strayed into the Inner Bay. Atlantic mackerel are also common outside of the Estuary during summer but were usually unavailable to our sampling gear, thus, the designation as "rare" is likely not representative of its true abundance.

Shorthorn sculpins (*Myoxocephalus scorpius*) were common in smelt nets set in the Inner Bay during autumn and winter but were seldom encountered at the Sheldrake Island site. They were rare between Chatham and the Outer Bay during other seasons. McKenzie (1959) reported that the grubby (*M. aeneus*) also was frequently taken in smelt nets during winter. This species was not present in our collections. One possibility is that the fish identified as *M. aeneus* by McKenzie (1959) were actually female shorthorn sculpin. Female shorthorn sculpin are also grey-brown in colour but are easily distinguished from the grubby, which has a slimmer body and a downward projecting spine on the preoperculum (Scott and Scott 1988). If McKenzie's identification was correct, however, the grubby has disappeared from the Miramichi Estuary. McKenzie (1959) also reported that most of the hake caught in the Estuary were squirrel hake (*Urophycis chuss*). The taxonomy of hakes was less well known at that time and the fish he collected were almost certainly white hake (*Urophycis tenuis*).

Butterfish (*Peprilus triacanthus*) were only caught during autumn 1992; apparently they are a periodic visitor to the Gulf of St. Lawrence (McKenzie 1959; Scott and Scott 1988). Three Arctic cod (*Boreogadus saida*) were caught during January 1992 from smelt nets set near Sheldrake Island. Arctic cod are seldom caught in the southern Gulf and represent strays from more northern waters (McKenzie 1953, 1959).

Seasonal patterns in abundance and distribution

No sampling was possible during March and April due to melting ice and high water levels. No samples were collected during December because ice formation begins during this month. Sampling during summer months was restricted to seining and trawling. The only commercial fishery during summer was for American eel and occurred in the same depths and sites covered by the seine surveys. The only source of winter samples was the bycatch from the smelt trapnet fishery. This restricted sampling to two sites: near Sheldrake Island and the Inner Bay. Thus, species that had moved further upstream or out to sea were not sampled.

There were clear seasonal patterns in the total abundance of all species (summed relative abundances) and numbers of species collected at the various sampling locations in the Estuary. The lowest relative abundances and species numbers were observed in the Inner Bay and Outer Bay locations whereas the Sheldrake Island location consistently showed the greatest numbers of fish and highest species number (Fig. 2). Although the greatest relative abundance and species number might be expected in the two locations furthest upstream (near the edge of the salt wedge), the Sheldrake Island location is in the area where the Estuary widens quickly and also receives freshwater from the Bartibog and Napan Rivers (Fig. 1). During winter, the water temperature is warmer at the Sheldrake Island location (-0.2 to -0.8) compared to the Inner Bay location (-1.0 to -1.5) (McKenzie 1959), which may explain the difference in fish abundance and number of species present during this season.

American eel were common in the Estuary, as evidenced by the commercial fishery for the

Table 2. List of species caught by trawling, seining, and as bycatch in commercial smelt and gaspereau trapnet fisheries in the Miramichi Estuary, including the shipping channel just outside of the Barrier Islands, from 1991 to 1993. Relative abundances are presented as abundant, common, scarce, and rare.

Species	Common name	Relative abundance
Petromyzontidae		
Petromyzon marinus	Sea Lamprey	Scarce
Rajidae		
Raja ocellata	Winter skate	Rare
Anguillidae		
Anguilla rostrata	American eel	Common
Clupeidae		
Alosa aestivalis	Blueback herring	Abundant
Alosa pseudoharengus	Alewife	Abundant
Alosa sapidissima	American shad	Abundant
Clupea harengus	Atlantic herring	Abundant
Salmonidae		
Salmo salar	Atlantic salmon	Common
Salvelinus fontinalis	Brook char	Scarce
Osmeridae		
Osmerus mordax	Rainbow smelt	Abundant
Cyprinidae		
Semotilus corporalis	Fallfish	Common
Catostomidae		
Catostomus commersoni	White sucker	Abundant
Ictaluridae		
Ameiurus nebulosus	Brown Bullhead	Rare
Gadidae		
Boreogadus saida	Arctic cod	Rare
Enchelyopus cimbrius	Fourbeard rockling	Rare
Gadus morhua	Atlantic cod	Rare
Gadus ogac	Greenland cod	Common
Microgadus tomcod	Atlantic tomcod	Abundant
Urophycis tenuis	White hake	Common
Zoarcidae		
Macrozoarces americanus	Ocean pout	Scarce
Cyprinidontidae		
Fundulus diaphanus	Banded killifish	Common
Fundulus heteroclitus	Mummichug	Common
Atherinidae		
Menidia menidia	Atlantic silverside	Abundant

Table 2. *(Concluded.)*

Species	Common name	Relative abundance
Gasterosteidae		
Apeltes quadracus	Fourspine stickleback	Common
Gasterosteus aculeatus	Threespine stickleback	Common
Gasterosteus wheatlandi	Blackspotted stickleback	Abundant
Pungitius pungitius	Ninespine stickleback	Scarce
Percithyidae		
Morone americana	White perch	Rare
Morone saxatilis	Striped bass	Abundant
Percidae		
Perca flavescens	Yellow perch	Rare
Labridae		
Tautogolabrus adspersus	Cunner	Scarce
Stichaeidae		
Lumpenus maculatus	Daubed shanny	Rare
Pholidae		
Pholis gunnelis	Rock gunnel	Rare
Cryptacanthodidae		
Cryptacathodes maculatus	Wrymouth	Rare
Ammodytidae		
Ammodytes americanus	American sand lance	Common
Scombridae		
Scomber scombrus	Atlantic mackerel	Rare
Stromateidae		
Peprilus triacanthus	Butterfish	Rare
Cottidae		
Hemitripterus americanus	Sea raven	Rare
Myoxocephalus octodecemspinosus	Longhorn sculpin	Scarce
Myoxocephalus scorpius	Shorthorn sculpin	Common
Cyclopteridae		
Cyclopterus lumpus	Lumpfish	Rare
Liparis atlanticus	Atlantic snailfish	Scarce
Bothidae		
Scophthalmus aquosus	Windowpane	Common
Pleuronectidae		
Hippoglossoides platessoides	American plaice	Rare
Pleuronectes americanus	Winter flounder	Abundant
Pleuronectes ferrugineus	Yellowtail flounder	Scarce
Pleuronectes putnami	Smooth flounder	Abundant

species from spring to autumn but our sampling gears did not catch large numbers of them except during spring (Fig. 3). American eels feed mostly at night and either bury themselves in the sediment or hide under rocks and logs during the day. This behaviour would have rendered eels largely immune to seines despite our seining in areas immediately adjacent to eel traps.

Adult American shad were abundant from near fresh water to Sheldrake Island during spring and were frequently collected as bycatch in gaspereau traps (Fig. 3). Similar to alewife and blueback herring, the abundance of American shad must have been high in the two downstream locations because this species also migrates into the Estuary from the Gulf of St. Lawrence in the

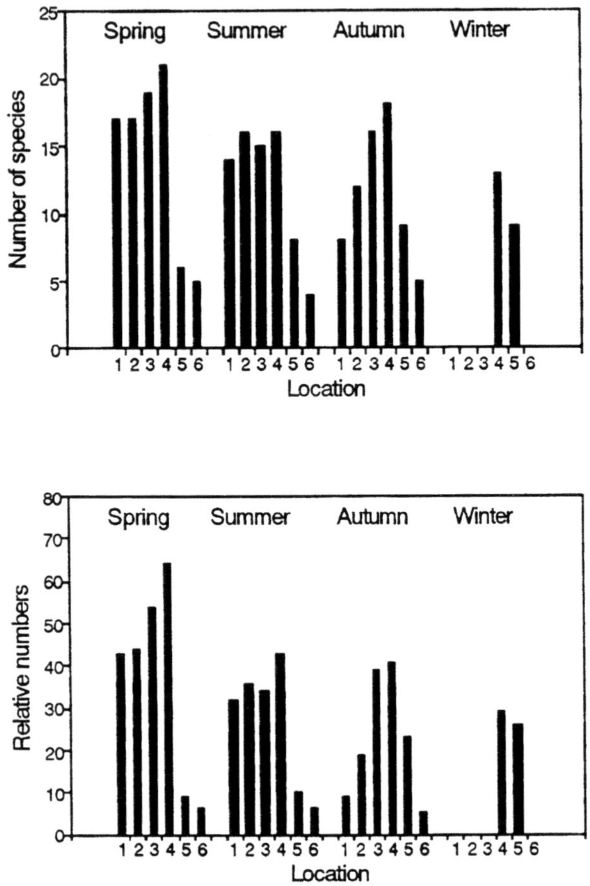

Fig. 2. Seasonal variation in the number of species captured and summed relative abundance of fish caught at six locations in the Miramichi Estuary:
near fresh water (1); Newcastle (2); Chatham (3); Sheldrake Island (4); Inner Bay (5); and Outer Bay (6).

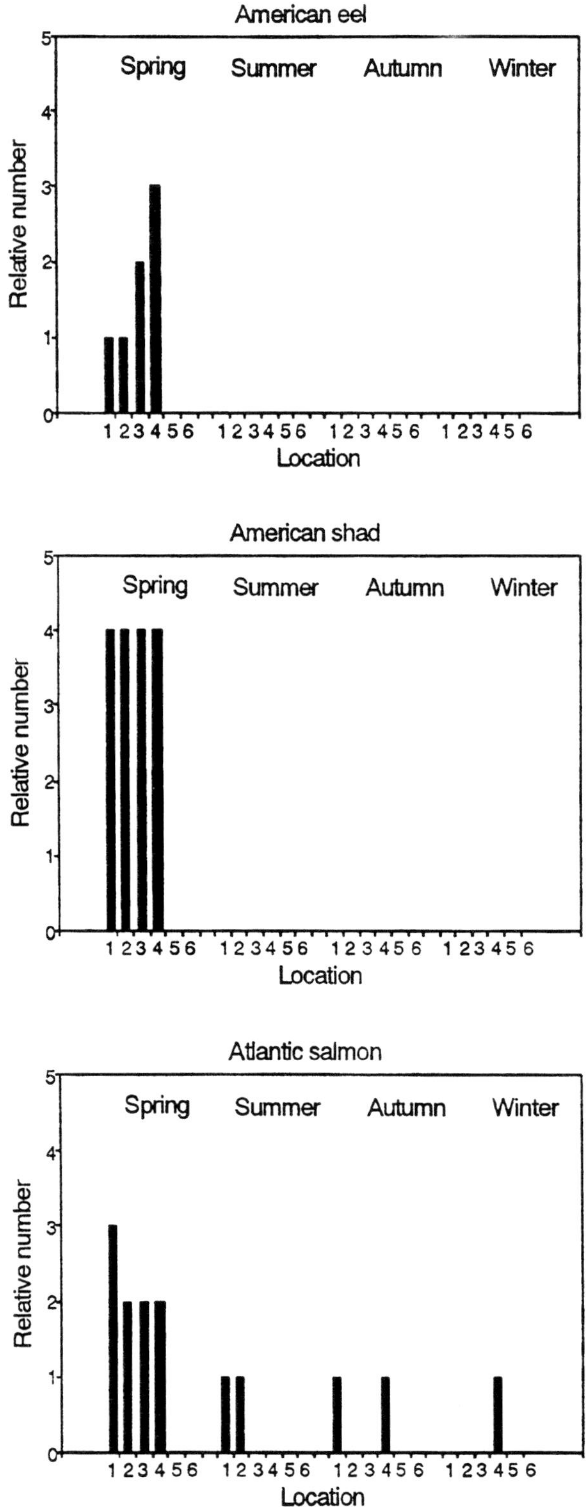

Fig. 3. Seasonal variation in the relative abundance of American eel, American shad, and Atlantic salmon caught at six locations in the Miramichi Estuary:
near fresh water (1); Newcastle (2); Chatham (3); Sheldrake Island (4); Inner Bay (5); and Outer Bay (6).

spring (McKenzie 1959; Chadwick and Claytor 1989; Chaput 1995). No adult American shad were caught during summer, autumn or winter. No juvenile American shad were caught during summer, suggesting that they remained in fresh water or were in the upper part of the water column in the offshore part of the Estuary. Larval and juvenile shad were not caught, however, in ichthyoplankton surveys conducted at two week intervals from 6 May to 25 September 1992 in an area from Redbank to the Outer Bay (Locke and Courtenay 1995a,b). No juvenile American shad were caught during autumn or winter in our seine, trawl, and smelt bycatch sampling.

Adult Atlantic salmon were only common at the most upstream site during spring (Fig. 3). The peak of the early run occurs from mid- to late-June (Chadwick and Claytor 1989; Chaput et al. 1994). Our sampling gear could not, however, catch adult Atlantic salmon during summer but a few parr were caught at the two most upstream sites. During autumn, the adult Atlantic salmon had largely finished their spawning migration into the River by 15 October (Chadwick and Claytor 1989; Chaput et al. 1994), which is the beginning of the smelt season; none were caught in smelt nets set from Chatham to the Inner Bay. The few Atlantic salmon caught during autumn were parr; those caught near Sheldrake Island may have been carried there by high freshwater runoff (possibly from the Bartibog or Napan Rivers). Similarly, the very few Atlantic salmon parr caught during winter only appeared at the Sheldrake Island location after several days of heavy rain.

Adult alewife and blueback herring were abundant in most locations during spring (Fig. 4). We did not capture these species in the Inner Bay or Outer Bay locations because no trapnets were present and we could not use gillnets to sample the fish community. Nonetheless, their abundance must have been high in the two downstream locations because these species migrate into the Estuary from the Gulf of St. Lawrence during May and June to spawn in fresh water (McKenzie 1959; Chadwick and Claytor 1989; Chaput 1995). Large numbers of blueback herring and alewife were caught during summer but all of these were young-

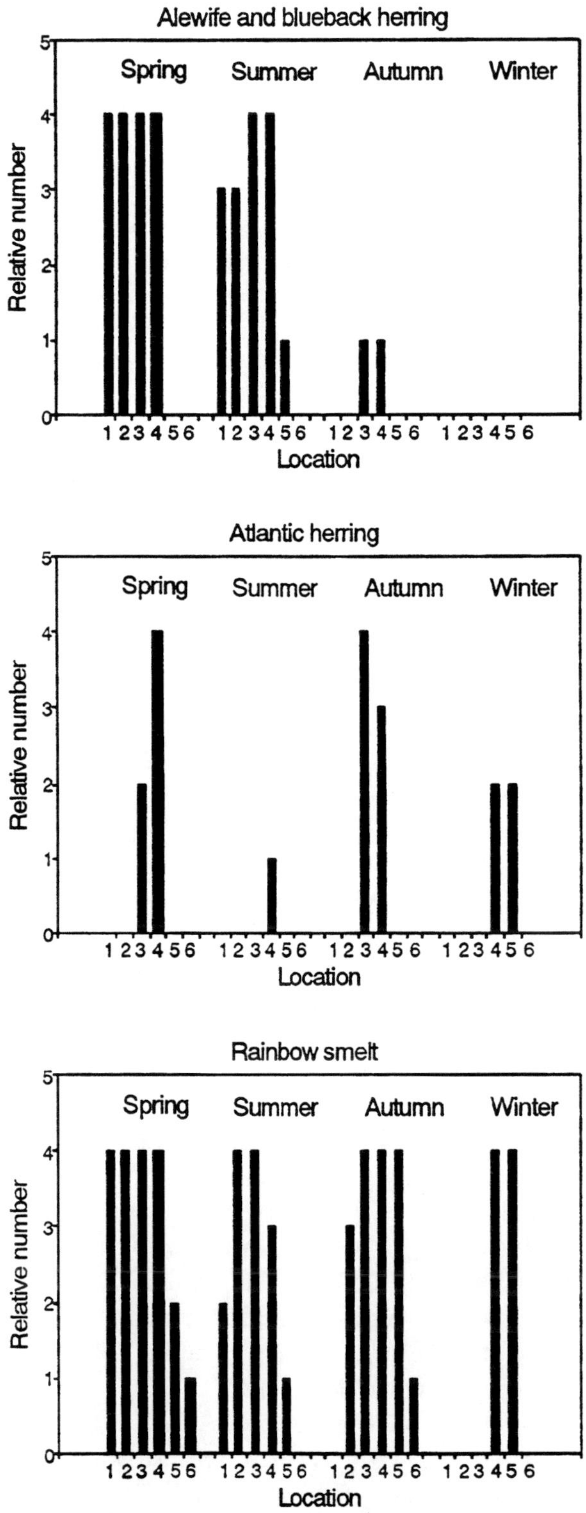

Fig. 4. Seasonal variation in the relative abundance of alewife and blueback herring, Atlantic herring, and rainbow smelt caught at six locations in the Miramichi Estuary: near fresh water (1); Newcastle (2); Chatham (3); Sheldrake Island (4); Inner Bay (5); and Outer Bay (6).

of-the-year. Nearly all of the juvenile blueback herring and juvenile alewife had left the Estuary by autumn and none were caught during winter.

Large numbers of adult herring occurred on a few occasions during spring in gaspereau traps set in the Sheldrake Island area (Fig. 4). A few small (< 8 cm long) herring were caught in seines set near Sheldrake Island during summer. Moderate numbers of juvenile herring appeared in smelt nets near Sheldrake Island during autumn and large numbers were caught in smelt nets at Chatham. Small numbers of juvenile herring appeared sporadically in smelt nets during winter.

Adult rainbow smelt overwintered in the Estuary and were abundant at the four upstream sites during spring. The low numbers (Fig. 4) collected from the most downstream sites during spring undoubtedly reflected the spawning migration to fresh water. Juvenile rainbow smelt were very common between Newcastle and Sheldrake Island locations during summer but few adults were caught. Large numbers of adults and juvenile rainbow smelt were caught between Newcastle and the Inner Bay during autumn, which confirms McKenzie's (1964) study indicating adult rainbow smelt began their return migration from the southern Gulf of St. Lawrence into the Estuary during early autumn. Rainbow smelt of all sizes were abundant at the two winter locations, which was expected given that we were sampling a commercial fishery directed at the species.

Fallfish (*Semotilus corporalis*) are primarily a freshwater species (Scott and Crossman 1973) but small (< 10 cm long) individuals occurred in low numbers at the most upstream site during spring (Fig. 5). Young fallfish were common at the most upstream site but absent from the four downstream sites during summer. Fallfish were rare even at the upstream locations during autumn and were not caught during winter.

Although white sucker (*Catostomus commersoni*) also live primarily in fresh water, large numbers of adult fish were caught in all but the two most downstream sites during spring (Fig. 5). The reason for this movement into the Estuary is unknown but likely was not related directly to spawning because white sucker spawn in freshwa-

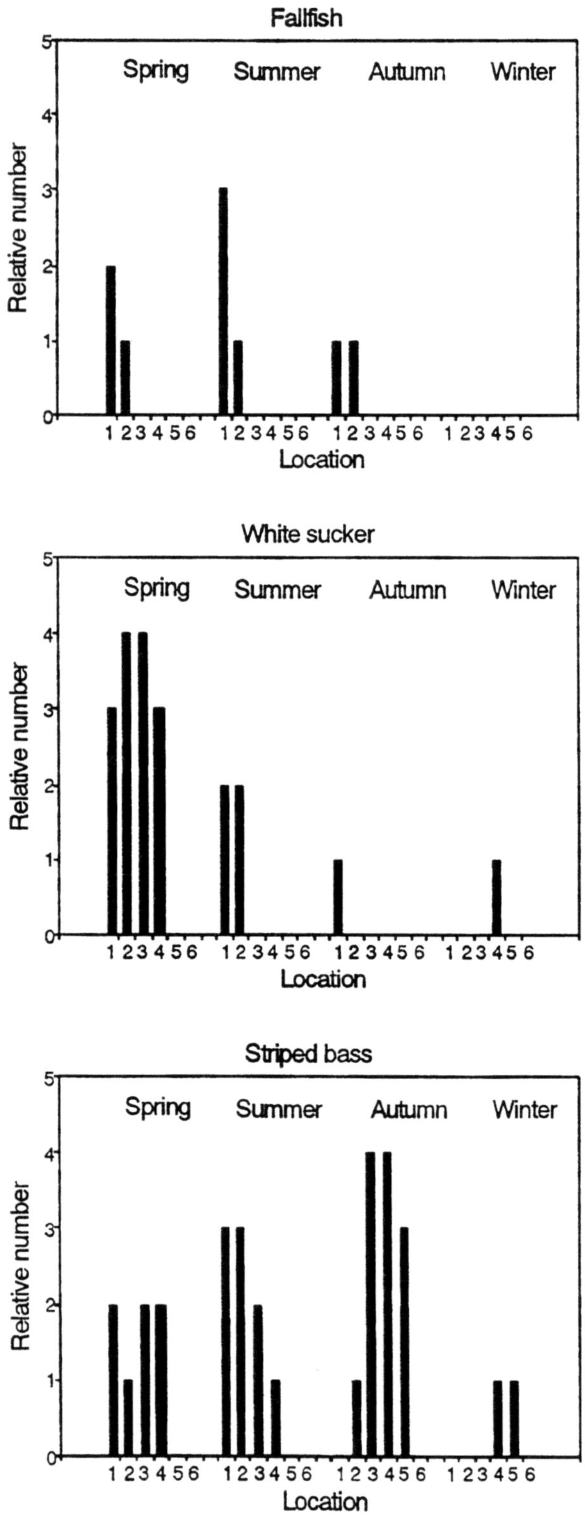

Fig. 5. Seasonal variation in the relative abundance of fallfish, white sucker, and striped bass caught at six locations in the Miramichi Estuary: near fresh water (1); Newcastle (2); Chatham (3); Sheldrake Island (4); Inner Bay (5); and Outer Bay (6).

ter (Scott and Crossman 1973). Adult white sucker were absent from the Estuary during summer; the few specimens captured at the two upstream sites were young-of-the year. Similarly, the only specimens caught during autumn were juveniles and at the most upstream site. Two specimens were caught in smelt nets at Sheldrake Island after several days of heavy rain during winter, probably washed out of nearby rivers.

Striped bass were not common at any location during spring (Fig. 5); the majority of the individuals captured were > 25 cm long. No adult striped bass were caught during summer but they would not have been available to our sampling gear. Young-of-the-year striped bass were common in the two upstream sampling areas during summer but rare or absent from the other four sites. Small numbers of adult striped bass (> 30 cm long) were caught in smelt nets and by trawling during autumn. In contrast, large numbers of juvenile striped bass were caught in smelt nets set between Chatham and the Inner Bay. During summer, juvenile striped bass were caught further upstream than during autumn. Very few striped bass were caught in smelt traps after the second week of November. The only striped bass collected during winter were caught after several days of heavy rain; unlike Atlantic salmon and white sucker, they were caught at both sampling locations.

Adult Atlantic tomcod were present in moderate numbers during spring between Newcastle and Sheldrake Island (Fig. 6). With the exception of a small number of adults caught near Sheldrake Island, the only Atlantic tomcod present in the Estuary during summer were young-of-the-year. These young tomcod were very common between Newcastle and Sheldrake Island and rare or absent elsewhere. Adult Atlantic tomcod returned to the Estuary during autumn and large numbers of adults and juveniles were caught by trawling and in smelt traps between Newcastle and the Inner Bay. All sizes of Atlantic tomcod were abundant at both winter sampling locations. Tomcod spawned in December and January - almost no fish caught during February and March still carried eggs (J. M. Hanson and S. C. Courtenay, unpublished data).

Only one white hake, a 52 cm long specimen,

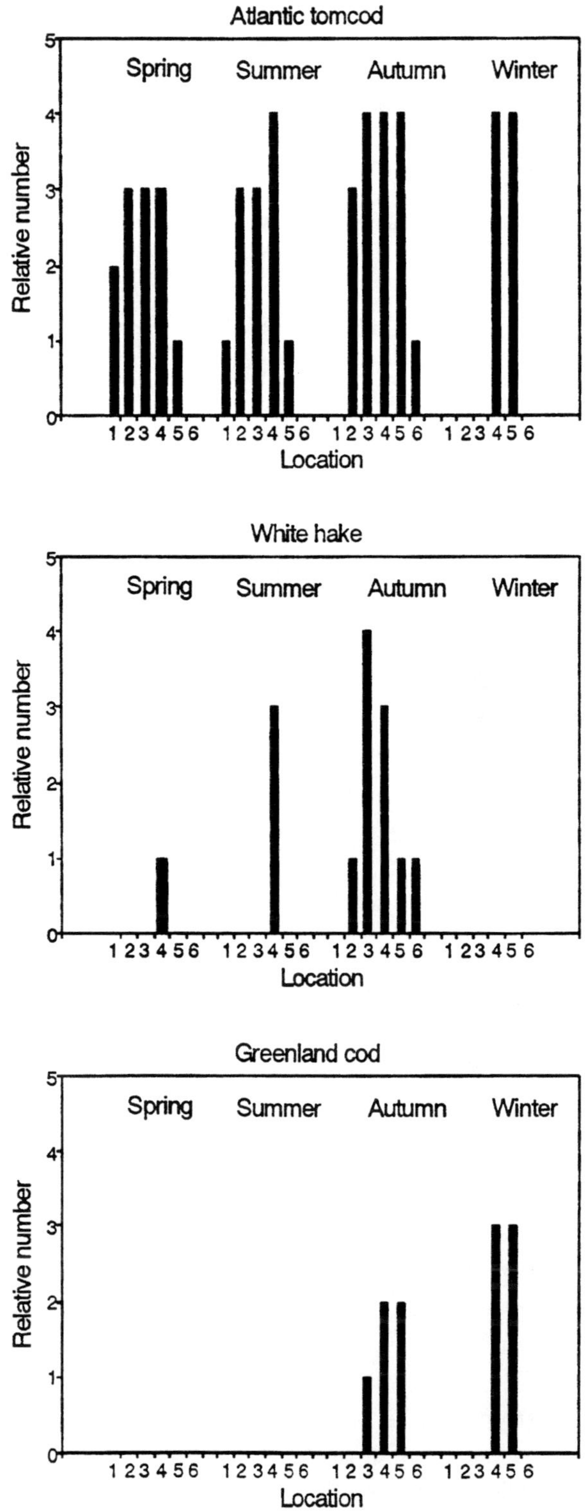

Fig. 6. Seasonal variation in the relative abundance of Atlantic tomcod, white hake, and Greenland cod caught at six locations in the Miramichi Estuary: near fresh water (1); Newcastle (2); Chatham (3); Sheldrake Island (4); Inner Bay (5); and Outer Bay (6).

was caught near Sheldrake Island during spring. Small (< 8 cm long) white hake were common near Sheldrake Island during summer but absent elsewhere (Fig. 6). Large numbers of juvenile white hake (mostly < 25 cm long) were caught in smelt nets during October and early November but, similar to striped bass, very few were caught after the second week of November. In contrast to autumn, no white hake were caught during winter months.

No Greenland cod (*Gadus ogac*) were present in the Estuary during spring or summer (Fig. 6). Small (< 15 cm long) Greenland cod were common in smelt nets and trawl surveys during October and November but larger specimens were rare. Greenland cod were common at both sites, especially during February. Unlike autumn, large Greenland cod (> 25 cm long) were caught during winter; those caught during February were in spawning condition and some caught in March clearly had just completed spawning. This confirms early speculation that Greenland cod spawn during February and March (Vladykov 1972; Scott and Scott 1988). The eggs or larvae appeared to be transported out of the Estuary in late winter or early spring because no larval Greenland cod were caught during May ichthyoplankton surveys (Locke and Courtenay 1995b).

Winter flounder were present in large numbers between Chatham and the Inner Bay but did not occur in the two upstream sites during spring (Fig. 7). Few winter flounder > 15 cm long were caught during spring. The very few larval winter flounder captured in ichthyoplankton surveys (< 0.1 tow-1) came from sites near the Barrier Islands during early June (Locke and Courtenay 1995b). With the exception of near Sheldrake Island, very few winter flounder of any size remained in the Estuary during summer months although they were common outside the Barrier Islands. Those remaining in the Estuary were mostly < 12 cm long (J.M. Hanson and S.C. Courtenay, unpublished data) but young-of-the-year (< 5 cm long) were only caught near the Barrier Islands. Winter flounder of all sizes (5 – 35 cm long) were very common during autumn between Chatham and the Inner Bay. Large winter flounder (> 20 cm long) were caught in moderate

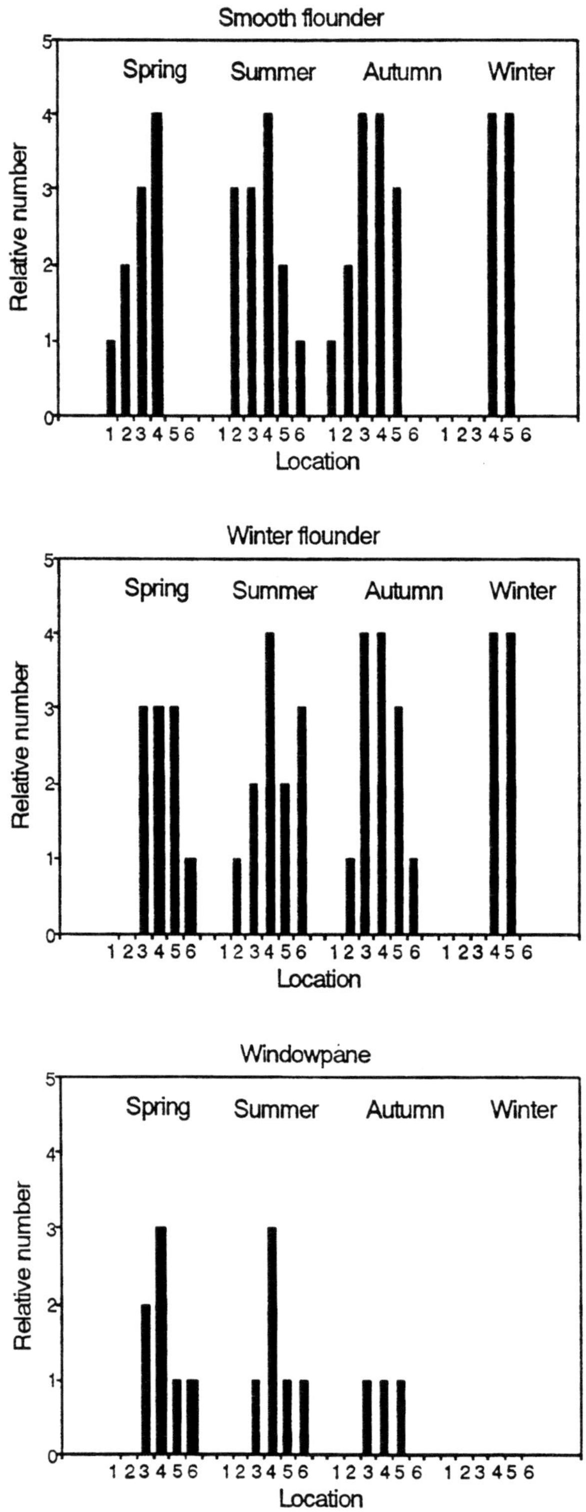

Fig. 7. Seasonal variation in the relative abundance of smooth flounder, winter flounder, and windowpane caught at six locations in the Miramichi Estuary: near fresh water (1); Newcastle (2); Chatham (3); Sheldrake Island (4); Inner Bay (5); and Outer Bay (6).

numbers in smelt traps during October and then in very large numbers during November. Winter flounder of all sizes were abundant at both sites during winter. Most of the fish > 18 cm long were marketed during both seasons whereas small fish were discarded.

Smooth flounder of all sizes excepting young-of-the-year were very common between Chatham and Sheldrake Island during spring but were not found in the Inner Bay and near the Barrier Islands (Fig. 7). Most age-0 smooth flounder were pelagic during spring; they were the only flatfish species caught during ichthyoplankton surveys conducted in the Estuary during May 1992 (Locke and Courtenay 1995b). In contrast to winter flounder, all sizes of smooth flounder were common to abundant between Newcastle and Sheldrake Island during summer and scarce in the Inner Bay. The young-of-the-year (< 5 cm long) were most common at the Newcastle site. Smooth flounder of all sizes were very common from Newcastle to Sheldrake Island. Unlike spring and summer, smooth flounder were also common in the Inner Bay during autumn. Smooth flounder of all sizes were abundant at both sampling locations during winter. During January, most of the adults (>15 cm) were in spawning condition and all were spent by early February (J.M. Hanson and S.C. Courtenay, unpublished data).

Small (10 to 20 cm long) windowpane (*Scopthalmus aquosus*) were only common near Sheldrake Island during spring and summer (Fig. 7). Windowpane were largely absent from the Estuary during autumn and absent from both sampling locations during winter.

Blackspotted (*Gasterosteus wheatlandi*) and fourspine (*Apeltes quadracus*) stickleback were common in shallow waters at the four upstream locations during spring while threespine stickleback (*G. aculeatus*) were only common near Sheldrake Island (Fig. 8). Fourspine stickleback were still common at the two upstream sites during summer but most of these were young-of-the-year. In contrast, most blackspotted and threespine stickleback had either left the Estuary during early summer or the young were pelagic and not available to our sampling gears. Very few specimens of

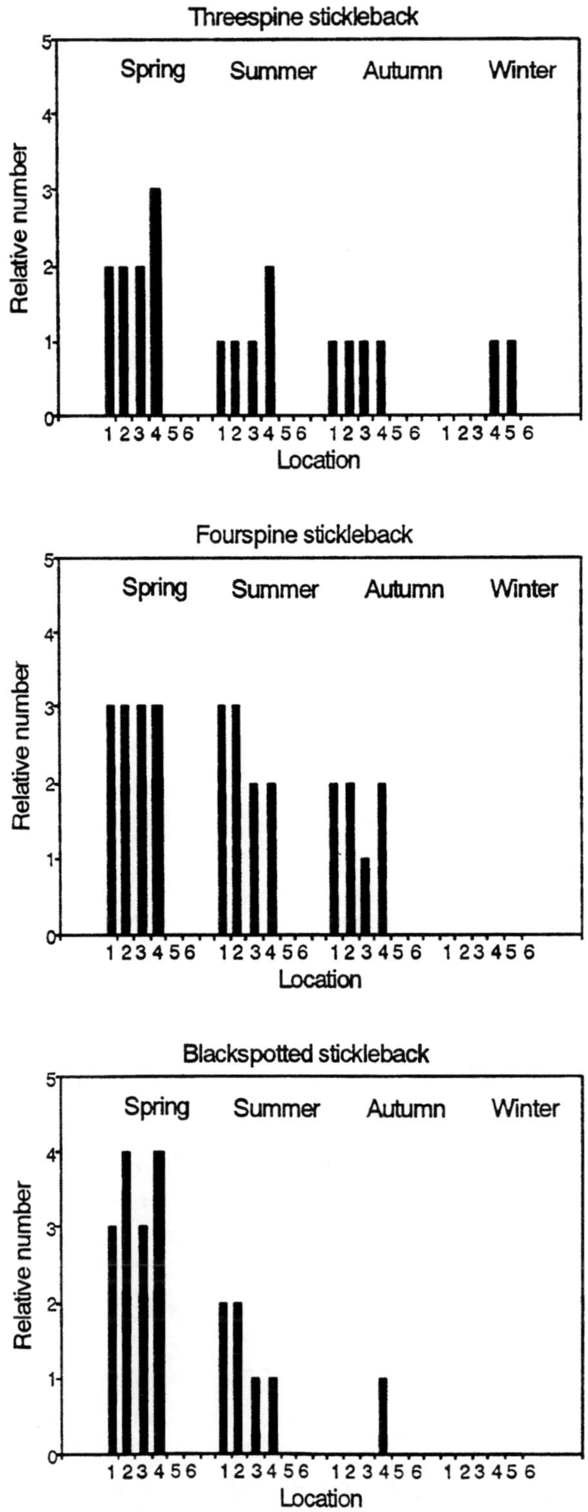

Fig. 8. Seasonal variation in the relative abundance of threespine, fourspine, and blackspotted stickleback caught at six locations in the Miramichi Estuary: near fresh water (1); Newcastle (2); Chatham (3); Sheldrake Island (4); Inner Bay (5); and Outer Bay (6).

either species were caught, however, during ichthyoplankton surveys conducted in the Estuary during July and August 1992 (Locke and Courtenay 1995a). The three species of stickleback were largely absent from the Estuary during autumn. No fourspine stickleback, or blackspotted stickleback were caught during winter and only a very few threespine stickleback were caught in the smelt fishery. Small numbers (1 or 2 per tow) of a fourth species, ninespine stickleback (*Pungitius pungitius*), were consistently caught from Newcastle to the Inner Bay during spring and summer.

Banded killifish (*Fundulus diaphanus*) and mummichug (*F. heteroclitus*) were present in moderate numbers near Chatham and Sheldrake Island during spring and at reduced numbers during summer (Fig. 9). Banded killifish and mummichug were seldom caught during autumn. No banded killifish and only a very few mummichug were caught in the winter smelt fishery.

Adult Atlantic silverside were common to abundant from Newcastle to Sheldrake Island during spring and both adults and young-of-the-year were abundant in shallow waters of the Estuary during summer (Fig. 9). The absence of specimens at the two downstream sites likely represented our inability to sample water < 4 m deep in these two areas. Atlantic silverside continued to be common in shallow water between Chatham and Sheldrake Island during autumn but had largely left the two upstream sites. Overall, autumn numbers of silverside caught were well below those seen during summer. Similar to striped bass and Atlantic parr, the few Atlantic silverside collected during winter were only caught after periods of heavy rain.

General Discussion

The Miramichi Estuary serves two functions for fish: as a nursery area for diadromous and estuarine species and an overwinter refuge for diadromous and marine species. The fish community in the Miramichi Estuary is unlike that of many estuaries studied elsewhere because the larval and juvenile fish originate from eggs laid in the River or Estuary. In most estuaries reported in the litera-

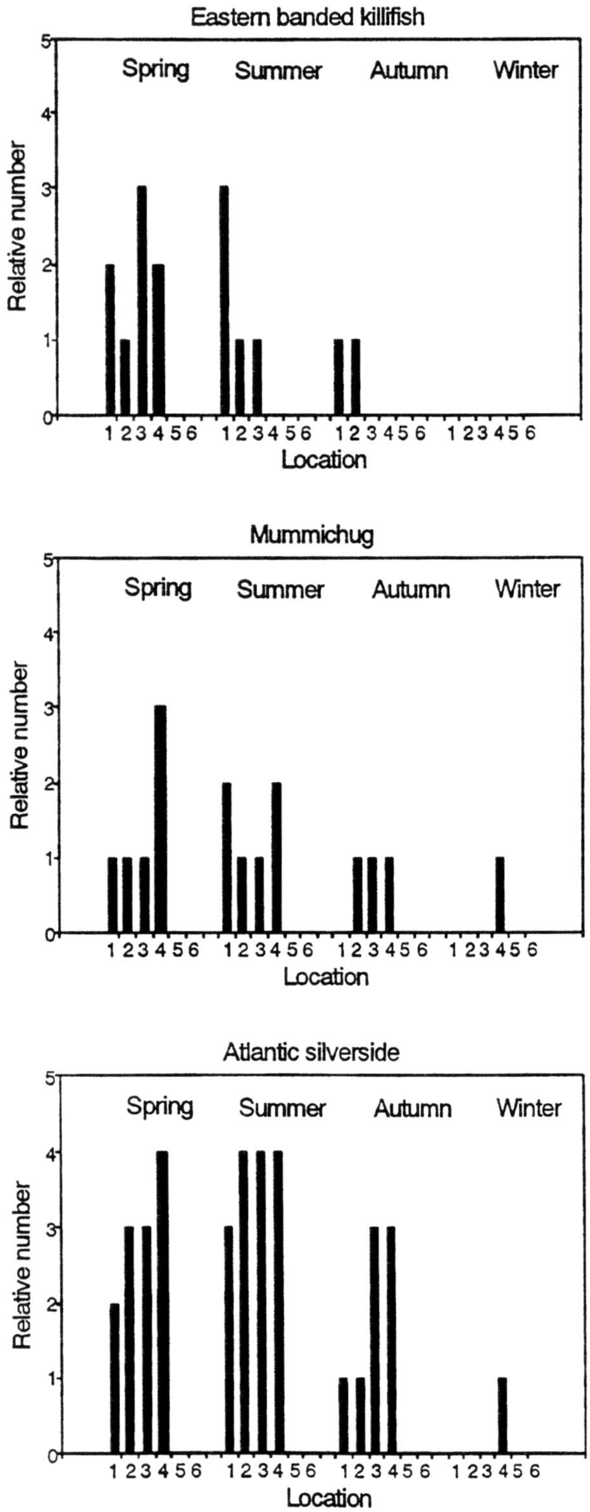

Fig. 9. Seasonal variation in the relative abundance of eastern banded killifish, mummichug, and Atlantic silverside caught at six locations in the Miramichi Estuary: near fresh water (1); Newcastle (2); Chatham (3); Sheldrake Island (4); Inner Bay (5); and Outer Bay (6).

ture, the dominant species are those in which the adults spawn at sea and the larvae drift into estuaries where they develop into juveniles and then move offshore (e.g., Norcross 1991; Roundtree and Able 1992; Deegan 1993).

Most of the juvenile fishes caught in the Miramichi Estuary originated from eggs laid during winter or spring. The winter spawning species were Atlantic tomcod and smooth flounder. The spring spawning species were: alewife, blueback herring, rainbow smelt, and striped bass. The adults of these fish species spent different amounts of time in the Estuary. Adult alewife and blueback herring entered the Estuary during May and June and apparently left soon after spawning in fresh water (this study; Chaput 1995). Striped bass spawned in the most upstream sites during late May or early June. Some adult striped bass appeared to leave the Estuary during early summer and to re-enter the Estuary during autumn (Bradford et al. 1995). Adult rainbow smelt entered the Estuary during autumn but did not spawn until the following spring. Greenland cod could be classified with these species because they appeared to spawn in the Estuary during winter, however, the larvae or pelagic young were not found in the estuary during spring (Locke and Courtenay 1995b).

The juveniles of the various fish species also spent different amounts of time in the Estuary. Large numbers of young-of-the-year alewife and blueback herring were present during summer but left the Estuary in late summer or early autumn. In contrast, juvenile American shad were never caught in the Estuary. Larval and juvenile American shad are primarily found in freshwater and migrate to salt water in early autumn (O'Leary and Kynard 1986; Stevens et al. 1987). This outward migration is rapid and the juveniles swim in the pelagic zone, which we could not sample. Juvenile striped bass first appeared at the near freshwater location in early July and spread downstream to the Inner Bay by early October. Juvenile striped bass apparently moved back upstream to overwinter and were occasionally washed downstream after heavy rainfall during winter. Young-of-year rainbow smelt and Atlantic tomcod were abundant throughout most of the Estuary from summer to autumn and were frequently caught in smelt nets set under the ice during winter. Thus these two species spent most of their first year in the Estuary before moving out of the Estuary the following spring.

The Miramichi Estuary is an important overwintering site for a number of fish species. Both adult and juvenile striped bass seemed to overwinter in the upper estuary or in fresh water. Adult rainbow smelt and Atlantic tomcod entered the Estuary during autumn and overwintered there. The presence of blood anti-freeze proteins in both species (Reisman et al. 1984; Ewart and Fletcher 1990) likely accounts for their ability to withstand the near freezing water of the Inner Bay during winter. Adults of both Atlantic tomcod and rainbow smelt fed in the Estuary during winter (J.M. Hanson and S.C. Courtenay; unpublished data). Atlantic tomcod spawned during winter but remained in the Estuary until late spring, possibly leaving in response to increasing water temperatures. Rainbow smelt did not spawn until spring and also left the Estuary by late spring.

Although it is a critical habitat for both, smooth and winter flounder differed in their use of the Estuary. Both species of flounder overwintered in the Estuary, both were abundant in the near-freezing water of the Inner Bay. The presence of blood antifreeze is well known for winter flounder (Fletcher and Smith 1980; Fletcher 1981) and antifreeze proteins are likely present in smooth flounder. Unlike Atlantic tomcod and rainbow smelt, however, neither flounder species fed during winter (J. M. Hanson and S. C. Courtenay; unpublished data). As found in Great Bay Estuary (Armstrong and Starr 1994), smooth flounder completed their entire life cycle in the Miramichi Estuary. Smooth flounder of all sizes were present in the Estuary all year. Smooth flounder can withstand higher water temperatures and a wider range of salinities than winter flounder (Scott and Scott 1988), hence they also occurred further upstream than winter flounder during spring and summer; they were found predominantly upstream of Sheldrake Island. Large numbers of adult smooth flounder were found in the Inner Bay area during autumn and winter, possibly a downstream migration to spawn. In contrast, winter flounder exhibited a pronounced seasonal migration into and out of the Estuary. Large numbers of winter flounder of all sizes moved into the Estuary from the southern Gulf of St. Lawrence during autumn to overwinter. Most winter flounder > 15 cm long left the Estuary during early spring, presumably to spawn and for summer feeding. The movement of winter flounder out of Estuaries appears to be triggered by warming water temperatures (McCracken 1963; Phelan 1992). Further evidence that winter flounder spawned outside of the Estuary was that we caught young-of-the-year fish outside but not inside of the Barrier Islands during summer. Some juvenile winter flounder, primarily individuals between 7 and 12 cm long, remained in the Estuary all year. Therefore, the Estuary is part of the inshore nursery for winter flounder during summer but not, apparently, for the young-of-the-year.

Watershed management programs and the development of habitat suitability indices require detailed knowledge of the degree of dependence of wildlife on specific habitats (e.g., Peters and Schaaf 1991; Rahel and Hubert 1991; Rutherford et al. 1992; Aadland 1993). That the Miramichi Estuary is a critical habitat for many fish species is now clear. Furthermore, recent declines in the abundances of commercially exploited fish species dependent upon the Estuary (e.g., rainbow smelt, Atlantic tomcod, American eel, striped bass, alewife, blueback herring; Chaput 1995) may be related to anthropogenic perturbation. Effluent from the bleached kraft mill and groundwood mill at Newcastle are sometimes toxic to fish and contain dioxins and other organic contaminants (Muir et al. 1992; N. Brodie, New Brunswick Department of the Environment, Fredericton, N.B., unpublished data). Indeed, high levels of dioxin have been found in lobster collected the Inner Miramichi Bay (Miramichi River Environmental Assessment Committee 1992; N. Brodie, unpublished data). Contaminants from the mills, a mine, and municipalities enter the Miramichi River and Estuary and many become concentrated in the Estuary and can be remobilized from the sediments (MacKnight 1995; Buckley 1995). These effluents discharge into a nursery area for rainbow smelt, alewife, blueback herring, striped bass, Atlantic tomcod, and smooth flounder (this study; Locke and Courtenay 1995a) possibly placing these species at risk.

There is mounting evidence that the degree of industrial contamination of estuaries is related to increased incidence of disease in fish (Stegeman et al. 1987; Mollers and Anders 1992; Wirgin et al. 1994) and failure of fish eggs and larvae to survive (Maurice et al. 1987; Perry et al. 1991; Hall et al. 1993). Mitigation efforts in the Miramichi River and Estuary are underway. N. Brodie (New Brunswick Department of the Environment, Fredericton, N.B. unpublished data) documents improved sewage treatment by municipalities, greatly reduced dioxin and furan emissions from the bleached kraft mill, and plans for secondary treatment of effluent from the groundwood mill. Monitoring of contaminant loads and their effects on fish and invertebrates is being carried out by the Department of Fisheries and Oceans under Canada's Green Plan for the Environment. It will be interesting to document the effects of decreasing contaminant loads on the abundances of fish that depend on the Miramichi Estuary as a nursery or overwinter refuge.

Acknowledgements

This study was part of the Miramichi Estuary Research Project and it would not have been possible without help from a large number of individuals. We are grateful to: the captains and crew of the M/V *Osborne*, CSS *Navicula*, and CSS *Opilio* for help during trawling operations; the numerous fishermen who permitted our sampling of their bycatch; and to L. Arsenault, J. Betts, L. Currie, C. Donahee, D. McCullough, C. Reid, K. Robichaud, A. St.-Hilaire, A. Vromans, and E. Wong, for technical support.

References

Aadland, L.P. 1993. Stream habitat types: their fish assemblages and relationship to flow. N. Amer. J. Fish. Manage. 13: 790-806.

Addison, R.F., Hansen, P.D., and Wright, E.C. 1991. Hepatic mono-oxygenase activities in American plaice (Hippoglossoides platessoides) from the Miramichi Estuary, NB. Can. Tech. Rep. Fish. Aquat. Sci. 1800: 18p.

Allen, D.M., Service, S.K., and Ogburn-Matthews, M.V. 1992. Factors influencing the collection efficiency of estuarine fishes. Trans. Amer. Fish. Soc. 121: 234-244.

Armstrong, M.P., and Starr, B.A. 1994. Reproductive biology of the smooth flounder in Great Bay Estuary, New Hampshire. Trans. Amer. Fish. Soc. 123: 112-114.

Boisclair, D., and Leggett, W.C. 1985. Rates of food exploitation by littoral fishes in a mesotrophic north-temperate lake. Can. J. Fish. Aquat. Sci. 42: 556-566.

Bradford, R.G., Robichaud, K.A., and Courtenay, S.C. 1995. By-catch in commercial fisheries as an indicator and regulator of striped bass (*Morone saxatilis*) abundance in the Miramichi River Estuary, p. 249–259. *In* E.M.P. Chadwick [*editor*]. Water, science, and the public: the Miramichi ecosystem. Can. Spec. Publ. Fish. Aquat. Sci. 123.

Buckley, D.E. 1995. Sediments and environmental quality of the Miramichi Estuary: new perspectives. p. 179–190. *In* E.M.P. Chadwick [*editor*]. Water, science, and the public: the Miramichi ecosystem. Can. Spec. Publ. Fish. Aquat. Sci. 123.

Buijse, A.D., Schaap, L.A., and Bult, T.P. 1992. Influence of water clarity on the catchability of six freshwater fish species in bottom trawls. Can. J. Fish. Aquat. Sci. 49: 885-893.

Chadwick, E.M.P., and Claytor, R.R. 1989. Run timing of pelagic fishes in Gulf of St Lawrence: area and species effects. J. Fish. Biol. 35(Suppl. A): 215-223.

Chaput, G. 1995. Temporal distribution, spatial distribution and abundance of diadromous fish in the Miramichi River watershed, p. 121–139. *In* E.M.P. Chadwick [*editor*]. Water, science, and the public: the Miramichi Ecosystem. Can. Spec. Publ. Fish. Aquat. Sci. 123.

Chaput, G.J., Moore, D., Biron, M., and Claytor, R. 1994. Stock status of Atlantic salmon (*Salmo salar*) in the Miramichi River, 1993. DFO Atl. Fish. Res. Doc. 94/20: 80p.

DeAlteris, J.T., Recksick, C.W., Fahfouhi, A., and Liuxiong, X. 1989. Comparison of the performance of two bottom sampling trawls. Trans. Amer. Fish. Soc. 118: 119-130.

Deegan, L.A. 1993. Nutrient and energy transport between estuaries and coastal marine ecosystems by fish migration. Can. J. Fish. Aquat. Sci. 50: 74-79.

Ewart, K.V., and Fletcher, G.L. 1990. Isolation and characterization of antifreeze protein from smelt (*Osmerus mordax*) and Atlantic herring (*Clupea harengus harengus*). Can. J. Zool. 68: 1652-1658.

Fletcher, G.L. 1981. Effects of temperature and photoperiod on the plasma freezing point depression, Cl- concentration, and protein "antifreeze" in winter flounder. Can. J. Zool. 59: 193-201.

Fletcher, G.L., and Smith, J.C. 1980. Evidence for permanent population differences in the annual cycle of plasma "antifreeze" levels of winter flounder. Can. J. Zool. 58: 507-512.

Hall, L.W. Jr., Ziegenfuss, M.C., Fischer, S.A., Sullivan, J.A., and Palmer, D.M. 1993. The influence of contaminant and water quality conditions on larval striped bass in the Potomac and Upper Chesapeake Bay in 1990: an in situ study. Arch. Environ. Contam. Toxicol. 24: 1-10.

Hanson, J.M., and Leggett, W.C. 1985. Experimental and field evidence for inter- and intraspecific competition in two freshwater fishes. Can. J. Fish. Aquat. Sci. 42: 280-286.

Lafleur, C., Pettigrew, B., St-Hilaire, A., Booth, D.A., and E.M.P. Chadwick. 1995. Seasonal and short-term variations in the estuarine structure of the Miramichi, p. 45-71. *In* E.M.P. Chadwick [*editor*]. Water, science, and the public: the Miramichi ecosystem. Can. Spec. Publ. Fish. Aquat. Sci. 123.

LeBlanc, C.H., and Chaput, G.J. 1991. Landings of estuarine fishes in the Gulf of St. Lawrence 1917-1988 / Débarquements de poissons estuariens dans le golfe du Saint-Laurent 1917-1988. Can. Data Rep. Fish. Aquat. Sci./Rapp. stat. can. sci. halieut. aquat. 842: 101 p.

Locke, A., and Courtenay, S.C. 1995a. Ichthyoplankton and invertebrate zooplankton of the Miramichi Estuary, p. 97–120. *In* E.M.P. Chadwick [*editor*]. Water, science, and the public: the Miramichi Ecosystem. Can. Spec. Publ. Fish. Aquat. Sci. 123.

Locke, A., and Courtenay, S.C. 1995b. Effects of environmental factors on ichthyoplankton communities in the Miramichi Estuary, Gulf of St. Lawrence. J. Plankton Res. 17: 333–349.

MacKnight, S. 1995. Contaminants in Miramichi sediments. p. 277. *In* E.M.P. Chadwick [*editor*]. Water, science, and the public: the Miramichi ecosystem. Can. Spec. Publ. Fish. Aquat. Sci. 123.

Maurice, K.R., Blye, R.W., Harmon, P.L., and Lake, D. 1987. Increased spawning by American shad coincident with improved dissolved oxygen in the tidal Delaware River, p. 79-88. *In* M.J. Dadswell, R.J. Klauda, C.M. Moffitt, R.L. Saunders, R.A. Rulifson, and J.E. Cooper [*editors*]. Common strategies of anadromous and catadromous fish. Amer. Fish. Soc. Symp. No. 1.

McCracken, F.D. 1963. Seasonal movements of the winter flounder, *Pseudopleuronectes americanus* (Walbaum), on the Atlantic coast. J. Fish. Res. Board Can. 20: 551-586.

McKenzie, R.A. 1953. Arctic or polar cod, *Boreogadus saida*, in Miramichi Bay, New Brunswick. Copeia 4: 238-239.

McKenzie, R.A. 1959. Marine and freshwater fishes of the Miramichi River and Estuary, New Brunswick. J. Fish. Res. Board Can. 16: 807-833.

McKenzie, R.A. 1964. Smelt life history and fishery in the Miramichi River, New Brunswick. Bull. Fish. Res. Board Can. 144: 77p.

Melvin, G.D. 1991. A review of striped bass (*Morone*

saxatilis) population biology in eastern Canada. p. 1-11. *In* R. H. Petersen [*editor*]. Proceedings of a workshop on biology and culture of striped bass (*Morone saxatilis*). Can. Tech. Rep. Fish. Aquat. Sci. No. 1832.

Miramichi River Environmental Assessment Committee. 1992. Summary and Final Report. Newcastle, N.B. 157p.

Mollers, H., and Anders, K. 1992. Epidemiology of fish diseases in the Wadden Sea. ICES J. Mar. Sci. 49: 199-208.

Moore, D.S., Chaput, G.J., and Pickard, R. 1995. The effect of fisheries on the biological characteristics and survival of mature Atlantic salmon (*Salmo salar*) from the Miramichi River, p. 229–247. *In* E.M.P. Chadwick [*editor*] Water, science, and the public: the Miramichi ecosystem. Can. Spec. Publ. Fish. Aquat. Sci. 123.

Mowbray, F., Chaput, G., and Courtenay, S. 1993. Assessment of the Miramichi River gaspereau fishery, 1991 and 1992. DFO Atl. Fish. Res. Doc. 93/51: 27p.

Muir, D.C.G., Fairchild, W.L., and Whittle, D.M. 1994. Predicting bioaccumulation of chlorinated dioxins and furans in fish near Canadian bleached kraft mills. Water Pollut. Res. J. Can. 27: 487-507.

Norcross, B.L. 1991. Estuarine recruitment mechanisms of larval Atlantic croakers. Trans. Amer. Fish. Soc. 120: 673-683.

O'Leary, J.A., and Kynard, B. 1986. Behaviour, length, and sex ratio of sea-ward migrating juvenile American shad and blueback herring in the Connecticut River. Trans. Amer. Fish. Soc. 115: 529-536.

Perry, D.M., Hughes, J.B., and Hebert, A.T. 1991. Sublethal abnormalities in embryos of winter flounder, *Pseudopleuronectes americanus*, from Long Island Sound. Estuaries 14: 306-317.

Peters, D.S., and Schaaf, W.E. 1991. Empirical model of the trophic basis for fishery yield in coastal waters of the eastern USA. Trans. Amer. Fish. Soc. 459-473.

Pierce, C.L., Rasmussen, J.B., and Leggett, W.C. 1990. Sampling littoral fish with a seine: corrections for variable capture efficiency. Can. J. Fish. Aquat. Sci. 47: 1004-1010.

Powles, P.M. 1965. Life history and ecology of American plaice (*Hippoglossoides platessoides* F.) in the Magdalen Shallows. J. Fish. Res. Board Can. 22: 566-598.

Rahel, F.J., and Hubert, W.A. 1991. Fish assemblages and habitat gradients in a Rocky Mountain-Great Plains stream: biotic zonation and additive patterns of community change. Trans. Amer. Fish. Soc. 120: 319-332.

Reisman, H.M., Kao, M.H., and Fletcher, G.L. 1984. Antifreeze glycoprotein in a "Southern" population of Atlantic tomcod. Comp. Biochem. Physiol. A, 78: 445-447.

Roundtree, R.A., and Able, K.W. 1992. Foraging habits, growth, and temporal patterns of salt-marsh creek habitat use by young-of-year summer flounder in New Jersey. Trans. Amer. Fish. Soc. 121: 765-776.

Rutherford, D.A., Echelle, A.A., and Maughan, O.E. 1992. Drainage-wide effects of timber harvesting on the structure of stream fish assemblages in southeastern Oklahoma. Trans. Amer. Fish. Soc. 121: 716-728.

Scott, W.B., and Crossman, E.J. 1973. Freshwater fishes of Canada. Bull. Fish. Res. Board Can. 184: 966p.

Scott, W.B., and Scott, M.G. 1988. Atlantic fishes of Canada. Can. Bull. Fish. Aquat. Sci. 219: 731p.

Stegeman, J.J., Teng, F.Y., and Snowberger, E.A. 1987. Induced cytochrome P450 in winter flounder (*Pseudopleuronectes americanus*) from coastal Massachusetts evaluated by catalytic assay and monoclonal antibody probes. Can. J. Fish. Aquat. Sci. 44: 1270-1277.

Stevens, D.E., Chadwick, H.K., and Painter, R.E. 1987. American shad and striped bass in California's Sacramento-San Joaquin River System, p. 66-78. *In* M.J. Dadswell, R.J. Klauda, C.M. Moffitt, R.L. Saunders, R.A. Rulifson, and J.E. Cooper [*editors*]. Common strategies of anadromous and catadromous fish. Amer. Fish. Soc. Symp. 1.

St-Hilaire, A., Booth, D., Bettignies, C., Chadwick, E.M.P., and Courtenay, S.C. 1995. Is the Miramichi a stratified estuary, p. 73–82. *In* E.M.P. Chadwick [*editor*]. Water, science, and the public: the Miramichi ecosystem. Can. Spec. Publ. Fish. Aquat. Sci. 123.

Swain, D.P., and Wade, E.J. 1993. Density-dependent geographic distribution of Atlantic cod (*Gadus morhua*) in the southern Gulf of St. Lawrence. Can. J. Fish. Aquat. Sci. 40: 725-733.

Vladykov, V.D. 1972. Morphological differences in male gonads among nine genera of Gadidae (Pisces). J. Fish. Res. Board Can. 29: 1709-1716.

Wirgin, I.I., Grunwald, C., Courtenay, S., Kreamer, G.-L., Reichert, W.L., and Stein, J.E. 1994. A biomarker approach in assessing xenobiotic exposure in cancer-prone Atlantic tomcod from the north American Atlantic coast. Environ. Health Perspect. 102(9): 764-770.

CHAPTER 10

Size and Growth of Striped Bass, *Morone saxatilis*, from the Miramichi River, Gulf of St. Lawrence, Canada

Gérald J. Chaput

Department of Fisheries and Oceans, Science Branch, P.O. Box 5030,
Moncton, New Brunswick, Canada E1C 9B6

and Kimberly A. Robichaud

Département de biologie, Université de Moncton,
Moncton, Nouveau-Brunswick, Canada E1A 3E9

Chaput, G.J., and Robichaud, K.A. 1995. Size and growth of striped bass, *Morone saxatilis*, from the Miramichi River, Gulf of St. Lawrence, Canada, p. 161-176. *In* E.M.P. Chadwick [*editor*]. Water, science, and the public: the Miramichi ecosystem. Can. Spec. Publ. Fish. Aquat. Sci. 123.

Abstract

The Miramichi River (Lat. 47°N) represents the northern limit of the known spawning distribution of striped bass (*Morone saxatilis* Walbaum). Analysis of the striped bass samples collected opportunistically over several years has provided the first description of age, size-at-age and growth-at-age of striped bass from the Miramichi River. Size and growth-at-age traits of bass from the Miramichi are similar to those of other populations of eastern Canada but bass from Canada are smaller at age than those of Chesapeake Bay and California. Length-at-age and annual length increments (growth) are similar for males and females up to age three but females are larger than males at older ages. Maximum growth in length occurred during the second year for both males and females. The size of young-of-the-year bass entering the first winter is highly variable within and between year-classes. Under the stringent winter conditions of the Miramichi, we would expect size-dependent overwinter survival to be an important factor regulating the recruitment strength of year-classes to the spawning population.

Résumé

La Miramichi (lat. 45 °N) représente la limite septentrionale de l'aire connue de reproduction du bar rayé (*Morone saxatilis* Walbaum). L'analyse d'échantillons de bar rayé recueillis à l'occasion pendant plusieurs années a permis d'établir la première description de l'âge, de la taille en fonction de l'âge et de la croissance en fonction de l'âge chez les bars rayés de la Miramichi. Les traits de la taille et de la croissance en fonction de la taille des bars rayés de la Miramichi sont similaires à ceux des autres populations de l'est du Canada, mais les bars du Canada sont plus petits selon l'âge que ceux de la baie Chesapeake et de la Californie. La longueur en fonction de l'âge et les intervalles annuels de longueur (croissance) sont similaires pour les mâles et les femelles jusqu'à l'âge de trois ans, mais les femelles sont par la suite plus grosses que les mâles. C'est la deuxième année que la croissance en longueur est maximale tant chez les mâles que chez les femelles. La taille des jeunes de l'année qui entament leur premier hiver est très variable au sein d'une classe d'âge et entre les classes d'âge. Étant donné les rudes conditions qui règnent en hiver sur la Miramichi, nous pouvons prévoir que la survie hivernale dépendante de la taille est un facteur important qui régle l'effectif du recrutement des classes d'âge dans la population reproductrice.

Introduction

The Miramichi River (Lat. 47°N) represents the northern limit of the known spawning distribution of striped bass (*Morone saxatilis* Walbaum) which extends south to the St. Johns River, Florida, U.S.A. (Scott and Scott 1988). The status of the striped bass resource of the Miramichi River is unknown but the population size is thought to be small (Chaput 1995; Chaput and Randall 1990). Striped bass from the Miramichi are genetically distinct from those of the Bay of Fundy and the eastern seaboard of the United States of America (Wirgin et al. 1993). The St. Lawrence River population has become extinct as a result of habitat degradation from river channel dredging in the vicinity of spawning areas, water pollution and intensive fishery exploitation (Beaulieu 1985; Beaulieu et al. 1990). Observed declines of striped bass populations in other rivers of Atlantic Canada have been partly attributed to excessive mortality of fertilized eggs and larvae caused by the interactive effects of low pH, aluminum, water hardness and salinity (Jessop 1990). Many southern populations along the eastern seaboard of U.S.A. have also declined as a result of fisheries exploitation and environmental degradation (Rago and Dorazio 1989).

The striped bass resource of the Miramichi River has been of minor economic importance. Reported landings of striped bass from commercial gear were 62 t in 1917 and have fluctuated between 2 and 48 tons between 1968 and 1992 (Chaput 1995). The abundance of striped bass in the Miramichi appears to be lower by several orders of magnitude, than in the centre of the species range where economically valuable commercial and recreational fisheries exist. In the U.S.A., striped bass commercial landings exceeded 6300 t in 1973 (Chaput and Randall 1990). The recreational fishery for striped bass in the U.S.A. has become more important than the commercial fishery; in 1992 the recreational harvest of 2200 t was three times larger than the commercial landing (Anon. 1993).

Recent enhanced interest in the striped bass population of the Miramichi River is the result of reductions and closures of commercial fisheries on other species, recreational fisheries promotion, interest in aquaculture of striped bass, and restoration programs for depleted river systems. Management initiatives introduced in 1993 to protect and enhance the recreational potential for striped bass in New Brunswick included the mandatory release of all bass greater than 38 cm total length (35.5 cm fork length) from commercial gear and release of all bass less than 68 cm total length in the recreational fishery (Dept. of Fisheries and Oceans 1993).

Much of what we know of striped bass biology come from studies of populations located within the centre of the species' distribution and not on studies from the Miramichi River. For example, somatic growth rates of male and female bass diverge at age of first maturity such that females are larger at age than males (Setzler et al. 1980). Size limits for hook and line fisheries could therefore potentially impart disproportionate harvests on males and females. Growth and age data specific to the Miramichi would therefore be of practical value in determining allowable keep sizes of bass in the fisheries in order to protect and enhance the resource. Such data would also be of ecological value. Population viability of many species of temperate fishes is dependent upon the ability of young-of-the-year fish to complete a minimum amount of growth before winter starvation, a condition which may be limited by the length of the growing season (Shuter and Post 1990). The determination of growth rates of different year-classes and sexes would provide insight into the viability of the striped bass population in the Miramichi River, especially when exposed to fisheries exploitation.

The main objective of this study is to describe the size, age and growth traits of striped bass from the Miramichi River. We determine if growth rates of male and female bass are different. We also compare growth rates at age of different year-classes. Length-at-age of bass from the Miramichi River is compared to that of other rivers in Atlantic Canada and the United States. Finally, we interpret these findings in the light of the striped bass population of the Miramichi which is at the northern limit of its distribution and assess how environmental factors may impact on the abundance and stability of the striped bass population.

Materials and Methods

Opportunistic sampling of striped bass for

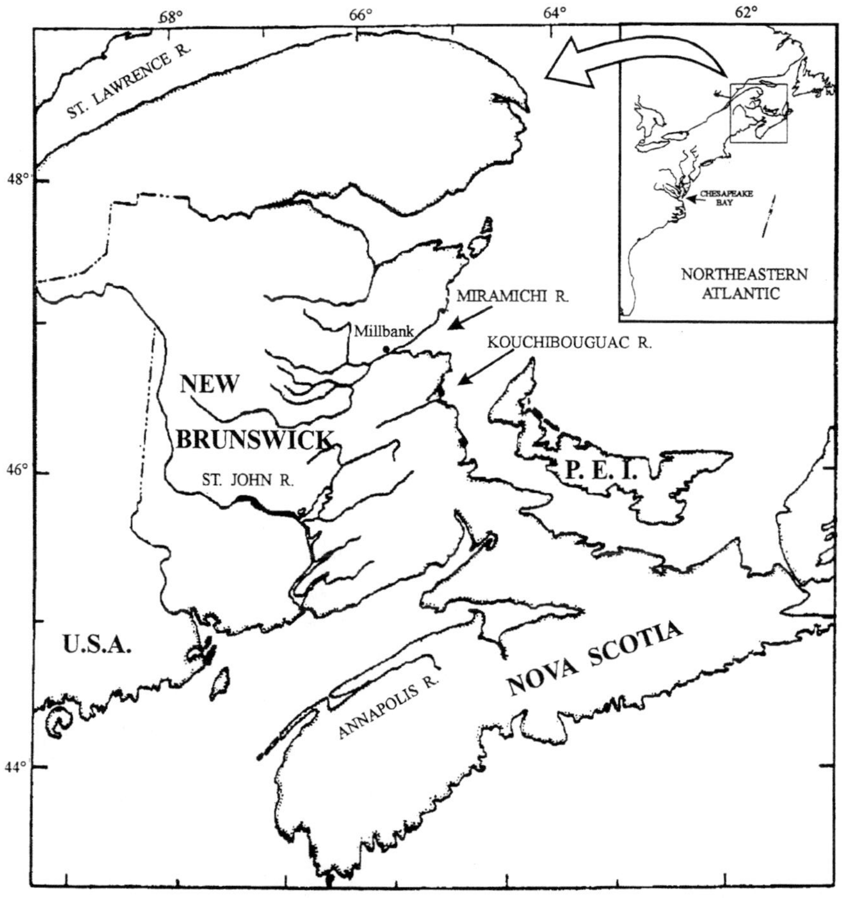

Fig. 1. Geographic location of the Millbank trapnet and the Miramichi River, New Brunswick, Canada.

fork length (0.1 cm), whole weight (50 g), scale samples and sex by internal examination were obtained from fish captured at the Millbank trapnet in the Miramichi River estuary (Fig. 1) for the years 1975 to 1977 and 1979 to 1982. The sampling rate over the season and for any year was not proportional to the catches of bass at the trapnet; most of the scale and biological samples were obtained in 1977 and the majority of the sampling effort occurred in May and June (Fig. 2). The length distribution of the sampled fish also varied with the month of sampling, most fish were greater than 35 cm fork length (Fig. 2). A sample of young-of-the-year bass was obtained from a smelt net in October, 1990.

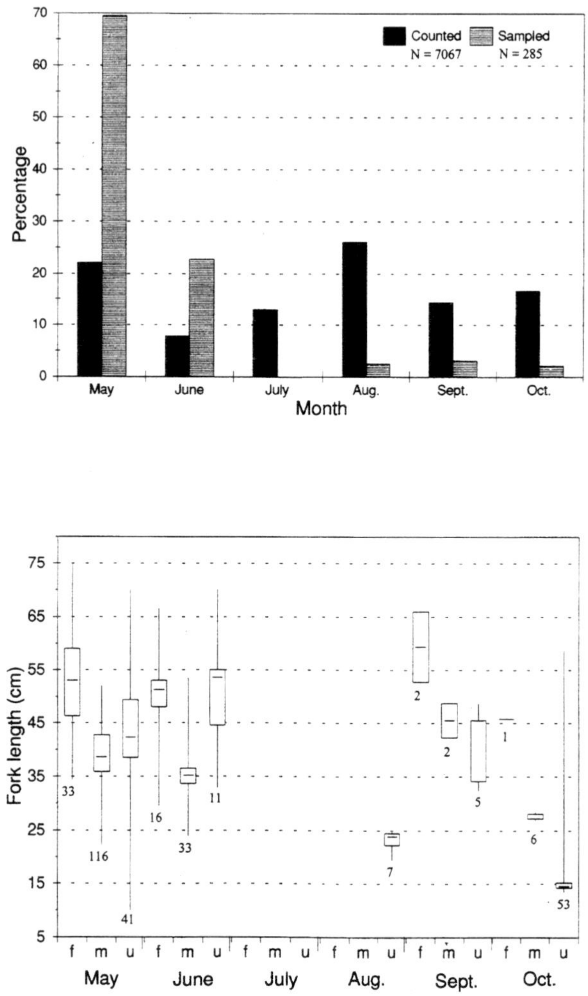

Fig. 2. Monthly sampling intensity (upper) and fork length distributions (lower) of (f=female, m=male, u=sex uknown) striped bass from the Millbank trapnet, 1975 to 1977 and 1979 to 1982. Length distribution for October includes the sample obtained in 1990 from the smelt fishery.

We tested for differences in the weight-length relationships of male and female striped bass from the Miramichi using analysis of covariance (Neter et al. 1983) after log transformations of the length and weight variables. The model fitted was:

log(Weight) = alpha + beta log(Length) + gamma*Sex + delta*Sex*log(Length)
where Sex = 0 for males, 1 for females

Scale samples were collected midway along the body, above the lateral line (Merriman 1941). Scales were cleaned in water and mounted between two glass slides. Age was determined according to criteria described by Merriman (1941). Scales are reliable structures for determining age up to a total length of 90 cm, at which point because of decreased growth, annuli cannot be reliably discerned (Welch et al. 1993). Detailed examination of scales collected from bass of the Miramichi River between May 15 and June 30 indicated that an annulus at the edge of the scale was generally visible after June 1. Bass sampled prior to June 1 whose scales showed plus growth but no visible annulus were designated as age equal to the number of annuli plus one. Fish sampled after June 1 with small plus growth were designated as age equal number of annuli +, the '+' meaning plus growth present.

Scales were measured from the focus to the anterior edge along the anterior-posterior axis of the scale and successively from the focus to each annulus. The measurements were obtained from a digitized image and processed with image analysis software. Image magnifications were calibrated with a micrometer ruler prior to scale measurements and for all changes in magnification.

A scale length to body length relationship was derived assuming the scale proportional hypothesis (SPH) of Francis (1990). Under this assumption, the proportion difference in the size of the scale from the expected size at a given length applies to the previous growth history of the fish. This linear regression of scale length on capture versus body length at capture was used to back-calculate the body length at each annulus using the following four step procedure (Francis 1990):

1) calculate mean scale radius for fish of length L using the SPH regression,
2) calculate ratio of the observed scale radius to mean scale radius of (1),
3) calculate expected scale radius for fish of length L using ratio (2),
4) calculate body length for expected scale radius (3).

We addressed concerns regarding Lee's phenomenon ("tendency for back-calculated lengths at any age to be smaller, the larger the fish from which they were calculated" (Ricker 1992:1024)) in the following manner. First, we plotted predicted size-at-age of different ages against total length of the fish used in the back-calculation and calculated the correlations by age group. Second, the presence of Lee's phenomenon was tested for an individual year-class for which samples were available from different sampling years. Since the size of the scale generally increases with size of fish sampled, under bias associated with Lee's phenomenon, we would expect the back-calculated size-at-age to decrease as progressively older fish of the year-class were used in the back-calculation. Because of limited sample sizes for individual cohorts over several sampling years, we only tested the 1973 cohort for back-calculated length-at-age. We used a one-way analysis of variance to test for differences in the predicted lengths of bass of the 1973 year-class sampled in 1976, 1977 and 1979. The significance level of the test statistic was determined by random permutations of the original data among all three sampling years (Edgington 1987).

Annual growth increment was calculated as the difference in back-calculated body length in year i from body length in year i-1. We tested three hypotheses regarding the annual increments in length of striped bass from the Miramichi. First, annual increments in length are the same for different year-classes (males and females separately). Second, annual increments in length are the same for male bass and female bass of the same year-class. Only year-classes with a minimum of five observations were included in the analyses. We tested the null hypotheses using one-way analysis of variance but with the significance level determined by randomization tests (Edgington 1987). Third, we determined when the maximum annual increment in length occurred using ANOVA method of log-transformed length increments at age against age (Proc GLM, SAS 1990).

For most of the statistical comparisons, we chose randomization procedures over classical significance tests (which are based on assumed distributions) for the following reasons: (1) the samples could not be considered as random samples from the bass population in any year, (2) generally were heteroscedastic, and (3) were not normally distributed. Under these constraints, the inferences we make refer to the fish in the sample and can only be extended to the striped bass population of the Miramichi River if the samples are considered as reasonable, although not random, samples of the population (Edgington 1987). Conclusions regarding the differences in growth rates between year-classes and sexes were based on the P-value of the randomization tests, the smaller the P-value, the greater the probability of the null hypothesis being false.

Results

Striped bass from the Miramichi River are between 30 and 50 cm fork length during their third year of growth and require at least 7 years to achieve fork lengths greater than 68 cm (Fig. 3). All the bass sampled from the Miramichi older than 7 years of age were female, and all male bass were of fork length less than 55 cm. The weight-length relationship was not significantly different for male and female bass (Fig. 4).

The scale length-body length relationship determined for the Miramichi River striped bass was strongly linear (Fig. 5). There was no evidence of "Lee's phenomenon" in the back-calculated length-at-age based on the SPH procedure (Fig. 6, Table 1). The opposite situation was observed; predicted length-at-age increased as the length of the fish when sampled increased. This indicates that the back-calculated length-at-age of the survivors at successive ages was greater than the average length of the year-class at previous ages. For the 1973 cohort, the back-calculated lengths-at-age of two of the three age groups were not significantly different between the years sampled. Although

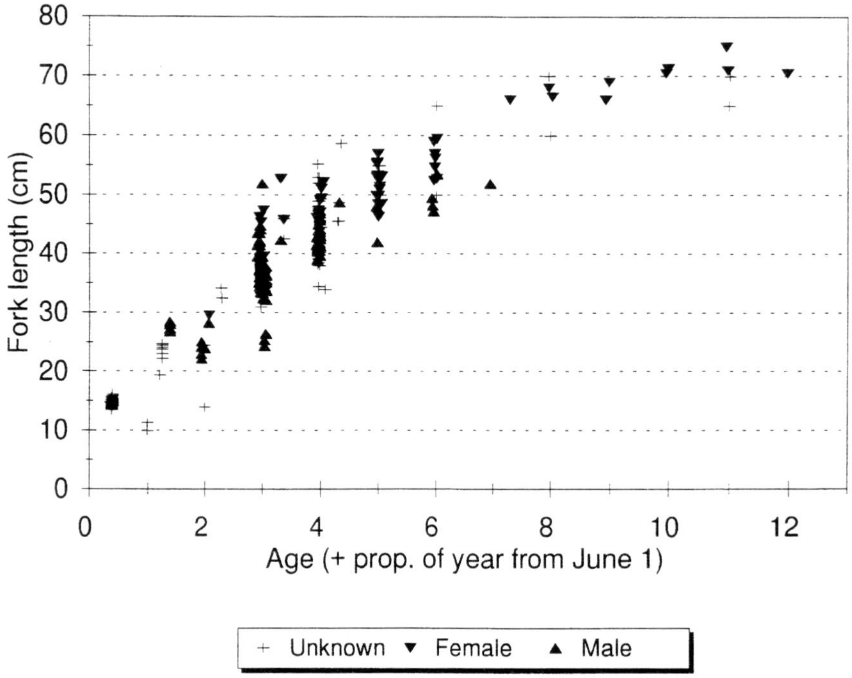

Fig. 3. Fork length-at-capture versus age of striped bass from the Miramichi River. Age axis corresponds to number of annuli plus proportion of year when sampled, assuming June 1 as birth date.

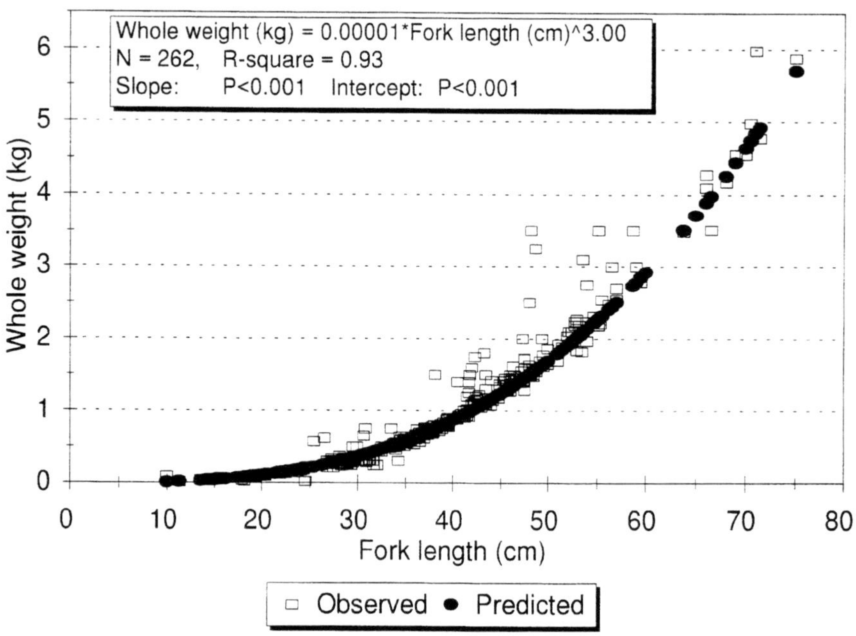

Fig. 4. Whole weight (kg) to fork length (cm) relationship of striped bass from the Miramichi River, sexes combined.

small sample size in the 1979 sampling year compromised the comparisons, we conclude that negative bias in the back-calculated length-at-age was not an important factor in our study (Table 2).

Striped bass from the Miramichi River are longer at age than those from the St. Lawrence River but similar to bass from the Annapolis River (Fig. 7). Northern populations of striped bass are

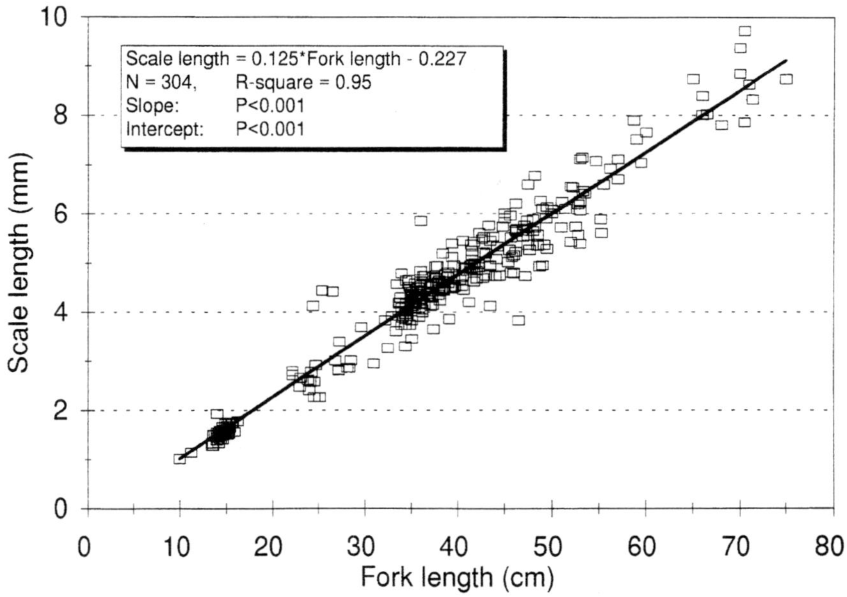

Fig. 5. Scale length (mm) to fork length (cm) relationship of striped bass from the Miramichi River.

Table 1. Relationship between predicted length-at-age versus length of fish when sampled for males, females, and sexes combined by age of back-calculation. *P*-value is the probability of the slope coefficient = 0. *N* is the sample size.

Sex	Age	Slope Coefficient	R^2	*P*-value	*N*
Combined	1	0.066	0.11	0.001	210
	2	0.095	0.09	0.001	197
	3	0.082	0.04	0.049	94
	4	0.152	0.18	0.007	38
	5	0.240	0.42	0.001	21
	6	0.324	0.45	0.035	10
Males	1	0.161	0.21	0.001	144
	2	0.279	0.26	0.001	139
	3	0.437	0.21	0.002	44
	4	0.657	0.53	0.042	8
	5	0.790	0.61	0.119	5
Females	1	0.006	0.01	0.821	48
	2	-0.046	0.03	0.269	48
	3	-0.059	0.02	0.375	40
	4	0.082	0.06	0.195	29
	5	0.124	0.14	0.160	16
	6	-0.198	0.13	0.347	9
	7	-0.356	0.30	0.124	9

of similar length-at-age, but all are of inferior size to the southern populations of Chesapeake Bay and California (Fig. 7). The differences between the Canadian populations (northern) and the bass from the southern areas are especially evident for both males and females older than 3 years of age.

Both sex and year-class contributed to the variability in length-at-age of striped bass from the Miramichi. With year-classes combined, female bass were significantly longer than males at all ages except age 3 (Table 3). When individual year-classes were considered, the length-at-age of 1-year-old

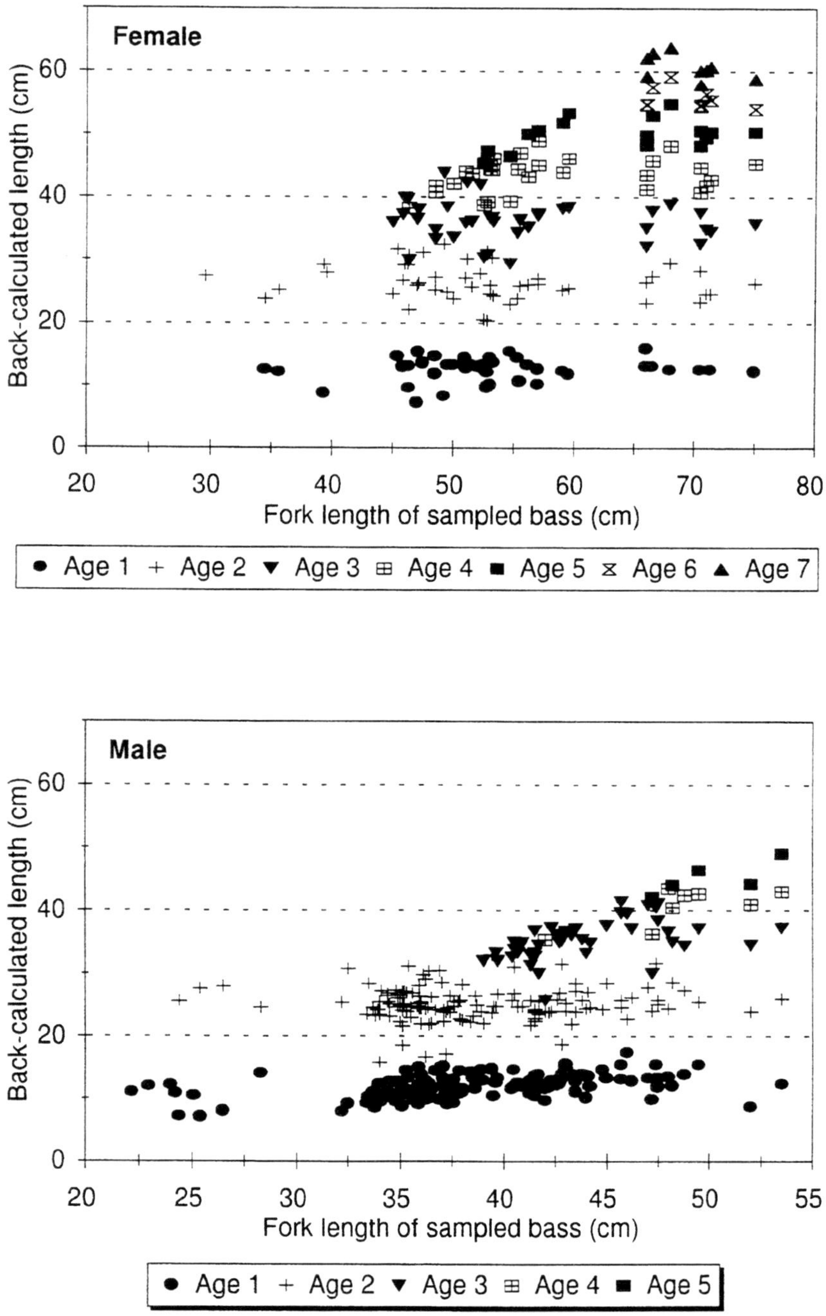

Fig. 6. Back-calculated fork length-at-age relative to the fork length of female (*upper*) and male (*lower*) striped bass when sampled.

Table 2. Analysis of possible bias associated with the back-calculation of length-at-age of bass of the 1973 cohort, all sexes combined, relative to the year samples were collected.

Year of Sampling	Length-at-age (cm)			
	Median	Min.	Max.	*N*
Age 1				
1976	13.8	10.2	15.3	26
1977	13.3	11.9	17.3	24
1979	12.5	12.0	15.4	4
H_0: $L_{1976} = L_{1977} = L_{1979}$; *P*-value = 0.239				
Age 2				
1976	26.4	23.9	31.8	26
1977	26.3	22.1	32.6	24
1979	24.2	23.0	27.2	4
H_0: $L_{1976} = L_{1977} = L_{1979}$; *P*-value = 0.039				
Age 3				
1976	37.4	37.3	45.3	3
1977	36.4	32.0	43.9	24
1979	36.1	29.4	37.3	4
H_0: $L_{1976} = L_{1977} = L_{1979}$; *P*-value = 0.152				

Table 3. Comparison of the back-calculated length-at-age of male and female striped bass from the Miramichi River between year-classes and differences between sexes by individual year-class and for year-classes combined. Distributions are shown in Fig. 8. *P*-value, on a total of 1000 randomizations, represents the probability of the observed size distributions if each is a sample from the same population.

Group	Age	Comparisons	*P*-value
Female	1	1967, 1969, 1973 to 1975	0.001
	2	1967, 1969, 1973 to 1975	0.028
	3	1967, 1969, 1973 to 1975	0.272
	4	1967, 1969, 1974, 1975	0.261
Male	1	1972 to 1976, 1978	0.001
	2	1972 to 1976, 1978	0.001
	3	1972, 1973, 1975	0.051
1973	1	Male versus female	0.294
	2	Male versus female	0.023
	3	Male versus female	0.032
1975	1	Male versus female	0.433
	2	Male versus female	0.270
	3	Male versus female	0.194
All year-classes sampled	1	Male versus female	0.006
	2	Male versus female	0.003
	3	Male versus female	0.168
	4	Male versus female	0.020
	5	Male versus female	0.002

male and female bass were not significantly different and in two of the three year-classes compared, length-at-age were also similar at age 2 and 3 (Table 3). Year-class effects were significant during at least the first three years in female bass and the first two years in male bass (Table 3, Fig. 8). With

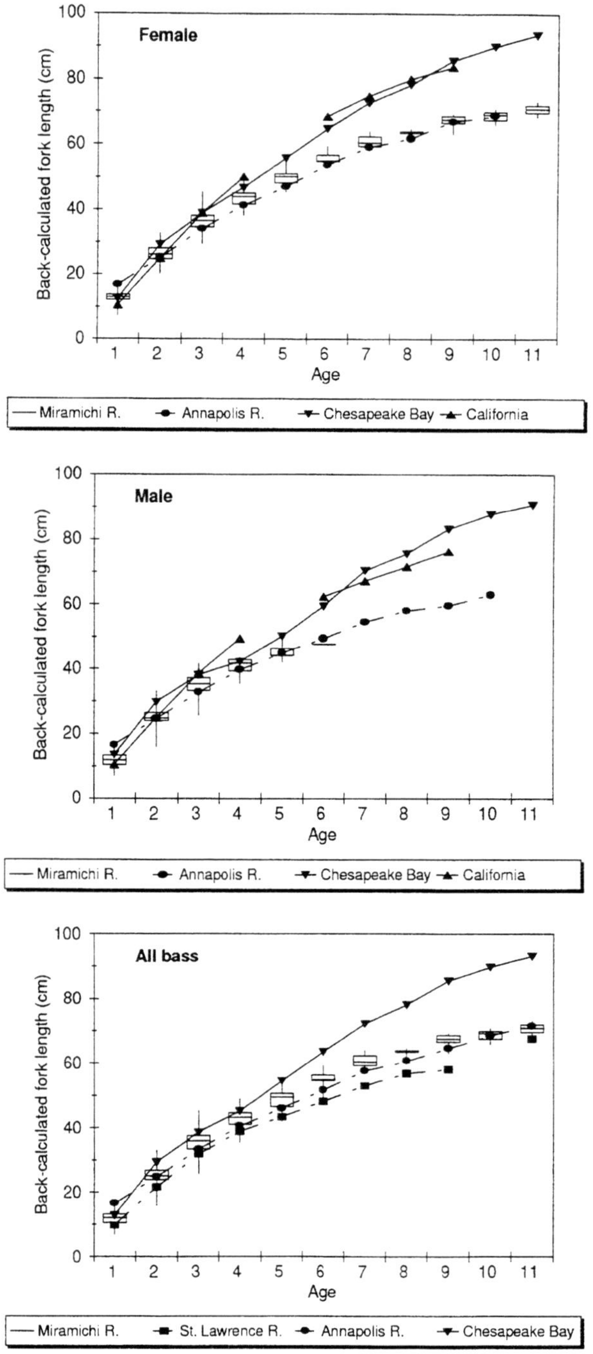

Fig. 7. Back-calculated fork length-at-age of female (*upper*), male (*middle*) and all striped bass (*lower*) from the Miramichi River compared to other rivers of eastern Canada and United States. Box plots for Miramichi River are interpreted as follows: single horizontal line represents the median fork length, rectangle defines the interquartile range, vertical lines represent minimum and maximum values. Length-at-age for other rivers are average length as reported from: St. Lawrence River (Magnin and Beaulieu 1967), Annapolis River (Williamson 1974), Chesapeake Bay (Mansueti 1961), and California (Robinson 1960).

fewer year-classes compared at older ages, differences in length-at-age were not significant.

As with length-at-age, annual length increment variability was greater between year-classes than among sex of the same year-class. Annual length increments were significantly different between year-classes for the first two years but growth during the third and fourth years was similar (Table 4, Fig 9). Growth increments of male as compared to female bass of the same year-class were generally similar during the first 3 years, the 1973 year-class was the only exception for which female bass had a greater annual length increment during the second year (Table 4, Fig. 9). With all the year-classes combined, the predicted annual length increments of male and female bass during the third to fifth years were not significantly different (Table 4). Although the growth increments of male bass in the fourth and fifth years (median values: 6.4 and 3.7 cm, respectively) were less than those of female bass (median values: 8.3 and 7.0 cm, respectively), the small sample sizes for male bass and the large variation in individual growth increments of both male and female bass for those years provided low statistical power for these comparisons (Fig. 10).

With year-classes and sexes combined, the annual growth in length was greatest in the second year (Fig. 10, Table 5). Growth in length in the first year was about 90% of that observed in the second year and growth in successive years decreased at a rate of 11% on average per year. The 1990 year-class had already attained fork lengths between 13 and 16 cm in October of 1990, large but within the

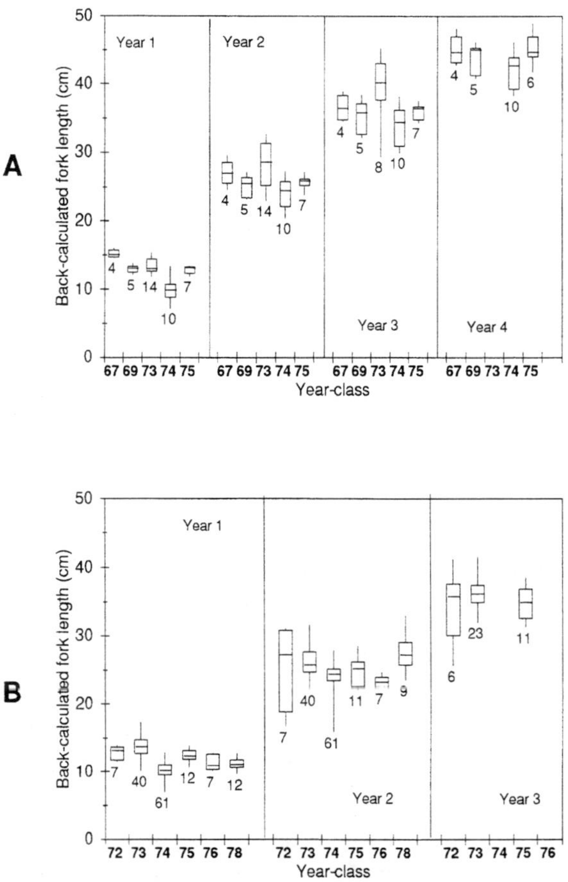

Fig. 8. Variation between year-classes in the back-calculated fork length-at-age for ages 1 to 4 years of female bass (*panel A*) and for ages 1 to 3 years for male bass (*panel B*). Box plots are interpreted as in Fig. 7. Number below each box plot is the sample size of the individual group.

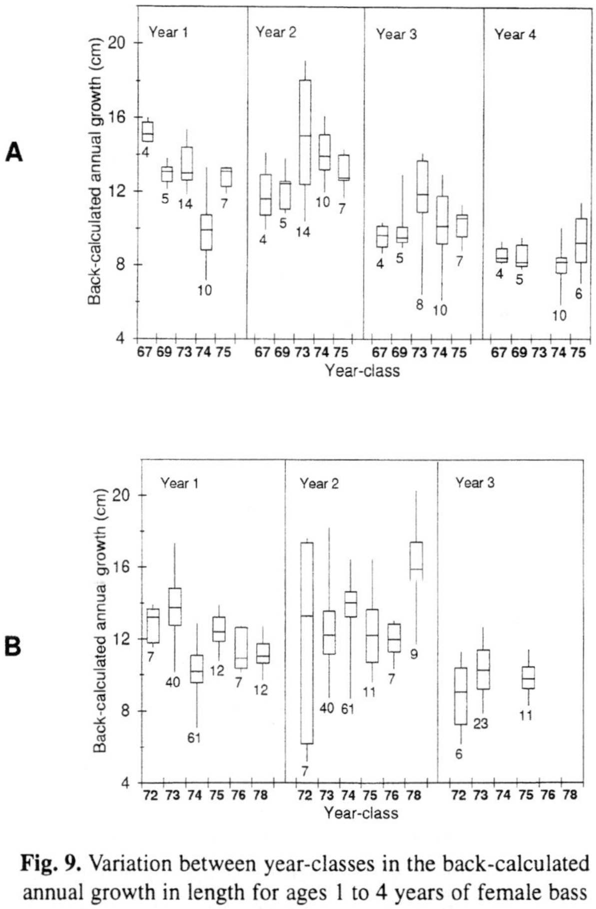

Fig. 9. Variation between year-classes in the back-calculated annual growth in length for ages 1 to 4 years of female bass (*panel A*) and for ages 1 to 3 years for male bass (*panel B*). Box plots are interpreted as in Fig. 7.

Table 5. Analysis of variance of the annual increment in length for successive ages of striped bass, sexes and year-classes combined, from the Miramichi River. Length increments were log-transformed before analysis. Year is treated as a categorical variable with year 2 as the reference level. Distributions of the annual length increments are in Fig. 10.

Model: Log_{10} (Length increment in cm) = $alpha_{Year}$ + Intercept

Source	df	Sum of Squares	*F*	*P*-value
Model	10	10.451	172.11	<0.001
Error	584	3.546		
Total	594	13.997		
R^2	0.747			

Parameter		Estimate	*P*-value H_0:Parameter = 0
Intercept		1.120	<0.001
Year	1	-0.048	<0.001
	2	0.000	
	3	-0.116	<0.001
	4	-0.220	<0.001
	5	-0.336	<0.001
	6	-0.432	<0.001
	7	-0.435	<0.001
	8	-0.552	<0.001
	9	-0.577	<0.001
	10	-0.667	<0.001
	11	-0.724	<0.001

range of back-calculated fork lengths observed for other year-classes during the first year (Fig. 10).

Discussion

Analysis of the striped bass samples collected opportunistically over several years has provided the first description of age, size-at-age and growth-at-age of striped bass from the Miramichi River. Size-at-age and growth-at-age traits of striped bass from the Miramichi River are similar to those of the Annapolis River population of eastern Canada (Williamson 1974). Size and growth traits of striped bass were similar for males and females, at least for the first 2 years but there were significant differences between year-classes.

Striped bass from the Miramichi River are genetically distinct from those of the Bay of Fundy (Wirgin et al. 1993) but the length-at-age is similar to other populations of eastern Canada. Bass from the southern areas, Chesapeake Bay and California, are longer at age than the more northern Canadian populations. Male and female striped bass from the Miramichi had a similar weight to length relationship, as was reported for the Annapolis River population (Jessop 1980). The samples in these studies were obtained during May to October. Significant differences between the sexes have been reported in the Chesapeake Bay population, in part as a result of using samples collected in a limited part of the year (Mansueti 1961). The Chesapeake Bay samples were collected in winter and spring and consisted mostly of maturing fish which could account for the females being heavier than males.

Length-at-age and annual length increments (growth) are similar for males and females up to age three but females are larger than males at older

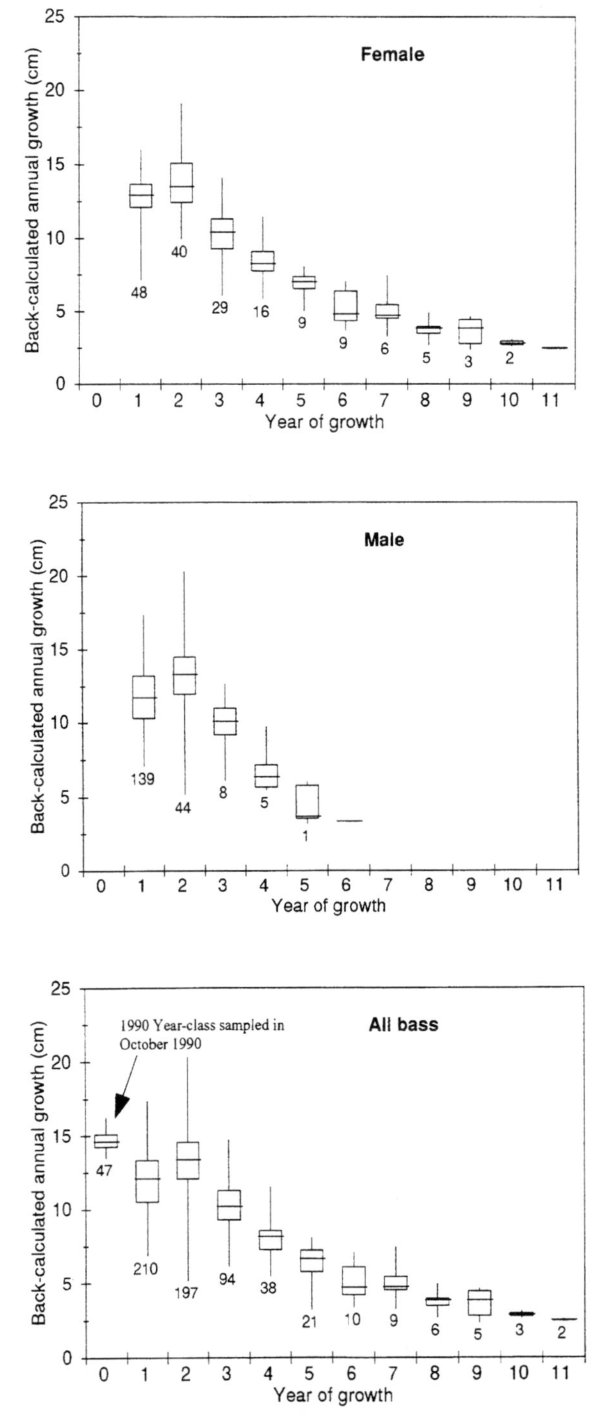

Fig. 10. Back-calculated annual growth in length by age for female (*upper*), male (*lower*) and all striped bass (*bottom*) from the Miramichi River, all year-classes combined. Box plots are interpreted as in Fig. 7. The fork length of the 1990 year-class sampled in October 1990 is included as an indication of the annual variation observed.

ages. Divergence in size-at-age which occurs after age three (Mansueti 1961; Robinson 1960; Scofield 1931) is probably related to the different maturation schedules of male and female bass. Male bass mature at younger ages, in southern populations as early as age 2 but more frequently at age three (Setzler et al. 1980). Mature male bass three years of age and mature female bass four years of age were reported from the Kouchibouguac River (Hogans and Melvin 1984). The larger length-at-age of females as compared to males would require that annual length increments be higher for females after age three. The absence of any detectable significant differences in annual length increments of male and female bass of the Miramichi was probably an artifact of small sample sizes for those older ages.

Maximum growth in length of the Miramichi bass occurred during the second year for both male and female bass, similar to populations from the St. Lawrence River (Magnin and Beaulieu 1967), from Chesapeake Bay (Mansueti 1961), and California (Robinson 1960). The maximum annual growth increment of 13.2 cm in the second year is higher than the 11.9 cm annual growth increment of the St. Lawrence population. The length increments from these northern areas are less than the annual increments of about 16.5 cm for the Chesapeake Bay bass and 14.5 for the California populations.

The observed inter-annual variability in the back-calculated fork length of one year old bass has important implications for the viability of the population as well as to the recruitment potential of striped bass to the Miramichi. Overwinter starvation mortality of young fish has been proposed as the critical factor limiting the northern distribution of yellow perch and smallmouth bass, two closely related species (Shuter and Post 1990). Overwinter survival in these species has been shown to be size-dependent, large fish can withstand winter starvation better than small fish because the ratio of energy stored to metabolic rate increases with body size (Shuter et al. 1980; Shuter and Post 1990). Winter duration in the Miramichi is long, freshwater lakes are frozen over during 125 to 155 days per year (Randall et al. 1989) and overwinter survival would be expected to be an important factor regulating recruitment.

Scales, as compared to otoliths, tend to underestimate the total age of striped bass but the discrepancy becomes important only in large and presumably older fish (Welch et al 1993). In fast growing populations, age determinations from scales provided 80% agreement with age determinations from otoliths for fish up to age four years; older fish were not examined (Heidinger and Clodfelter 1987). More importantly, there were no significant differences between the back-calculated total lengths from scales and the measured total lengths at known age (Heidinger and Clodfelter 1987). In the Miramichi, the maximum age interpreted from the scales was 12 years but the majority (94%) of the bass analyzed for age, size and growth traits from the Miramichi were less than 6 years old. Bass of 19 years of age have been reported from the St. Lawrence River (Magnin and Beaulieu 1967) while 20-year-old bass have been reported from the Annapolis River (Jessop 1980).

We consider the back-calculated length-at-age and growth-at-age values reported in our study to be unbiased estimates of the true traits of this population for the following reasons. First, we assumed that the radius of the scale mark is equivalent to the radius of the scale when the mark was formed (Francis 1990). This is a valid assumption because scale erosion was not evident in the samples which we examined and there is no evidence from previous studies that the shape of a scale at younger ages would change with age. Second, we determined when the annulus was formed by examination of a large number of scales during the period of annulus formation. We estimated that the time of annulus formation was in the vicinity of June 1 based on scales of young bass, less than five years of age, sampled in late May and June. It is important to sample young fast growing fish for which annual growth of the scale is greater than the width of the annulus. Finally, we used a back-calculation method which relates, without bias, the scale radius and body size for each fish (SPH method of Francis 1990). The scale to body length sample covers the full size range of the population of interest, (10 – 75 cm fork length, Fig. 5) and under this condition, the slope of the regression line should be closest to the true slope (Francis 1990). As in several other studies (Magnin and Beaulieu 1967; Mansueti 1961; Robinson 1960), we found that the scale length to body length relationship for striped bass from the Miramichi River was strongly linear. Although Lee's phenomenon was not evident in our study, this by itself is insufficient evidence for the accuracy of the back-calculation method. In our study, the opposite situation occurred; generally, back-calculated size-at-age of male bass, but not female bass, increased as successively older fish were used. Such a phenomenon could result from non-representative sampling, or from size-related survival. The striped bass samples which we used can not be considered as random samples from the population, because they were collected opportunistically, but we have no reason to suspect that they were collected in a biased manner. It would be impossible for field personnel to select bass for sampling on the basis of whether they were fast growing or slow growing individuals of a year-class. Size-dependent survival could account for increased predicted length at young ages as successively older fish are used. The only valid way of addressing size-dependent survival is to sample the same year-classes over several years, including during the late fall in the first year of growth.

In spite of being at the northern limit of its species distribution, striped bass in the Miramichi can grow as quickly as bass from other populations in eastern Canada. Male and female bass have similar growth rates during the first year but the size of young-of-the-year bass entering the first winter is highly variable. Under the stringent winter conditions of the Miramichi, we would expect size-dependent survival to be an important factor regulating recruitment to the spawning population. Winter conditions may ultimately determine the carrying capacity and the minimum viable population (Shuter and Post 1990). This could explain the lower abundance of striped bass in the Miramichi River compared to the abundance of the southern populations of the eastern United States.

Acknowledgements

The study was partly funded through a National Sciences and Engineering Research Council grant to Dr. Peter Beninger, University of Moncton. We thank Mr. Brian Jessop for providing guidelines for ageing the striped bass scales. Thanks to Rod Bradford for his keen interest and numerous suggestions for improving an earlier draft of this manuscript.

References

Anon. 1993. Status of fishery resources off the Northeastern United States for 1993. By Conservation and Utilization Division, Northeast Fisheries Sciences Centre. October 1993. 140 pp. NOAA Technical Memorandum NMFS-F/NEC-101.

Beaulieu, H. 1985. Rapport sur la situation du bar rayé (*Morone saxatilis*). Faune et flore à protéger au Québec. Association des biologistes du Québec. Publ. No. 7. 53 p.

Beaulieu, H., Trepanier, S., et Robitaille, J.A. 1990. Statut des populations indigènes de bar rayé (*Morone saxatilis*) au Canada. Ministère du Loisir, de la Pêche, Direction générale des espèces et des habitats. Rapp. tech. xii + 50p.

Chaput, G.J. 1995. Temporal distribution, spatial distribution, and abundance of diadromous fish in the Miramichi River watershed, p. 121–139. *In* E.M.P. Chadwick [*editor*]. Water, science, and the public: The Miramichi ecosystem. Can. Spec. Publ. Fish. Aquat. Sci. 123.

Chaput, G.J., and Randall, R.G. 1990. Striped bass, *Morone saxatilis*, from the Gulf of St. Lawrence. Can. Atl. Fish. Sci. Adv. Comm. Res. Doc. 90/71: 29p.

Dept. of Fisheries and Oceans. 1993. Striped bass management plan for New Brunswick: major elements. 5p.

Edgington, E.S. 1987. Randomization tests (2nd ed.). Statistics, Textbooks and Monographs, Vol. 77. Marcel Dekker, Inc. New York, NY.

Francis, C. 1990. Back-calculation of fish length: a critical review. J. Fish. Biol. 36: 883-902.

Heidinger, R.C., and Clodfelter, K. 1987. Validity of the otolith for determining age and growth of walleye, striped bass, and smallmouth bass in power plant cooling ponds. Pages 241-251. *In* R.C. Summerfelt and G.E. Hall [*editors*]. Age and growth of fish. Iowa State University Press, Ames, IA.

Hogans, W., and Melvin, G.D. 1984. Kouchibouguac National Park striped bass, *Morone saxatilis* (Walbaum), fishery survey. 91 p. Available from Environment Canada, Kouchibouguac National Park, Kouchibouguac, New Brunswick E0A 2A0, Canada.

Jessop, B.M. 1980. Creel survey and biological study of the striped bass fishery of the Annapolis River, 1978. Can. Manusc. Rep. Fish. Aquat. Sci. No. 1566: x + 20p.

Jessop, B.M. 1990. The status of striped bass in the Scotia-Fundy Region. Can. Atl. Fish. Sci. Adv. Comm. Res. Doc. 90/36: 22p.

Magnin, E., and Beaulieu, G. 1967. Le bar, *Roccus saxatilis* (Walbaum), du fleuve Saint-Laurent. Nat. Can. 94: 539-555.

Mansueti, R. 1961. Age, growth and movements of the striped bass, *Roccus saxatilis*, taken in size selective fishing gear in Maryland. Chesapeake Sci. 2: 9-36.

Merriman, D. 1941. Studies on the striped bass (*Roccus saxatilis*) of the Atlantic coast. U.S. Fish Wildl. Serv. Fish Bull. 50(35): 77p.

Neter, J., Wasserman, W., and Kutner, M.H. 1983. Applied Linear Regression Models. Irwin, IL.

Randall, R.G., O'Connell, M.F., and Chadwick, E.M.P. 1989. Fish production in two large Atlantic coast rivers: Miramichi and Exploits, p. 292-308. *In* D.P. Dodge [*editor*] Proceedings of the International Large River Symposium. Can. Spec. Publ. Fish. Aquat. Sci. 106.

Rago, P.J., and Dorazio, R.M. 1989. Emergency striped bass research study report for 1988. Dept. of Commerce, National Marine Fisheries Service. 55 p.

Ricker, W.E. 1992. Back-calculation of fish lengths based on proportionality between scale and length increments. Can. J. Fish. Aquat. Sci. 49: 1018-1026.

Robinson, J.C. 1960. The age and growth of striped bass (*Roccus saxatilis*) in California. Calif. Fish Game 46: 279-290.

SAS. 1990. SAS User's Guide: Statistics. SAS Institute Inc. Cary, North Carolina.

Scofield, E.C. 1931. The striped bass of California (*Roccus saxatilis*). Calif. Div. Fish Game, Fish Bull. 29: 1-84.

Scott, W.B., and Scott, M.G. 1988. Atlantic Fishes of Canada. Can. Bull. Fish. Aquat. Sci. No. 219: 731 p.

Setzler, E.M., Boyton, W.R., Wood, K.V., Zion, H.H., Lubbers, L., Mountford, N.L., Frere, P., Tucker, L., and Mihursky, J.A. 1980. Synopsis of biological data on striped bass, *Morone saxatilis* (Walbaum). NOAA Tech. Rep. NMFS Circ. 433: 68p.

Shuter, B.J., MacLean, J.A., Fry, F.E., and Regier, H.A. 1980. Stochastic simulation of temperature effects on first-year survival of smallmouth bass. Trans. Amer. Fish. Soc. 109: 1-34.

Shuter, B.J., and Post, J.R. 1990. Climate, population viability, and the zoogeography of temperate fishes. Trans. Amer. Fish. Soc. 119: 314-336.

Welch, T.J., Van Den Avyle, M.J., Betsill, R.K., and Driebe, E.M. 1993. Precision and relative accuracy of striped bass age estimates from otoliths, scales, and anal fin rays and spines. N.Amer. J. Fish. Manage. 13: 616-620.

Williamson, F.A. 1974. Population studies of striped bass (*Morone saxatilis*) in the Saint John and Annapolis rivers. M.Sc. thesis, Acadia Univ., Wolfville, NS, Canada. 60p.

Wirgin, I.I., Ong, T.-L., Maceda, L., Waldman, J.R., Moore, D., and Courtenay, S. 1993. Mitochondrial DNA variation in striped bass (*Morone saxatilis*) from Canadian rivers. Can. J. Fish. Aquat. Sci. 50: 80-87.

EVALUATING HUMAN IMPACTS

"The operative word must be we, as in we've got to work together and protect our river."

Premier Frank McKenna
April 13, 1994, Miramichi Environmental Workshop

Smelt is the most abundant fish species in the Miramichi River. Part of the commercial catch is taken in winter by box traps set under the ice. These traps provide information on the winter distribution and abundance of many fish species in the estuary.

CHAPTER 11

Sediments and Environmental Quality of the Miramichi Estuary: New Perspectives

Dale E. Buckley

Atlantic Geoscience Centre, Bedford Institute of Oceanography, Dartmouth, Nova Scotia, Canada, B2Y 4A2

Buckley, D.E. 1995. Sediments and environmental quality of the Miramichi Estuary: new perspectives, p. 179-190. *In* E.M.P. Chadwick [*editor*], Water, science, and the public: the Miramichi ecosystem. Can. Spec. Publ. Fish. Aquat. Sci. 123.

Abstract

The Miramichi Estuary has been one of the most thoroughly studied estuaries in Canada. Previous studies have included fisheries biology, physical oceanography, geochemistry, sedimentology, and engineering feasibility for the dredging of the shipping channel. The Miramichi River and Estuary system is highly dynamic, with respect to seasonal variability, as well as longer term changes as the coastal environment has evolved. The spring floods provide a five-fold increase in fresh water input to the estuary over an average of 40 days and discharge 80% of the annual supply of sediments (95 600 tonnes per year), including as much as 20 000 tonnes per year of organic matter from the forest industries during the decades between 1940 and 1980. Sedimentation rates in the upper estuary are up to 4 cm per year. Contaminant metals, including zinc, lead, copper, cadmium, and mercury are carried from the watershed and are predominantly deposited in the estuarine sediments during the spring flood period. Some of these contaminants are then remobilized from the sediments over the rest of the year, with 90 tonnes of zinc being discharged to the Gulf of St. Lawrence each year. Dredging of the shipping channel through the estuary during the 1980's resulted in the displacement of 6 000 000 m^3 of sediment to dump sites within the system. The long-term stability of some of these dump sites is unknown.

In recent years new techniques for evaluating environmental quality of other estuarine systems have improved. New techniques of conducting high resolution acoustic surveys of geomorphology and geology can be used to identify processes of erosion and deposition of sediments. New methods of chemical analyses and statistical interpretation for estuarine sediments have been developed so that discrete sources of contamination, such as sewer outfalls, industrial waste, and surface drainage can be identified. Precise isotopic dating of sediment cores can be used to compile historical records of contamination, and determine the significance of changes in major sources of contamination. These new techniques could now be applied to the Miramichi system so that rational management of environmental quality could be obtained.

Résumé

L'estuaire de la Miramichi est l'un des estuaires canadiens qui a fait l'objet des études les plus exhaustives : biologie des pêches, océanographie physique, géochimie, sédimentologie, et faisabilité technique du dragage de la voie de navigation. Le système de la Miramichi et de son estuaire est extrêmement dynamique en ce qui concerne la variabilité saisonnière, et a subi des changements à long terme à mesure qu'évoluait le milieu côtier. Pendant les crues du printemps, l'apport d'eau douce dans l'estuaire est multiplié par cinq sur une période de 40 jours en moyenne, avec décharge de 80 % de l'apport annuel de sédiments (95 600 tonnes par an), dont 20 000 tonnes par an de matière organique produite par les industries forestières entre 1940 et 1980. Le taux de sédimentation dans la partie supérieure de l'estuaire va jusqu'à 4 cm par an. Les métaux polluants, notamment le zinc, le plomb, le cuivre, le cadmium et le mercure descendent du bassin versant et se déposent principalement dans les sédiments estuariens pendant la crue printanière. Certains de ces polluants sont alors remis en suspension pendant le reste de l'année, avec un rejet de 90 tonnes de zinc par an dans le golfe du Saint-Laurent. Le dragage de la voie de navigation dans l'estuaire dans les années 80 a causé le déplacement de 6 000 000 m^3 de sédiments vers les zones de décharge du système. On ne connaît pas la stabilité à long terme de certaines de ces zones de décharge.

Ces dernières années, on a amélioré des techniques d'évaluation de la qualité environnementale de certains autres systèmes estuariens. De nouvelles techniques faisant appel à des relevés acoustiques à haute résolution de la géomorphologie et de la géologie peuvent servir à repérer les processus d'érosion et de dépôt des sédiments. De nouvelles méthodes d'analyse chimique et d'interprétation statistique ont été mises au point pour les sédiments estuariens, de sorte qu'il est possible d'identifier les sources discrètes de pollution, notamment les décharges d'eaux usées, les déchets industriels et le drainage des eaux superficielles. La datation isotopique précise des carottes de sédiments peut servir à établir l'histoire de la pollution et à déterminer la signification des modifications observées dans les principales sources de pollution. Ces nouvelles techniques pourraient maintenant être appliquées au système de la Miramichi, ce qui permettrait de rationaliser la gestion de la qualité de l'environnement.

Introduction

The Miramichi Estuary has been one of the most thoroughly studied estuaries in Canada. In a compiled bibliography of technical and scientific reports on the Miramichi Estuary (Acadia Centre for Estuarine Research 1990), 390 references were listed. Some of the earliest studies were related to marine biology, such as the distribution of diatoms (Fritz 1918), salmon (Kerr 1929), oysters (Medcof 1940), and smelt (McKenzie 1943). Other early studies included a significant study of the tidal circulation in Miramichi Bay (Fothergill 1953), the physical characteristics of the Miramichi Estuary (Bousfield 1955), and an analysis of sedimentary environments from studies of foraminifera (Bartlett 1966).

Some of these studies and reports provided a basis for assessing potential and actual changes in environmental quality of the system that occurred during the 20th century. During the 1970's many studies and research projects were carried out in relation to proposed dredging of the shipping channel through the estuary (Miramichi River Navigation Channel Study 1974). Studies related to this major public works project represented one of the most significant attempts to assess the environmental impact of modifications to a marine system in eastern Canada. These studies provided information on water and sediment impacts (Ambler 1976; Philpott and Duncan 1977), as well as some basic understanding of characteristics and processes in the estuarine system. These latter studies included evaluation of the relationship of barrier island stability to Miramichi Bay (Owens 1974), analyses of the geological history of sediment deposition in Miramichi Bay (Howells and McKay 1977; Wagner and Schafer 1980; Schafer and Smith 1983), interpretations of the relationship between physical estuarine processes and sediment distributions (Reinson 1976, 1977, 1980; Schafer et al. 1977), the geochemistry of surficial sediments in the estuary (Rashid and Reinson 1979; Willey and Fitzgerald 1980), the physical oceanography of the estuary (Vilks and Krauel 1982), the dynamics of suspended sediment transport through the estuary (Winters 1981, 1983), and assessment of the annual geochemical transport through the estuarine system (Buckley and Winters 1983).

After this period of intensive study a major capital dredging project between 1981 and 1983 moved about 6.5 x 10^6 m^3 of sediment from the main shipping channel to three designated dump sites in the estuarine system. This project was monitored for perturbations to the suspended sediment characteristics and other environmental impacts (Unpublished contract reports by MacLaren Plansearch between 1981 and 1985; Macgregor and Packman 1983; Krank and Milligan 1989). Later reviews of the impact of this capital dredging project have also been published (Lewis and Buckley 1990; Buckley 1990; Chadwick 1990; MacKnight 1990; Sephton 1990; Solomon 1990). These reviews pointed out the difficulty of identifying specific impacts from short-term changes in water quality, suspended sediment loading, and transitory deposition patterns.

Results of estuarine studies over the past two decades have identified new techniques and approaches to assessing environmental quality, and point the way to developing means for environmental management of these valuable coastal systems. Knowledge of the significant environmental quality impacts are reviewed in this paper, and new techniques and approaches will be highlighted with suggestions on their application to the Miramichi estuarine system.

Significant Environmental Impacts

Organic Matter

The influence of human activity on the Miramichi River and Estuary probably became significant about 300 years ago with the development of the forest product industry. Initially this would have been related to the harvesting of mature timber in the watershed areas of the Miramichi Rivers. The initial impact may have been increased erosion of forest soils, with the result that more silt was probably discharged to the upper estuary channels. Most of this discharge, rising to about 95 600 tonnes per year by the mid 20th century, would occur during the spring floods of the drainage system (Buckley and Winters 1983).

About 200 years ago saw mill operations became a significant industry in the watershed, and about 100 years ago pulp mills began operations at Newcastle. These industries contributed an increasing amount of organic waste to the drainage system and the upper estuary. An estimate of the average pulp waste production from the Newcastle mill between 1890 and 1950 is about 2 500 tonnes per year (Philpot and Duncan 1977). Thus, during this time about 2.6% of the total solids discharge of the Miramichi River consisted of organic matter. Between 1950 and 1977 the loss of bark, wood fibre, and biosolids rose to an annual average of 20 000 tonnes. During this time the percentage of organic matter in the discharge of total solids rose to 20%. After 1977 there may have been a reduction in this type of waste loss, due to the introduction of more stringent environmental controls.

The impact of forest industry waste on the estuary was significant. Suspended particulate matter in the water column of the estuary became organic-rich, increasing surface water turbidity and chemical oxygen demand. When the organic rich sediments were eventually deposited on the bottom of the estuary they created a sediment layer with a very high chemical oxygen demand. Samples of mud from the upper estuary taken during the mid 1970's confirm that very high organic contents were present (up to 23% organic carbon, with an average of 6.7% organic carbon, Willey and Fitzgerald 1980). Sediments from Miramichi Bay also show the impact of the 20th century anthropogenic input of organic matter, with an average organic carbon content of 2.3% (Fig. 1), as compared with pre 20th century deposits that contain less than 1.0% organic carbon. Isotopic carbon analyses of sediments from throughout the estuary confirmed that the source of most of the recent carbon was from the forest products plants (Rashid and Reinson 1979).

Examination of sediment cores taken from

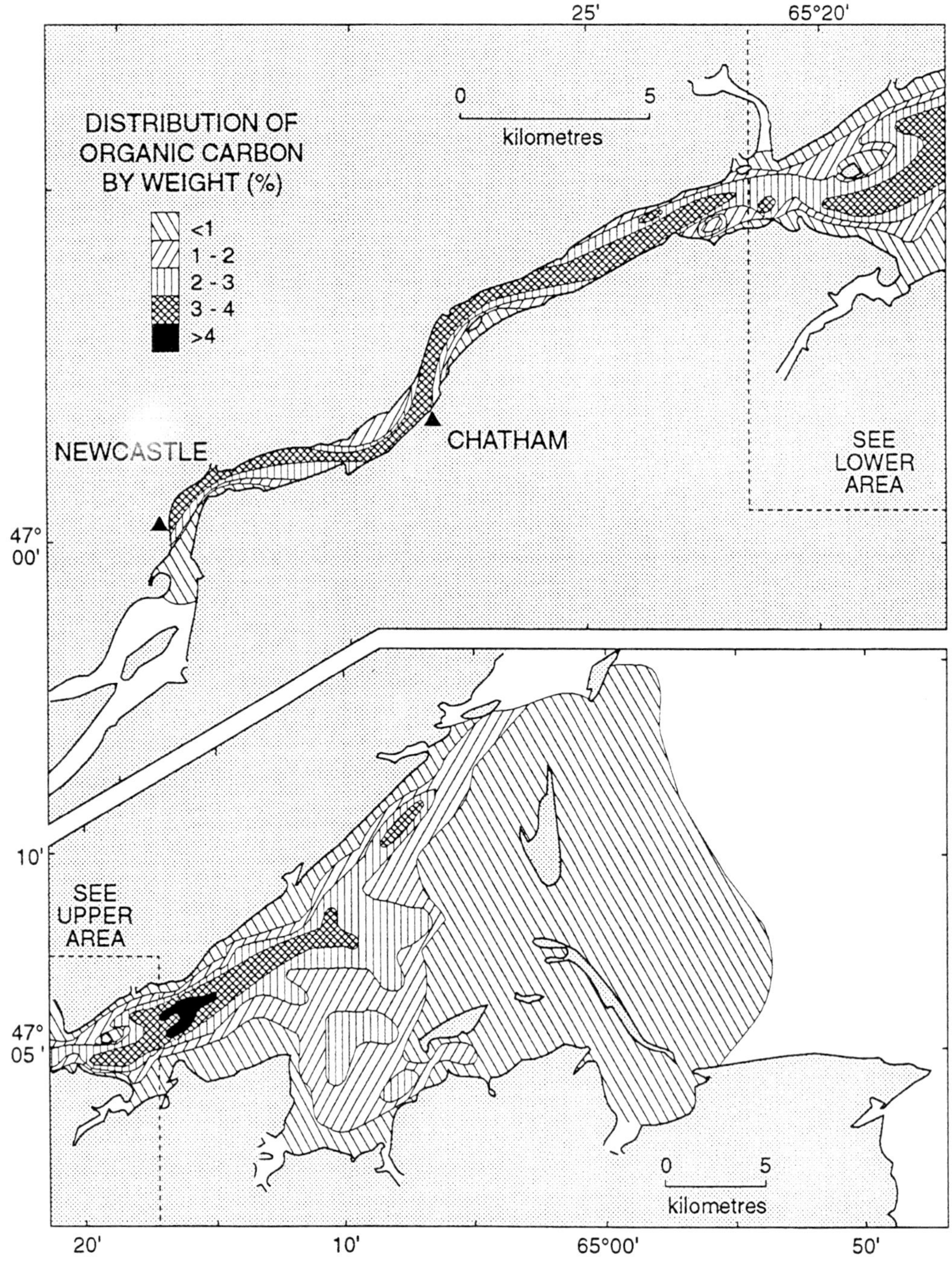

Fig. 1. Distribution of organic carbon in surficial sediments of Miramichi Estuary (after Rashid and Reinson 1979).

the upper estuary found wood chips buried to depths of 20 cm (Philpott and Duncan 1977), indicating that these sediments may have been deposited during the past 100 to 200 years. A sediment core taken in the lower river section at the limit of tidal influence found extremely enriched organic matter only to a depth of 6 cm (Winters at al. 1978). However, other cores taken from the estuary between Newcastle and east of Chatham found this extremely organic-rich layer extended from the surface mud to depths ranging from 97 cm to 22 cm. Similar organic-rich sediments were even found in the central part of the Inner Miramichi Bay, where the thickness of the layer was about 2 cm. This organic-rich layer probably marks the interval of time between 1949 and 1977, during which increased pulp production took place at the Newcastle plant. If this is true, then the indicated rates of sediment deposition varied from about 0.1 cm per year in the lower Miramichi River, to high rates of about 4 cm per year in the estuary near Newcastle, and 1 cm per year around Chatham. The sedimentation rates in the Inner Bay appear to have been the lowest, less than 0.1 cm per year.

Sediments with very high organic matter are notable indicators of the environmental quality of the aquatic system in which they have been deposited. Systems that have a high rate of supply of organic matter, such as the Miramichi, and high sediment accumulation rates, such as occurs in much of the Miramichi Estuary, preserve chemically reduced bottom sediments (Cranston 1994). Typically these sediments are black, and contain high concentrations of metals in a chemically reduced state. Shallow subsurface layers of these estuarine sediments often contain elevated concentrations of ammonium, and reduced quantities of sulfate as a result of sulfate reduction under anoxic conditions. This type of sediment generally creates unfavourable habitat for many estuarine benthic fauna, often leading to low diversity biological communities (Schafer 1973).

The loading of organic matter in water and sediments of the Miramichi Estuary probably has the most significant impact on the environmental quality of the system. This observation can be evaluated by comparing the organic carbon loading of the Miramichi Estuary with the organic carbon loading in other marine systems. The relatively high sediment accumulation rates between 0.1 and 4 cm per year and the high organic carbon content in sediments, between 2.6 and up to 23%, indicates that the Miramichi system is being highly loaded with organic matter. In a compilation of sedimentation rates and organic carbon content measured in about 100 other marine systems, Cranston (1994) demonstrated that only about 10% of those systems have values as high as have been found in the Miramichi, Fig. 2. Also by comparison, the total organic carbon accumulation in the Miramichi system is similar to that found in Halifax Harbour; a system that has been identified as being highly contaminated (Buckley and Winters 1992).

Metal Contamination

Other environmental quality concerns for the Miramichi system are related to the accumulation of potentially toxic metals. These metals include copper, lead, zinc, mercury, cadmium, and arsenic. Studies of the concentration of some of these met-

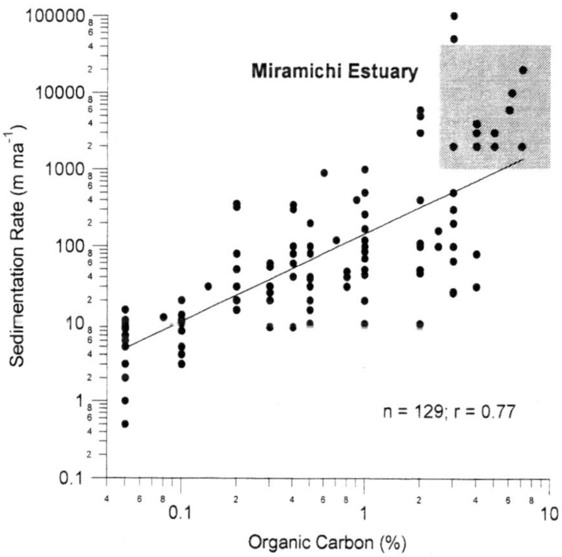

Fig. 2. Relationship between deposition rate and organic carbon preserved in marine sediments from a variety of marine systems. Variability in sedimentation rates and organic carbon for the Miramichi Estuary is depicted by shaded area (after Cranston 1994).

als in bottom sediments (Willey and Fitzgerald 1980), as well as in the water column and suspended sediments (Buckley and Winters 1983) demonstrated that there are probable sources of these metals from mining activities along the upper reaches of the Miramichi River, from the pulp and wood products plants at Newcastle and Chatham, and from the discharge of sewage and storm drainage outfalls of the towns of Newcastle and Chatham. Some of these metals are sequestered in the fine-grained and organic-rich muds of the upper estuary, while others are transported through the system in the dissolved state or in association with the suspended particulate matter. For example, an estimated 90 tonnes of zinc are exported to the Gulf of St. Lawrence from the Miramichi Estuary each year (Fig. 3). About 45% of this metal is associated with the 71 000 tonnes of suspended particulate matter exported from the estuary each year (Buckley and Winters 1983).

In spite of the large net annual exports of metals from the Miramichi Estuary there are indications that significant accumulation of some metals is occurring in the bottom sediments within the estuary. Estimates of about 16 tonnes of zinc, and 5 tonnes of lead have accumulated in the bottom sediments every year (Buckley and Winters 1983). This sedimentary accumulation is much less than is occurring in sediments in Halifax Harbour where 36 tonnes of zinc and 35 tonnes of lead are deposited each year (Buckley and Winters 1992). This difference is due to a greater degree of containment of contaminants in Halifax Harbour. Most of the metal accumulated during the 20th century in both the Miramichi and Halifax Harbour is associated with organic-rich muds.

Evidence of the history of metal contamination can be obtained from analyses of sediment core samples. In the upper reaches of the estuary above Newcastle sediment core data indicate that contamination extended to 8—10 cm depths in the sediment (Willey and Fitzgerald 1980). This depth is slightly greater than the horizon where very high organic contamination began, suggesting that zinc contamination began at a time that precedes the increased production of the pulp mill at Newcastle in 1949. Core samples from the estuarine sections east of Newcastle and Chatham show that high concentrations of copper, lead, zinc, and cadmium begin at depths ranging from 10 to 30 cm below the surface muds (Willey and Fitzgerald 1980; MacLaren Marex 1978). These measurements suggest that significant accumulation of metal contamination began in this area during the 1940's and 1950's. In sediments from the Inner Miramichi Bay the metal contamination layer averages about 4 cm thickness suggesting the beginning of contamination in about the 1930's. Some of this metal contamination has been attributed to operations of the Heath Steel lead-zinc mine in the watershed of

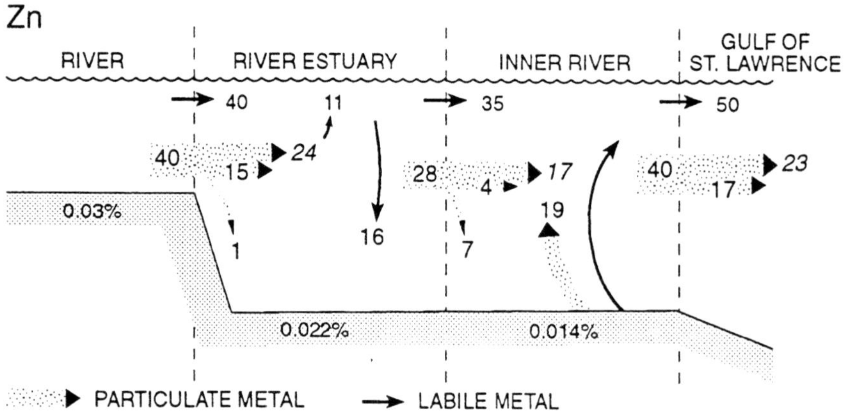

Fig. 3. Model of geochemical transport of zinc metal through the Miramichi Estuary. Transports are in tonnes per year. Concentration of zinc in fine-grained bottom sediments are shown in percentages. Italized numbers in particulates represents zinc metal that is potentially reactive or labile (after Buckley and Winters 1983).

the Northwest Branch of the Miramichi River (Willey and Fitzgerald 1980). However, this mining operation was closed for several years in the early 1980's (MacKnight 1990), with subsequent ore concentrates being shipped by rail through northern New Brunswick. Metal analyses of surficial sediments collected after 1983 indicate that there was little change in the distribution or average concentration of metals after the closure of the mining operations (MacKnight 1990). This suggests that there are other sources of metal contamination or that there is a hysteresis effect from previous contaminants in the watershed.

Organic Contaminants

There are only limited data on organic contaminants in the Miramichi system. These compounds generally include PCB compounds, DDT pesticides, and hydrocarbon compounds such as PAHs. Most of these contaminants were found in the surficial and near surface muds in the upper estuary near Newcastle and Chatham (MacLaren Atlantic 1978). The sharp decrease in concentration below 5 cm depth in the sediments suggests that these contaminants are a relatively recent addition to the system, possibly after the 1960's. The main sources of contamination are probably from the wood preservation plant near Newcastle, from the power generating station at Chatham, and from hydrocarbon residues in surface land drainage.

Dredging

Concerns for management of environmental quality in the Miramichi system have been most acute with respect to the impact of dredging operations to maintain or improve the shipping channel. The impact of the main capital dredging project carried out in 1981 and 1982 was not expected to significantly change the overall characteristics of the estuary because this dredging would only increase the cross-sectional area by 0.7% (Willis 1990). However, concerns have centred on the impact of increased turbidity in the water during dredging operations and the dispersion of contaminants as a result of dredge spoils disposal at the dump sites. As a result of limited monitoring studies conducted during the dredging and disposal operations, MacLaren Plansearch (1985) found that suspended particulate matter increased by up to an order of magnitude at some locations in the middle estuary. After the main capital dredging operations were completed in 1982 the level of suspended sediment decreased to the range of values found in studies conducted before the project began. This period of increased turbidity was found to be coincident with observations of increased body burdens of cadmium in some benthic fauna during 1981 and 1982 and a return to pre-dredging levels in 1983 (MacLaren Plansearch 1985).

The stability of the dredge spoil dump sites ("A" in the upper estuary; "B" in the Inner Bay; "C" outside the barrier islands in Outer Miramichi Bay) have been of considerable concern. Dredge spoils deposited at dump site "A" totalled 1 439 000 m^3, and were deposited in a secondary channel of the upper estuary between Newcastle and Chatham. The material dumped here was dredged from the main upper estuary channel in the vicinity of Newcastle and Chatham, and east of Chatham to Logieville. This dump site appears to have been relatively stable and may contain most of the spoils deposited there. Dump site "B" in the central part of the Inner Bay received 4 928 000 m^3 of sediments dredged mainly from the eastern part of the inner estuary and the western part of the Inner Bay. Dump site "C" received 1 349 000 m^3 of sediments from the main channel in the eastern part of the Inner Bay.

Most questions (see for example, Review of Impacts of Channel Dredging on Biological Resources of the Miramichi System, Acadia Centre for Estuarine Research 1990) have been raised regarding the stability of dump site "B". An estimated average thickness of dredge spoils in the 21 km^2 dump site should have been 23 cm, however, Krank and Milligan (1989) found that most measurements from core samples indicated thickness much less than 23 cm. These data suggest that material has been dispersed from the dump site after the end of the capital dredging project. One study (Krank 1989), estimated that 4 000 000 m^3 of dredge spoils may have been lost from the dump site between 1983 and 1988. If this estimate was accurate an assessment of environmental impact could be made by comparing the volume of "normal" sediment that would have been dispersed into the Inner Bay from the upper estuary of the Miramichi, with the suggested volume of remobilized dredge spoils. The annual discharge of about 100 000 m^3 would yield a 5-year volume of 500 000 m^3 of sediment in the Inner Bay. Krank's estimate of 4 000 000 m^3 of remobilized dredge spoils over 5 years would thus be a very significant redistribution of sediments! The suggested remobilized dredge spoils would equal about 40 years of sediment discharge from the upper estuary. Such a large and relatively rapid remobilization of sediment should have had a significant impact on benthic communities in Inner Miramichi Bay. It is now suggested that Krank's (1989) estimates of lost dredge spoils were excessive.

The long-term impacts of the deepening of the shipping channel were not expected to be profound. The change in penetration of salt water into the upper reaches of the estuary was predicted to be insignificant after dredging (Willis et al. 1977). Similarly, changes in tides and currents were not expected to be significant.

Some significant historical changes in the morphology of the barrier islands at the eastern end of Inner Miramichi Bay may have been in response to previous dredging (up to 1964) of about 2 000 000 m^3 of sediment from the shipping channel in the Inner Bay. The size and orientation of Portage Island has changed very notably over the past 150 years (Owens 1974; Reinson 1976, 1977, 1980). Portage Gully on the north end of the island has moved 3 km south between 1945 and 1974. The main channel through the barrier islands, Portage Channel, increased in cross sectional area by about 10% between 1954 and 1975. Fox Channel at the southern end of Fox Island closed completely in 1974.

Need for New Perspectives

Previous studies of the Miramichi Estuary have provided a tremendous storehouse of information on aspects of the environmental quality of the system and have identified key processes that are responsible for controlling the environmental quality of this important coastal system. Many of these previous studies could be enhanced with application of new technology. However, if a decision is made to supplement previous studies, or to undertake new surveys, consideration should be given to the focus of assessment. What are the important issues that should be addressed?

The interested community should determine the relative importance and urgency of three types of issues: (1) Are the aesthetics of the Miramichi Estuary and its resources important, considering the quality of sight, smell, and taste of the physical environment and the living resources? (2) What are the important habitats for fish and wildlife that must be restored or protected? (3) What are the risks to human health from exposure to contaminants or consumption of toxins in fish?

During the time since many of the environmental studies were conducted in the Miramichi system in the 1970's and 1980's there have been important scientific and technological advances made that could now be applied to a new environmental assessment of the Miramichi. Some of these advances have built on the experience that was gained in studies of other Atlantic Canada estuaries and inlets, such as Malpeque Bay, Prince Edward Island (Buckley 1969); the LaHave River and Estuary (Cranston and Buckley 1972); the Strait of Canso and Chedabucto Bay (Buckley et al. 1974); the Bay of Chaleur (Schafer 1973); and Halifax Harbour (Buckley and Winters 1992).

One of the technological aspects that has advanced most significantly is that of geological surveys using acoustic methods (Stea et al. 1994). This technology includes high resolution digital swath bathymetry, side scan sonar mosaics, and acoustic reflection profiling. These survey tools, coupled with high resolution positioning systems, allow small scale morphological features to be identified and mapped so that interpretations of natural and anthropogenic features can be made much more accurately than was possible 10 years ago. Acoustic surveys of estuarine systems, such as Halifax Harbour (Fader et al. 1991), allow one to identify important dynamic processes, such as deposition and erosion of sediments in estuaries. Accurate characterization of anthropogenic features, such as dredge and dump sites, waste deposits, and sewage banks can be obtained with this modern technology. Areas of disturbed seabed caused by shipping activity and fishing can be readily identified. Maps produced by these techniques are essential guides for the planning of sampling programs that may later supplement the acoustic property maps.

High resolution acoustic reflection profiles provide a cross section of sedimentary strata that may be used to interpret the history of sedimentation in estuaries. In some cases acoustic profiles may be used to identify important geochemical processes, such as the biogenic production of methane in chemically reduced subsurface sediments.

Application of these new acoustic survey techniques could solve some of the outstanding problems in the Miramichi Estuary. Because of the high resolution of the sidescan sonar and swath bathymetry, such surveys should be able to determine where spoils were originally deposited on dump site "B", and determine where material has been subsequently dispersed. These survey techniques should also be able to precisely determine where sediments are presently accumulating in the shipping channel so that appropriate maintenance dredging could be managed. New techniques of monitoring the stability of deposited sediments from dumping of dredge spoils have been developed that will allow assessments to be made of potential damage to habitats in the Miramichi Estuary (Amos et al. 1994).

Although extensive chemical and geochemical surveys were previously carried out in the Miramichi Estuary some new techniques have been developed to augment interpretation of these data. Research carried out in Halifax Harbour (Buckley and Winters 1992) determined that results of total and labile metal analyses, together with determinations of organic carbon and sediment texture, could be used in a statistical analysis to determine source and process factors that identified how contaminants entered the harbour sediments. This factor analysis technique (Winters and Buckley 1992) is not entirely new, but its application to environmental assessment in complex estuarine systems has only recently been demonstrated.

Application of the environmental factor analysis technique to data that already exists for the Miramichi may be possible. Results of this analysis may resolve sources of potentially toxic metals and organic compounds. With these results it may be possible to identify that recent contamination by metals such as lead, zinc, copper, mercury, and cadmium is from more than a single source, and not solely a result of industrial activity. This factor analysis technique can also be applied to sediment core data so that historical environmental factors can be identified (Buckley et al. 1995). Resolution of these historical factors is important because it demonstrates how changes in industrial activity, waste treatment, and resource developments have impacted the estuarine environment (Gearing et al. 1991). An understanding of the relationship between human activities and environmental impacts provides a means for developing an effective environmental quality management strategy.

With the wealth of environmental data that already exists for water, sediment, and biological communities in the Miramichi system it is possible to develop and apply local water and sediment environmental quality standards so that industrial planning and development can be carried out with minimal future impacts. This would reduce the risks to future aesthetic qualities, habitat destruction, and human health. Recently, Cranston (1994) proposed a method for estimating the carrying capacities of coastal inlets for development of aquaculture. This method relies on estimating rates of burial of carbon and production of ammonium from the sediments. These essential data would have to be obtained from a new set of sediment cores from the Miramichi. Although this method was developed to assess potential development of aquaculture it may also be applied to assessing the "health" of inlet systems for indigenous aquatic stock.

The relationship between environmental quality of sediments and biological communities is a fundamental principle that is now being recognized throughout the world. Benthic biological community diversity and abundance often indicate effects of sediment toxicity. Studies of benthic biological communities are especially useful in understanding the effects of overall environmental stress because benthic communities are sessile and respond to an integrated impact of water and sediment qualities. Analyses of past ecological conditions is possible by studying organisms that leave a fossil imprint of the community. Such organisms include foraminifera, ostracods, and molluscs. The numbers and diversity of these animals, as preserved in fossil remains in buried sediments, will often indicate the impact of changing environmental conditions (Buckley et al. 1974). Some of this type of analysis was done in the Miramichi Estuary during the 1970's (Schafer et al. 1977; Schafer and Smith 1983), however, there is a need for it to be expanded and enhanced.

Acknowledgements

Many of the references cited in this review identify research contributions by colleagues who participated in a multidisciplinary study of the Miramichi during the 1970's. The many contributions of these friends of the author are greatly appreciated. A draft version of this paper was reviewed by Dr. Steve Solomon and Mr. Gary Winters. The manuscript was critically reviewed for publication by Dr. E.M.P. Chadwick, Charles Gobeil, and Mike Brylinsky. This paper is Geological Survey of Canada contribution number 48593.

References

Acadia Centre for Estuarine Research. 1990. Bibliography of the Miramichi. Acadia Centre for Estuarine Research, Acadia University, Wolfville, N.S. 27 p.

Ambler, D.C. 1976. Surface water sediment investigation, Miramichi River Navigation Channel Study, Water Survey of Canada, Environ. Can., Atl. Div., Halifax. 71 p.

Amos, C.L., Brylinski, M., Christian, H.A., and Daborn, G.R. 1994. Seabed stability, liquefaction and the development of fluid mud during dredging and dumping, Inner Miramichi Bay. Report of contract SSC # E0225-3-0020/01-OSC, Public Works Canada, Architectural and Engineering Services, St. John, New Brunswick, January, 1994, also Acadia Centre for Estuarine Research Publication No. 32.

Bartlett, G.A. 1966. Distribution and abundance of foraminifera and thecamoebina in Miramichi River and bay. Can. Comm. Oceanogr. Bedford Inst. Oceanogr. Rep. 66-2: 107 p.

Bousfield, E.L. 1955. Some physical features of the Miramichi Estuary. J. Fish. Res. Board Can. 12: 342-361.

Buckley, D.E. 1969. Sedimentological studies at Malpeque Bay, Prince Edward Island. Atl. Oceanogr. Lab., Bedford Inst., Rep. A.O.L. 69-3: 60 p.

Buckley, D.E. 1990. Sedimentary geology of the Miramichi Estuary system in relation to dredging and dumping activities: a review, p. 2.1-2.10. *In* Review of Impacts of Channel Dredging on Biological Resources of the Miramichi System. Acadia Centre for Estuarine Research, Acadia University, Wolfville, NS.

Buckley, D.E., and Winters, G.V. 1983. Geochemical transport through the Miramichi Estuary. Can. J. Fish. Aquat. Sci. 40 (Suppl. 2): 162-182.

Buckley, D.E., and Winters, G.V. 1992. Geochemical characteristics of contaminated surficial sediments in Halifax Harbour: impact of waste discharge. Can. J. Earth Sci. 29: 2617-2639.

Buckley, D.E., Smith, J.N., and Winters, G.V. 1995. Accumulation of contaminant metals in marine sediments of Halifax Harbour, Nova Scotia: environmental factors and historical trends. Appl. Geochem. v.10. (in press).

Buckley, D.E., Owens, E.H., Schafer, C.T., Vilks, G., Cranston, R.E., Rashid, M.A., Wagner, F.J.E., and Walker, D.A. 1974. Canso Strait and Chedabucto Bay: a multidisciplinary study of the impact of man on the marine environment, p. 133-160. *In* Pelletier [*editor*], Offshore Geology of Eastern Canada, , Geol. Surv. Can. Pap. 74-30, v.1.

Chadwick, M. 1990. Review of Miramichi Fisheries Data, p. 7.1-7.3. *In* Review of Impacts of Channel Dredging on Biological Resources of the Miramichi System. Acadia Centre for Estuarine Research, Acadia University, Wolfville, N.S.

Cranston, R.E. 1994. Dissolved ammonium and sulfate gradients in surficial sediment pore water as a measure of organic carbon burial rate, p. 39-65. *In* Hargrave, B., Silvert, W., Sowles, J., Churchill, L., and Cranston, R. [*editors*], Chapter IV, Modelling Benthic Impacts of Organic Enrichment from Marine Aquaculture. Fish. Aquat. Sci. Tech. Rep., Department of Fisheries and Oceans, Canada.

Cranston, R.E., and Buckley, D.E. 1972. Mercury pathways in a river and estuary. Environ. Sci. Tech. 6: 274-278.

Fader, G.B.J., Miller, R.O., and Pecore, S.S. 1991. The marine geology of Halifax Harbour and adjacent areas. Geol. Surv. Can. Open File Rep. 2384, 1, and 2: 23p.

Fothergill, N.O. 1953. Tidal circulation in the Miramichi Bay. Canada Department of Mines and Technical Surveys, Hydrographic Service, Ottawa, 18 p.

Fritz, C.W. 1918. Distribution of diatoms at the mouth of the Miramichi River, New Brunswick. Atl. Biol. Stn., Original Manuscr. Rep. No. 416.

Gearing, J.N., Buckley, D.E., and Smith, J.N. 1991. Hydrocarbon and metal contents in a sediment core from Halifax Harbour: a chronology of contamination. Can. J. Fish. Aquat. Sci. 48: 2344-2354.

Howells, K., and McKay, A.G. 1977. Seismic profiling in Miramichi Bay, New Brunswick. Can. J. Earth Sci. 14: 2909-2927.

Kerr, R.B. 1929. Investigations on the Atlantic salmon (Salmo salar) of the Miramichi River, New Brunswick, Canada. Biol. Board Can. Manuscr. Rep. Biol. Stn. No. 43: 10p.

Krank, K. 1989. Cores collected in 1988 from the vicinity of dumpsite B, Miramichi Bay, New Brunswick. Physical and Chemical Sciences Branch, Scotia Fundy Region, Department of Fisheries and Oceans. (unpublished, Bedford Institute of Oceanography).

Krank, K., and Milligan, T.G. 1989. Effects of a major dredging program on the sedimentary environment of the Miramichi Bay, New Brunswick. Can. Tech. Rep. Hydrogr. Ocean Sci. No. 112: 61p.

Lewis, C.F.M., and Buckley, D.E. 1990. The Miramichi: shipping channel design and industrial pollution, p. 763-767. *In* C.F.M. Lewis and M.J. Keen [*editors*], Constraints to Development, Chapter 14; Geology of the Continental Margin of Eastern Canada, M.J. Keen and G.L. Williams [*editors*]; Geol. Survey Can., Geology of Canada, no. 2 (also Geol. Soc. Amer. The Geology of North America, v. 1-1).

MacGregor, C., and Packman, G.A. 1983. Winter water quality studies in the Miramichi Estuary, p. 48-56. *In* Report on Ocean Dumping Research and Development Fund, Atlantic region, 1981-1982. Proceedings of workshop January 18, 1983, St. Andrews Biol. Stn., N.B., Surveillance Rep. EPS-5-AR-83-9.

MacKnight, S. 1990. Contaminants in the Miramichi Estuary system, p. 3.1-3.4. *In* Review of Impacts of Channel Dredging on Biological Resources of the Miramichi System. Acadia Centre for Estuarine Research, Acadia University, Wolfville, N.S.

MacLaren Atlantic Ltd. 1978. Miramichi channel study supplementary environmental studies and addenda. Report to District Office, Public Works Canada, Saint John, N.B., (unpublished).

MacLaren Marex Inc. 1978. Commentary on stability of disposal site "A". Public Works Canada. (unpublished report).

MacLaren Plansearch Ltd. 1981. Environmental monitoring of the dredging of the Miramichi River, Test disposal at site "B". Public Works Canada.

MacLaren Plansearch Ltd. 1982. Environmental monitoring of the dredging of the Miramichi River, summary of 1981 operations. Public Works Canada.

MacLaren Plansearch Ltd. 1983. Environmental monitoring of the dredging of the Miramichi River, volume 1, summary of 1982 operations. Final report, Public Works Can. 168 p.

MacLaren Plansearch Ltd. 1985. Environmental monitoring of the dredging of the Miramichi River. Summary Report 1981-1984, 74p.

McKenzie, R.A. 1943. Smelt movements, 1943-44 tagging, Miramichi Bay. Fish. Res. Board Can. Manuscr. Rep. Biol. Stn. No. 369, Part 2: 7p.

Medcof, J.C. 1940. Oyster investigations in 1940. Fish. Res. Board Can. Rep., 99p.

Miramichi River Navigation Channel Study. 1974. Miramichi River navigation channel study. Working Group, Department of Public Works, Report April 30, 1974.

Owens, E.H. 1974. Barrier beaches and sediment transport in the southern Gulf of St. Lawrence. Proceedings of 14th Conference on Coastal Engineering, Copenhagen, V. II: 1173-1193.

Philpott, K.L., and Duncan, G.R. 1977. Miramichi Channel Study. Initial Environmental Evaluation, Environment Canada.

Rashid, M.A., and Reinson, G.E. 1979. Organic matter in surficial sediments of the Miramichi Estuary, New Brunswick, Canada. Estuarine Coastal Mar. Sci. 8: 23-26.

Reinson, G.E. 1976. Surficial sediment distribution in the Miramichi Estuary, New Brunswick, p. 41-44. In Geol. Survey Can. Report of Activities, Part C, Pap. 76-1C.

Reinson, G.E. 1977. Tidal-current control of submarine geomorphology at the mouth of the Miramichi Estuary, New Brunswick. Can. J. Earth Sci. 14: 2524-2532.

Reinson, G.E. 1980. Variations in tidal current morphology and stability, northeast New Brunswick, p. 23-39. *In* S.B. McCann [*editor*], The Coastline of Canada, Littoral Processes and Shore Morphology. Geol. Survey Can. Pap. 80-10.

Schafer, C.T. 1973. Distribution of foraminifera near pollution sources in Chaleur Bay. Water, Air, Soil Pollut. 2: 219-233.

Schafer, C.T., and Smith, J.N. 1983. River discharge, sedimentation, and benthic environmental variations in Miramichi Inner Bay, New Brunswick. Can. J. Earth Sci. 20: 388-398.

Schafer, C.T., Cole, F.E., and Wagner, F.J.E. 1977. Relationship of foraminifera distribution patterns to sedimentary processes in the Miramichi Estuary, New Brunswick. Geol. Survey Can., Report of Activities, Part C, Pap. 77-1C: 1-7.

Sephton, T.W. 1990. Review of impacts of Miramichi dredging on fisheries resources; review of benthos and shelfish literature, p. 6.1-6.3. *In* Review of Impacts of Channel Dredging on Biological Resources of the Miramichi System. Acadia Centre for Estuarine Research, Acadia University, Wolfville, N.S.

Solomon, S. 1990. Suspended particulate material in the Miramichi Estuary, New Brunswick: summary of pre- and post-dredging data, p. 3.1-3.22. *In* Review of Impacts of Channel Dredging on Biological Resources of the Miramichi System. Acadia Centre for Estuarine Research, Acadia University, Wolfville, N.S.

Stea, R.R., Boyd, R., Fader, G.B.J., Courtney, R.C., Scott, D.B., and Pecore, S.S. 1994. Morphology and seismic stratigraphy of the inner continental shelf off Nova Scotia, Canada: evidence for a -65 m lowstand between 11650 and 11250 C^{14} yr B.P. Mar. Geol. 118: 1-20.

Vilks, G., and Krauel, D.P. 1982. Environmental geology of the Miramichi Estuary: physical oceanography. Geol. Surv. Can. Pap. 81-24: 53p.

Wagner, F.J.E., and Schafer, C.T. 1980. Upper Holocene paleoceanography of Inner Miramichi Bay. Marit. Sed. 16: 5-10.

Willey, J.D., and Fitzgerald, R.A. 1980. Trace metal geochemistry in sediments from the Miramichi Estuary, New Brunswick. Can. J. Earth Sci. 17: 254-265.

Willis, D.H. 1990. Miramichi Estuary – Physical Oceanography, p. 1.1-1.10. *In* Review of Impacts of Channel Dredging on Biological Resources of the Miramichi System. Acadia Centre for Estuarine Research, Acadia University, Wolfville, N.S.

Willis, D.H., Crookshank, N.L., and Johnson, R.R. 1977. Miramichi channel study, hydraulic investigation. NRC, Div. Mechanic. Eng., Hydraulics Lab. Tech. Rep. No. LTR-HY-56.

Winters, G.V. 1981. Environmental geology of the Miramichi Estuary: suspended sediment transport. Geol. Surv. Can. Pap. 81-16: 12p.

Winters, G.V. 1983. Modelling suspended sediment dynamics of the Miramichi Estuary, New Brunswick, Canada. Can. J. Fish. Aquat. Sci. 40 (Suppl. 1): 105-116.

Winters, G.V., and Buckley, D.E. 1992. Factor analyses as a method of evaluating sediment environmental quality in Halifax Harbour, Nova Scotia, p. 165-171. In Current Research, Part D; Geol. Surv. Can. Pap. 92-1D.

Winters, G.V., Fitzgerald, R.A., and Buckley, D.E. 1978. Analyses of water column and bottom sediment samples from the Miramichi Estuary, New Brunswick. Bedford Inst. Oceanogr. Data Ser. BI-D-78-8, December 1978, 81p.

CHAPTER 12

Addressing Forestry Impacts in the Catamaran Brook Basin: An Overview of the Pre-Logging Phase[1]

Richard A. Cunjak

Department of Fisheries and Oceans, Science Branch, P.O. Box 5030, Moncton, New Brunswick, Canada E1C 9B6

Cunjak, R.A. 1995. Addressing forestry impacts in the Catamaran Brook basin: an overview of the pre-logging phase, p.191-210. *In* E.M.P. Chadwick *[editor]*. Water, science, and the public: the Miramichi ecosystem. Can. Spec. Publ. Fish. Aquat. Sci. 123.

Abstract

There has been much concern in eastern Canada about the potential impacts of forestry activity on aquatic resources, particularly Atlantic salmon (*Salmo salar*). However, until the Catamaran Brook project began in 1990, no detailed studies have addressed this issue in the Miramichi. Research at Catamaran Brook has adopted a multi-disciplinary approach for quantifying the potential impacts of clear-cut timber harvest on fish production and aquatic habitat. The study is divided into three 5-year phases: pre-logging (1990-94); logging (1995-1999); and, a post-logging phase (2000-2004) which will incorporate experimental manipulations. This paper examines some of the major findings of the pre-logging phase, specifically those environmental variables and ecosystem processes most likely to be affected by deforestation - i.e. water temperature, water quality, sediment loading, and hydrologic regime. These findings are then discussed in relation to previously published information of forestry impacts elsewhere in North America. Finally, the population dynamics of Atlantic salmon in Catamaran Brook between 1990-1994 is briefly described, as is ongoing research on other biota in the drainage basin.

Studies are also being carried out away from the stream proper to determine the inputs and export of materials via the stream channel. Meteorological data monitoring, precipitation interception in the tree canopy, soil nutrient loss, and surficial geochemistry are some additional studies presently underway. Effectiveness of the riparian zone (30–60 m width) to buffer the effects in adjacent timber harvest blocks is also being tested through studies of understory plant diversity and amphibian abundance.

[1] Contribution No. 12 to the Catamaran Brook Habitat Research Project.

Résumé

On s'inquiète beaucoup dans l'est du Canada des impacts potentiels de la foresterie sur les ressources aquatiques, particulièrement le saumon atlantique (*Salmo salar*). Toutefois, jusqu'au lancement du projet du ruisseau Catamaran, en 1990, aucune étude détaillée n'avait abordé ce sujet pour la Miramichi. Les recherches menées au ruisseau Catamaran ont adopté une approche pluridisciplinaire pour quantifier les impacts potentiels de la coupe à blanc sur la production de poisson et l'habitat aquatique. L'étude se divise en trois phases de cinq ans : avant l'exploitation (1990-1994); la phase d'exploitation (1995-1999); après l'exploitation (2 000-2 004), avec des travaux expérimentaux. Nous examinons ici certains des principaux résultats de la phase avant exploitation, et plus précisément les variables environnementales et les processus écosystémiques qui risquent le plus d'être touchés par la déforestation - température de l'eau, qualité de l'eau, charge en sédiments et régime hydrologique. Nous mettons ensuite ces résultats en rapport avec des données déjà publiées sur les impacts de la foresterie observés ailleurs en Amérique du Nord. Enfin, nous décrivons brièvement la dynamique de la population de saumon atlantique dans le ruisseau Catamaran entre 1990 et 1994, ainsi que des recherches en cours sur d'autres biotes du bassin.

Certaines études sont également réalisées hors du cours d'eau lui-même en vue de mesurer les apports et les exportations de matériaux par le lit du cours d'eau. La surveillance météorologique, l'interception des précipitations par le couvert forestier, les pertes de matières nutritives du sol et la géochimie superficielle sont d'autres domaines où des recherches sont en cours. On mesure aussi l'efficacité de la zone riveraine (de 30–60 m de largeur) pour contrer les effets de la coupe effectuée dans les parcelles adjacentes en étudiant la diversité végétale de l'étage inférieur et l'abondance des amphibiens.

Introduction

The commercial harvesting of trees has been an important activity in the Miramichi region since the 18th century when lumber, particularly white pine, was exported throughout North America and Europe where it was used to build masts for ships in the British Navy (Wright 1944; Johnson 1945). Today, mill and woodland operations in the Miramichi Region directly employ 3500 people in an industry worth an estimated $500M annually to the local economy (J. O'Neill, Miramichi Pulp and Paper, Inc., pers. comm.).

Concern over the possible impacts of forestry on the environment of the Miramichi River basin has steadily increased in the past two decades. For example, the use of pesticides such as DDT and fenitrothion have been widely criticized because of the toxicity to non-target aquatic biota (e.g., Elson 1967, 1974; Symons 1977). Elson (1974) reviewed the impacts of forestry and other anthropogenic activities (e.g., base metal mining, urbanization) on the stocks of Atlantic salmon in the Northwest Miramichi River. He suggested that extensive cutting of forests resulted in the river becoming wider and shallower, and that the best habitats for juvenile salmon were lost, lowering production. However, these habitat changes were largely speculative and no cause-effect relationship was possible because of the lack of specific data on habitat. More recently, a report published by the Miramichi River Environmental Assessment Committee (MREAC 1992) summarized the concerns of residents about environmental degradation in the Miramichi River and made recommendations to improve the situation for the future. For example, changes in the hydrological regime as a result of forestry practices was suggested as a direction for future investigation.

In 1990, the Department of Fisheries and Oceans (Gulf Region) initiated a long-term research project at Catamaran Brook, a tributary of the Little Southwest Miramichi River, with the aim of quantifying the impacts of forestry activity on fish habitat in a small (52 km^2) stream basin (Cunjak et al. 1990). Although the effects of deforestation in stream catchments have been extensively studied in the past, much of this research has been carried out on the West Coast of North America (e.g. Carnation Creek) or in steep, small-order streams draining deciduous forests (Table 1) where climate, hydrology, forest-type, and fish populations are quite different from those found in Atlantic Canada, and specifically in the Miramichi.

The inherent complexity (spatial and temporal) of aquatic ecosystems required that the Catamaran Brook project be long-term (Schindler 1987) and with a 'pre-treatment' period to permit a distinction between patterns of natural variability and changes resulting from land-use activity (Rinne 1990). The adopted format followed that used in the Carnation Creek study (see review by Hartman and Scrivener 1990), probably the most comprehensive study of forestry-fishery interactions to date. Briefly, the Catamaran Brook project comprises three phases: a 5-year, *pre-logging phase* (1990-1994) to collect background data will serve as the basis for comparison with the *logging phase* (1995-1999). Within the logging (or "treatment") phase, studies will involve investigation of the effects of road construction in 1995 and timber harvest in pre-determined harvest blocks (1996-1999) which constitute approximately 7% of the stream basin (Fig. 1). The third phase, or *post-logging phase* (2000-2004), will serve as a post-treatment assessment of habitat changes and a period for testing hypotheses developed during the previous years.

The purpose of this paper is to summarize some of the major findings of the Catamaran Brook project during the pre-logging phase, specifically those environmental variables and ecosystem processes most likely to be affected by deforestation - i.e., water temperature, water quality, sedimentation, and hydrologic regime. These findings are then discussed in relation to previously published information of forestry impacts elsewhere in North America. Finally, the population dynamics of Atlantic salmon in Catamaran Brook between 1990-1994 is briefly described, as is ongoing research on other biota in the drainage basin; where

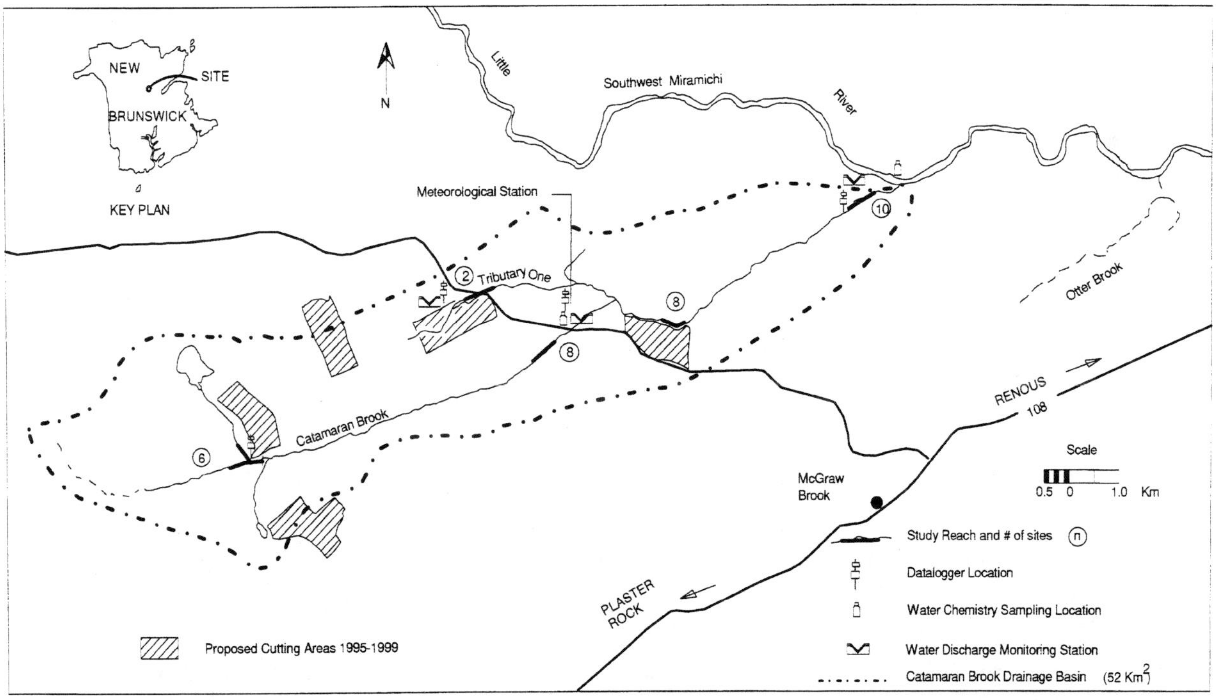

Fig. 1. Map of the Catamaran Brook drainage basin showing location of scheduled timber harvest blocks, data-loggers, sampling/monitoring stations, and the experimental stream reaches.

applicable, potential impacts are predicted.

Study Design

Details concerning the drainage basin, experimental design, and historic land-use in the area were provided by Cunjak et al. (1990); the various studies underway at Catamaran Brook, and preliminary data on the physical, chemical, and biological conditions measured between 1990-1992 were reported in Cunjak et al. (1993). Field research is carried out at various temporal and spatial scales. In total, 34 sites involving 4 habitat-types (riffles, runs, flats and pools) are sampled within 4 stream reaches from headwaters to stream mouth (Fig. 1). Electro-fishing and physical habitat surveys are carried out at all sites 2 times each year; water chemistry (>25 parameters analyzed including pH, DOC, conductivity, alkalinity, major ions, trace metals, and turbidity) is determined monthly from 2 locations in the brook (Fig. 1) and also from the Little Southwest Miramichi River upstream of Catamaran Brook; suspended sediment concentrations are determined for selected storm events; woody debris is quantified annually within the experimental reaches; snow depth and water equivalency are measured semi-monthly during winter in designated harvest blocks and in 'control' plots; water temperature, stream discharge, and meteorological data are measured hourly and recorded using data-loggers located throughout the basin (Fig. 1).

Water Temperature

With the removal of canopy cover and consequent exposure of the stream and soil surface in the catchment, the average summer water temperature and diel temperature fluctuations increase following deforestation (Campbell and Doeg 1989; Hartman and Scrivener 1990; Garman and Moring 1991). The extent and timing of the

Table 1. Selected long-term studies of forest ecosystems and forestry impacts in North America.

Project	Location and Forest Type	Focus	Key References
Coweeta 1939-	North Carolina (s. Appalachia) Eastern Deciduous Forest	hydrology & water quality changes ; ecosystem processes	Swank & Crossley (1988) Webster et al. (1983)
Carnation Creek 1970-	Vancouver Island Pacific Coastal Rainforest	effects of different forestry activities on vegetation, soil, water, fish	Hartman & Scrivener (1990)
Clearwater River 1972-	Olympic Peninsula, Washington Pacific Coastal Rainforest	impacts of clear-cutting and road construction on water quality and anadromous fish production	Cederholm and Reid (1987)
Hubbard Brook 1963-	New Hampshire (White Mtns) Northern Hardwood Forest	study of forest and stream ecosystems; catchment management problems	Bormann & Likens (1979) Pierce & Siccama (1986)
Nashwaak River 1971-82	Central New Brunswick Northern Hardwood - Mixed Forest	effect of clear-cutting, fertilization, insecticide spraying on hydrology, water quality, nutrient cycling	Dickison (1988) Powell (1983)
Alsea 1958-73	mid-Oregon Coast Pacific Coastal Rainforest	impact of clear-cutting (with and without buffer strips) on water quality & anadromous fishes	Hall et al. (1987) Moring & Lantz (1975)

temperature increase within a stream basin are dependent on many factors including size (area) and slope orientation of the harvest block, previous cutting history, groundwater influence, and whether riparian buffer strips are present. For example, Brown and Krieger (1970) found little or no change in stream temperature, following logging, where buffer strips were retained along tributaries of the Alsea basin (Oregon). In contrast, Garman and Moring (1991) measured significant increases in mean daily water temperature and thermal regime just one year following the clear-cutting of a Maine river catchment. In the Catamaran Brook basin, riparian buffer-strips will be retained according to the provincial forestry guidelines. Therefore, for those timber harvest blocks adjacent to tributaries and headwater streams (see Fig. 1), buffer strip width will be 30 m; for the block along the main channel (3rd order stream), the buffer will be 60 m wide.

The implications of a water temperature increase are many and complex. Following logging in the Carnation Creek catchment, an increase in water temperature during winter had the most profound impacts on coho (*Oncorhynchus kisutch*) and chum (*O. keta*) salmon populations (Hartman and Scrivener 1990). The higher winter temperature resulted in a shorter incubation time for developing alevins and an earlier emergence time than was the case prior to logging (Holtby 1988). This change had a negative impact on chum salmon which emigrated to the ocean immediately after emerging. Initially, the winter temperature increase appeared to benefit coho salmon which experienced a longer growing season such that a portion of the population were able to smoltify and move seaward at a younger age. However, earlier emergence exposed juvenile coho to strong spring flows which displaced some seaward, presumably to their demise. Also, the younger (and smaller) smolts emigrated seaward earlier than normal but had poorer marine survival than was the case prior to logging. Hartman et al. (1987) used the temperature change example to underline the need for long-term research and multi-disciplinary investigation of both the freshwater and marine phases of anadromous salmonids to understand the overall

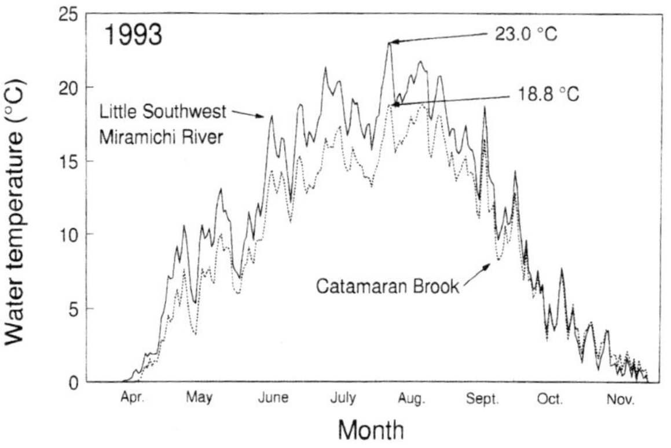

Fig. 2. Mean daily water temperature profiles for Catamaran Brook (dotted line) and the Little Southwest Miramichi River (solid line) between April and November, 1993. Maximum mean daily water temperatures, recorded August 4, were 23.0°C (Little Southwest) compared with 18.8°C (Catamaran Brook).

effects of land-use activity.

In Catamaran Brook, Atlantic salmon smolts leaving the system (1990-1994) were predominately age-3 which is typical for the Miramichi River system and northern New Brunswick rivers but older than smolts from more southerly Maritime rivers such as the Pollet and Margaree (Table 2) where growing seasons are longer. This positive latitudinal cline in age at smoltification is well known for Atlantic salmon (Randall et al. 1987). The timing, size, and age of smolts will continue to be monitored following timber harvest (1996) to determine if any changes have occurred. Adult returns are also monitored at the counting-fence near the mouth of the brook so that marine survival can be calculated. Similar to Pacific salmon (Peterman 1981), marine survival of Atlantic salmon has been related to smolt size (Chadwick 1987). Timing, feeding ecology and distribution of recently-emerged salmonid fry is presently being studied. These data, together with historic data on salmonid emergence in Catamaran Brook (Randall 1982) and the calculated number of degree-days during egg incubation will be compared with data from post-impact years.

Water temperature also affects biota other than fish and such relationships are being investigated in Catamaran Brook. Garnett (1994) recently reported on the emergence patterns of aquatic insects from different stream reaches and habitat-types in Catamaran Brook. She noted that the emergence timing of several species of adult Plecoptera (stoneflies) during the spring of 1993 was 2-3 weeks later than expected and probably reflected the cooler water temperatures compared with 1991 and 1992.

The high summer water temperatures experienced in the main branches of the Miramichi River are speculated to be the result of historic logging practices in the basin (e.g., see Elson 1974). Many residents of the Miramichi suggest that such high temperatures were rare in the past prior to the clear-cutting of vast tracts of forested lands. Unfortunately, dependable water temperature records quantifying such changes are rare. Water temperature has been continuously recorded in the Little Southwest Miramichi River since 1990 as part of the regular data monitoring network in Catamaran Brook (Fig. 1). Maximum daily stream temperatures > 25°C are measured each year for a few days during summer, low-flow conditions. Differences in the temperature regimes of the Little Southwest Miramichi River and Catamaran Brook are greatest during this period of the summer (Fig. 2) and are probably related to the surface exposure and greater heat-retention of the Little Southwest Miramichi River which is 5-6 times wider with less canopy cover and more exposed boulders compared with Catamaran Brook. Coincident with these high temperature events, significant numbers of brook trout (*Salvelinus fontinalis*) and salmon parr move into Catamaran Brook (Cunjak et al. 1993) which remains 3-4°C cooler (Fig. 2). That this fish movement is repeated each year attests to the importance of tributary streams like Catamaran Brook to serve as thermal refugia for fishes experiencing high temperature stress. If water temperature increases after timber harvest, as hypothesized, will the temperature difference between Catamaran Brook and the Little Southwest Miramichi River be maintained during the critical period of thermal stress? Such questions can certainly be addressed with the data being gathered from (i) the network of thermistors scattered throughout the river system, (ii) previous research (El-Kourdahi 1993) and current modelling of the thermal regime in Catamaran Brook, and (iii) from the monitoring of daily fish movements through the counting-fence.

Water Quality

Following deforestation, water quality in the stream can change, often a result of increased runoff and sub-surface throughflow in the basin which leach soil nutrients and ions en route to the stream. In Carnation Creek following logging, conductivity (which reflected total ion concentrations)

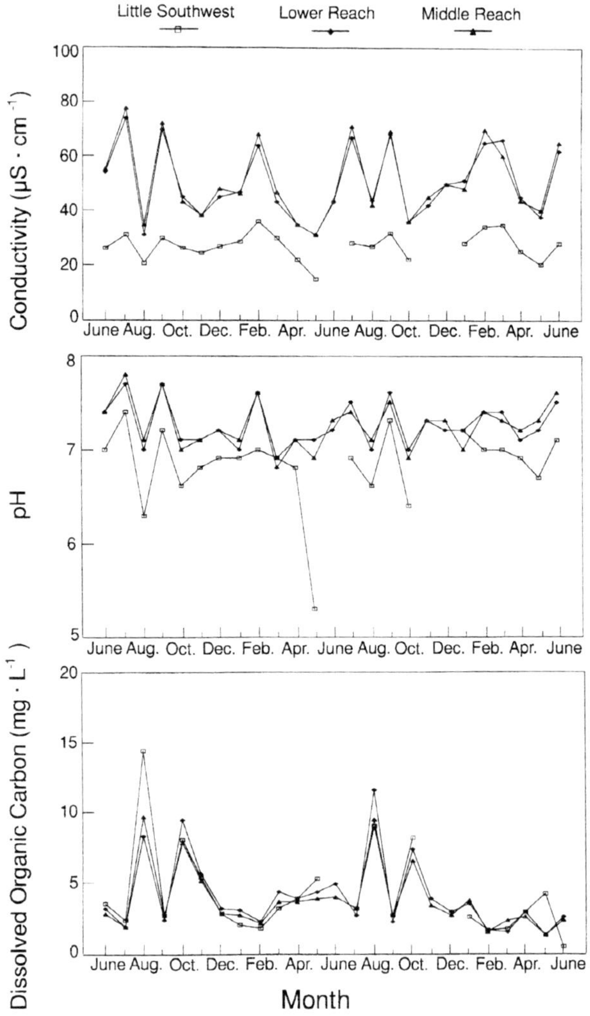

Fig. 3. Monthly changes in conductivity, pH, and dissolved organic carbon as measured for water samples taken from the Little Southwest Miramichi River and Catamaran Brook (Middle and Lower Reaches) between June 1990 and June 1992

and nitrate concentrations increased and were most evident at high stream discharges (Hartman and Scrivener 1990). Similarly, Likens et al. (1970) measured large increases in most major ions following deforestation of a Hubbard Brook subbasin. Meyer et al. (1988) measured lower DOC concentrations in Coweeta stream water 1-2 years following clear-cutting of the basin; 6 years after disturbance, DOC concentrations had increased, but with seasonal discrepancies, which suggested varying biotic processes were important causal agents for increasing the DOC in the disturbed basin. In Catamaran Brook prior to deforestation, DOC concentrations are similarly low in both stream reaches and in the Little Southwest Miramichi River for much of the year (Fig. 3). The seasonal peaks in late-summer (August) and autumn (October) are likely associated with storm events which flush in-stream algae/diatoms and terrestrially-derived carbon (i.e., leaf litter), respectively, through the system.

Water chemistry is analyzed from monthly water samples collected in mid-basin, at the mouth of the brook and in the Little Southwest Miramichi River (Fig. 1). Data from 1990-1993 indicate that Catamaran Brook is a well-buffered, neutral pH (average = 7.26), soft-water stream with a seasonal pattern of dissolved ions typical of groundwater baseflow interrupted by storm events and spring snowmelt. Water chemistry is similar throughout the brook but differs markedly from the Little Southwest Miramichi River which is more dilute (note low conductivities compared with Catamaran Brook, Fig. 3) and subject to pH depressions below 5.5, most obvious during spring snowmelt events (Fig. 3). During a significant snowmelt freshet in May, 1991, a pH of 5.3 was measured in the Little Southwest Miramichi River, the lowest value recorded during our study; in comparison, pH in the Catamaran Brook samples remained between 6.9 and 7.1 (Fig. 3). Catamaran Brook is sufficiently well-buffered that acidifying events are unable to reduce stream alkalinity to zero, as occurred in the Little Southwest Miramichi River during the May, 1991 snowmelt event; alkalinity in the Catamaran Brook reaches has ranged from 9-32.3 mg L^{-1} between 1990-1992 (Cunjak et al. 1993).

Nutrient and major ion concentrations (e.g., NO_3, Cl^-, Mg^{2+}, K^+, PO_4^{3-}) and dissolved organic carbon (DOC) were low or close to the laboratory equipment's detection limits (Cunjak et al. 1993). These results are indicative of a stream draining a "healthy" forested catchment and suggest that, during baseflow-dominated periods, most of the ions and DOC are being removed from the groundwater or retained in the soil and surficial deposits before entering the stream channel (Bormann and Likens 1979). Continuation of the water chemistry monitoring program through the years of catchment disturbance (i.e., 1995-1997) will permit us to distinguish any 'leaking' of nutrients and major ions. For example, autumnal DOC concentrations are expected to decrease following tree removal although late summer DOC concentrations may rise if an increase in water temperature and light penetration to the stream promote autochthonous carbon production. Conductivity and nitrate concentrations may also rise after clear-cutting in the Catamaran Brook basin.

As part of a study to investigate groundwater changes in harvested versus control plots, scientists from the University of New Brunswick (Civil Engineering) are monitoring groundwater dynamics and chemistry for samples taken from 4 wells and 10 mini-piezometers (which measure hydraulic head differential) situated throughout the basin (see Jones and Bray 1994 for details). Preliminary results show that groundwater chemistry would classify the water as bicarbonate-carbonate based; water levels in the wells show strong seasonal trends with major recharge periods occurring in the spring and autumn, but mainly in the latter; and, hydraulic gradients indicate that groundwater is discharging into the stream, but at variable strengths depending on location in the basin. Monitoring of groundwater will continue following timber harvest.

Soil temperature and soil solution chemistry are being evaluated at different depths in the soil horizon with a network of lysimeters (Titus et al. 1994) placed in future harvest blocks and in control blocks. This project, initiated in 1993, is being co-ordinated by scientists from the Canadian

Forest Service in Fredericton. Finally, surficial geochemistry changes due to timber harvest and soil disturbance and the potential loss of ions to the stream are being studied along Tributary One (Fig. 1) by researchers from the Université du Québec à Montréal (Geography).

Sediment Input

Forestry activity inevitably causes soil disturbance and exposure of the land surface which can lead to increased erosional processes and runoff. In reviewing the phenomenon, Everest et al. (1987) noted that quantifying cause and effect of increased sediment loading in a stream catchment is a complex problem depending on many factors at the levels of macro-scale (e.g. geomorphology, meteorology, hydrology) and micro-scale (e.g. gradient, channel morphology, hydrograph shape). As the sediments settle onto the stream substrate, imbedding bed particles, the stream biota are affected in various ways.

One of the major, short-term impacts on aquatic biota from disturbance in a drainage basin is an increase in sediment input to the stream (Campbell and Doeg 1989; Rinne 1990). For the purposes of this paper, I will be referring to the impacts of fines (i.e., <2-3 mm in diameter) when referring to sediments entering the stream. Sediments in suspension will be treated separately from those fines settling onto the stream bottom (i.e., sedimentation) because the impacts to aquatic biota and their habitat can be different.

For macroinvertebrates (see review in Campbell and Doeg 1989), suspended sediments probably have their greatest effect on filter-feeders as filtering mechanisms and nets become clogged and damaged. Also, invertebrate drift increases as a response to the stressful conditions and, during high, turbid flows, some benthic invertebrates are displaced. In general, sedimentation often changes the invertebrate community by reducing biomass and species diversity. In Carnation Creek, for example, species abundance and diversity were reduced in substrate of smaller diameter, and where there was less detritus accumulated, which in turn was negatively affected by logging activity (Hartman and Scrivener 1990).

Fine sediments in the stream can directly affect fish in numerous ways. Egg-to-fry survival of salmonids can be affected because of: (1) suffocation of eggs and alevins in redds; (2) reduced intra-gravel water flow and dissolved oxygen; (3) a physical barrier to emergence (Everest et al. 1987). Grant et al. (1986) carried out a post-treatment assessment of logging impacts for streams in Nova Scotia and New Brunswick (including streams in the Miramichi). They found that

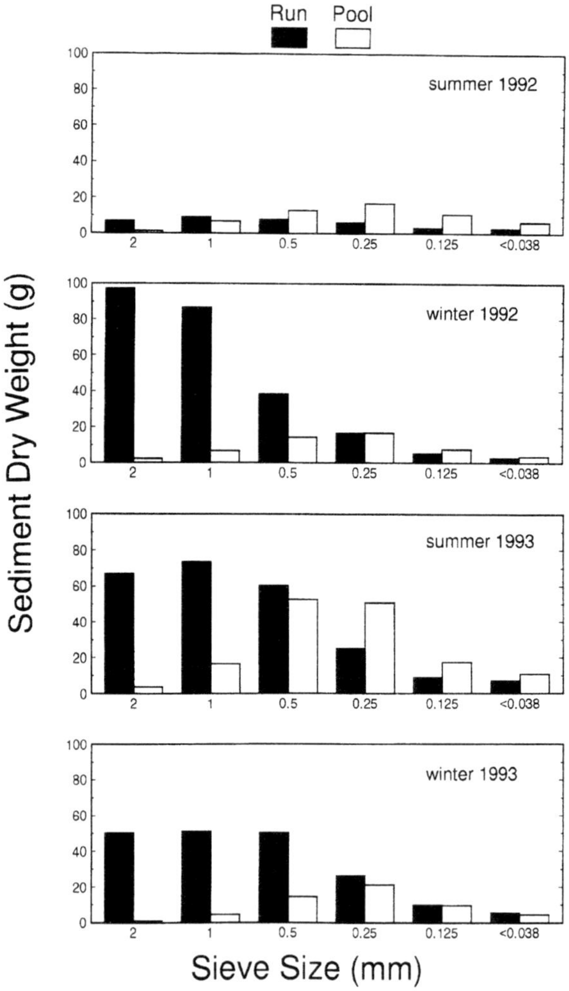

Fig. 4. Dry weight distribution of sediment particle sizes collected in Whitlock-Vibert boxes placed in runs and pools in the Upper Reach of Catamaran Brook during two years. Sedimentation was estimated separately for "winter" (November - May) and "summer" (May - November). For example, winter - 92 refers to the period from November, 1992 to May, 1993.

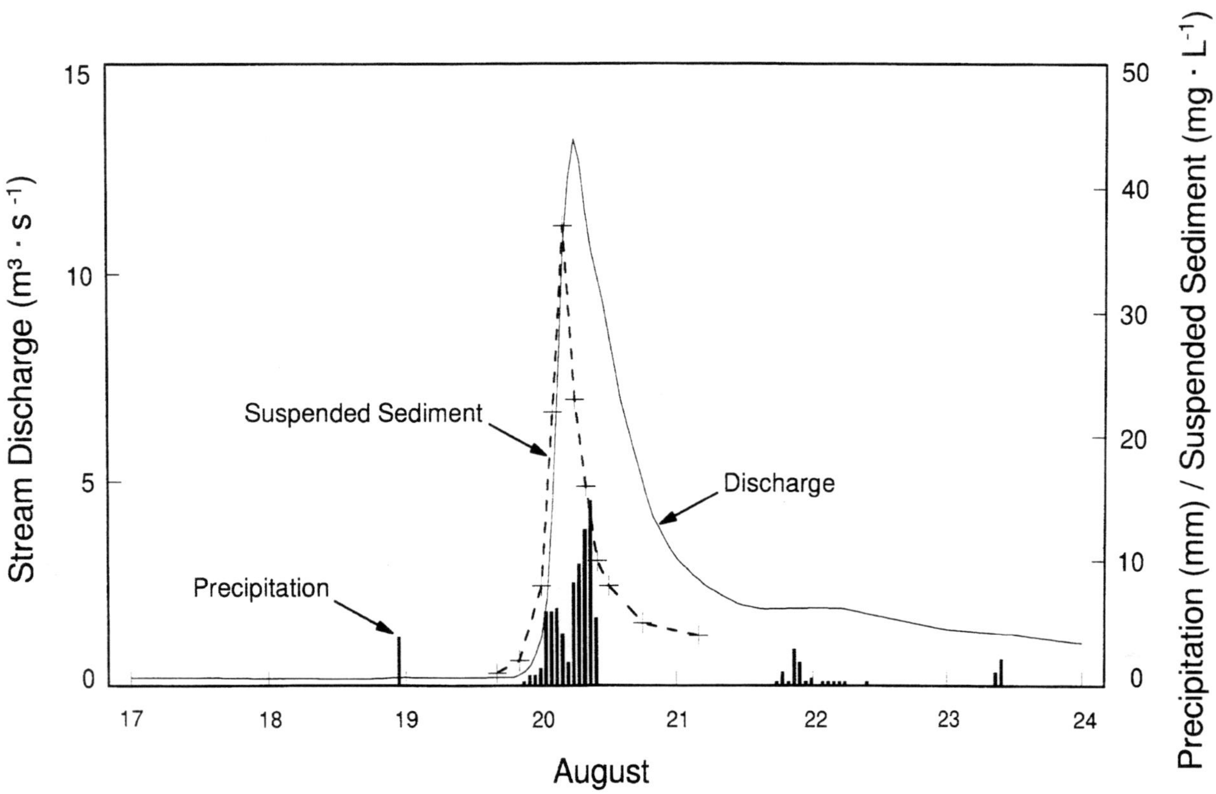

Fig. 5. Response of stream discharge ($m^3 \cdot s^{-1}$) and suspended sediments ($mg \cdot L^{-1}$) in the Lower Reach of Catamaran Brook during an extreme storm event (Hurricane Bob), August 19-21, 1991. Vertical bars represent hourly precipitation (mm).

salmonid biomass decreased significantly in two streams, downstream of logging roads, and speculated that the reason was increased siltation.

Imbeddedness (i.e., compaction) of stream substrate caused by the settling of fines can reduce microhabitat heterogeneity and preclude access beneath stones that stream fishes use as cover. For young Atlantic salmon, sedimentation can alter habitat suitability particularly in winter (Cunjak 1988) resulting in movement to other habitats (Cunjak and Randall 1993). Recent winter studies on movement and survival of juvenile salmon in Catamaran Brook indicate that movement over winter is extensive for some parr (>500 m) with mortality as high as 50-60% in some habitat-types during severe winters (Cunjak, unpubl data). Consequently, future estimates of salmon survival and winter habitat suitability after timber harvest will be compared with the pre-impact results.

In Catamaran Brook, the hypothesized impacts of fine sediments are being evaluated from two sources: timber harvest (1996-99); and, also from road construction (1995). The contribution to increased soil erosion from road construction has been found to be significant in Coweeta (Swift 1988) and Carnation Creek (Hartman and Scrivener 1990). Sedimentation is being directly measured with Whitlock-Vibert boxes (Wesche et al. 1989) placed in different habitat-types in each of the four experimental reaches (Fig. 1), and sampled semi-annually. Preliminary results indicate that relatively more larger particles (i.e., >1 mm diam.) are settling in run-type habitats than in pools (Fig. 4) which are more depositional. Proportionally little of the accumulated sediment was made up of silt-type particles (i.e., <0.125 mm diam., Fig. 4). The accumulation of such fines are often associated with land-use disturbance and these can be the most detrimental to stream biota. Similar trends in sedimentation were found throughout the length of stream (Cunjak et al. 1993) with more large particles accumulating in the fast-flow habitats (riffles, runs) than in slow-flow habitats (pools, flats) and greater loads deposited in the lower sections of river compared with the headwaters, although site-specific and temporal variability was high (Cunjak, unpubl. data). For example, sediment-size distribution was relatively similar for the two "winter" samples in each habitat-type (Fig. 4). However, there were marked differences between the two "summers" with relatively little sedimentation in 1992 compared with 1993 (Fig. 4). The effectiveness of this technique for quantifying sedimentation is being evaluated in Catamaran Brook by researchers from McGill University (Geography). Field experiments to quantify the impacts of fines on egg survival and emergence success of Atlantic salmon in Catamaran Brook are planned for 1994/95 using a modification of the trap design of Bardonnet et al. (1993).

Suspended sediments are being analyzed, primarily in association with storm events during the open water season. The working hypothesis is that increased runoff and erosion following forestry disturbance will be measurable as an increase in suspended sediment concentrations relative to the measurements taken during the pre-logging years. Consequently, suspended sediments have been measured since 1990 for storm events of varying magnitude. Figure 5 shows suspended sediment concentrations measured near the mouth of Catamaran Brook during a major storm in 1991, Hurricane Bob, when 108 mm of rain fell in only 14 hours. The suspended sediment concentration peaked approximately 2-3 hours before peak discharge which is similar to the lag time recorded for

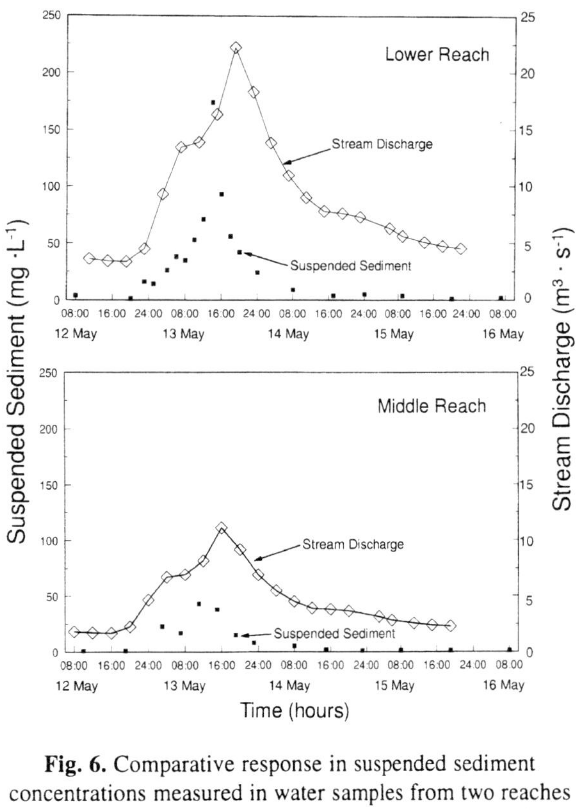

Fig. 6. Comparative response in suspended sediment concentrations measured in water samples from two reaches of Catamaran Brook in relation to stream discharge during a storm event (May 12-15, 1994) when 62 mm of rain fell in the basin.

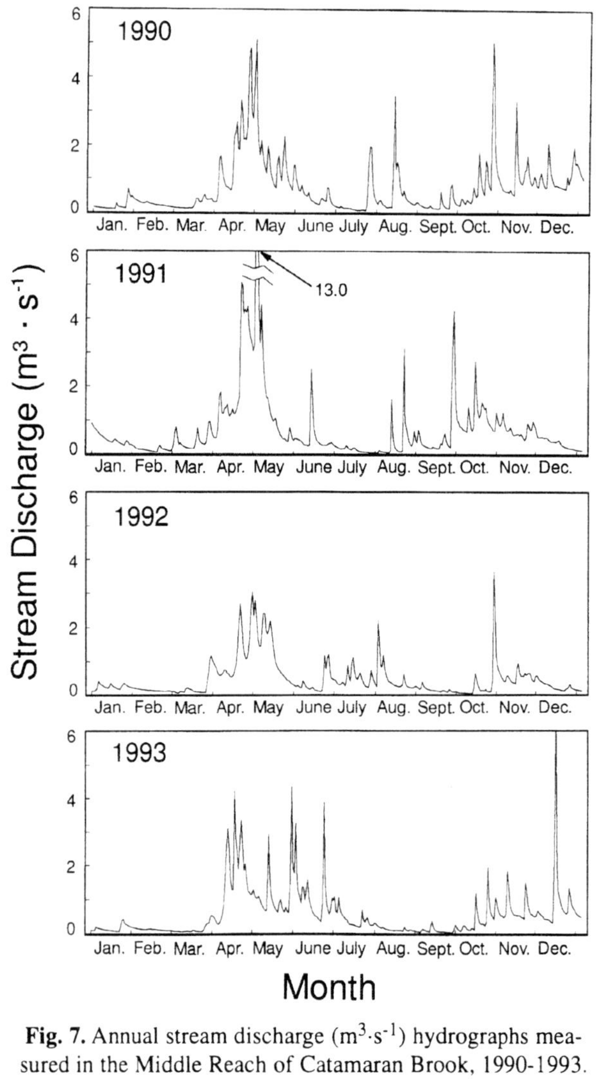

Fig. 7. Annual stream discharge ($m^3 \cdot s^{-1}$) hydrographs measured in the Middle Reach of Catamaran Brook, 1990-1993. Note that water discharge at the mouth of Catamaran Brook is approximately twice that measured at the Middle Reach hydrometric station.

other storm events in Catamaran Brook (Cunjak et al. 1993).

Beginning in 1994, suspended sediment in Catamaran Brook samples is being measured simultaneously at the stream mouth (Lower Reach) and in the Middle Reach for comparison of sediment response curves for the two sub-basins. Results from a significant storm event in May, 1994 (Fig. 6) indicated that suspended sediment concentrations in the upper sub-basin peaked at a much lower concentration (43 mg$\cdot$ L^{-1}) than would be have been expected based on the discharge-response curve derived from near the stream mouth. That is, peak suspended sediment concentration at the stream mouth (174 mg$\cdot L^{-1}$) was 4 times the peak concentration measured in the Middle Reach although stream discharge was only 2 times greater at the mouth (Fig. 6). The likely explanation is that much of the suspended sediment is entering the stream via bridge crossings and road runoff, conditions which are relatively scarce in the upper sub-basin. This situation will change with the beginning of forestry activity in 1995 and may result in increases in suspended sediment concentrations compared with pre-impact years. As for the Hurricane Bob event in 1991, the May 1994 peak in suspended sediment concentration lagged behind the peak in stream discharge (in this case, by 3-4 hours) and the situation was the same in both reaches (Fig. 6).

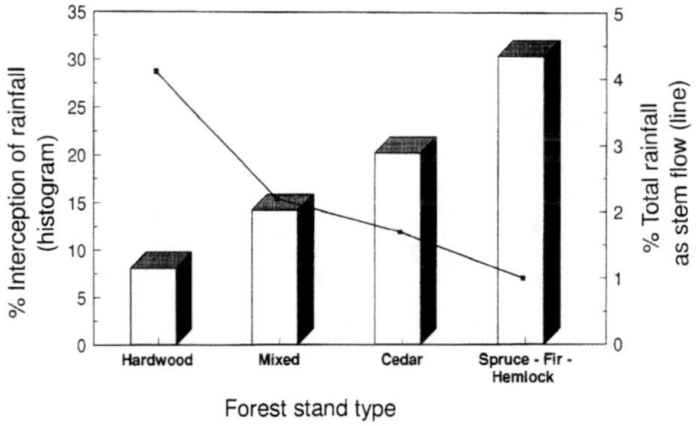

Fig. 8. Percent of rainfall intercepted in the canopy or measured as stemflow within four different forest stand-types in the Catamaran Brook basin. Results are for data collected between June and September of 1993 (Barry and Coté 1994).

Hydrologic Regime

Stream hydrology can be markedly influenced by forestry activities in the catchment (see Powell 1983; Swank and Crossley 1988; Hartman and Scrivener 1990). Most commonly, there is an increase in water yield to the stream and greater variability in the hydrograph after disturbance in the catchment. That is, for a given storm event, discharge changes faster, reaching a higher peak than was the case prior to the disturbance. Consequences include flooding, greater erosional forces, bedload movement, and scouring. The hydrologic modelling is further complicated in regions such as the Miramichi where snowpack accumulates differently within, and outside of, clear-cuts. Investigating these hypothesized changes in Catamaran Brook is critical to an assessment of habitat impacts and changes in fish production after the start of forestry activities. Detecting changes in water yield at the scale of whole basin may be difficult with just 7% of the Catamaran Brook basin being clear-cut. Bosch and Hewlett (1982) reviewed 94 studies and indicated that >20% of a basin had to be clear-cut in order to detect change in water yield. However, by monitoring hydrologic variables in the sub-basins immediately downstream of forestry activity, such changes may be more readily measured.

In Catamaran Brook, hydrology and physical habitat are being studied (Caissie and El-Jabi 1995). The program involves three main components: (1) calculation of the hydrologic budget of the Catamaran Brook basin which is largely dependent on multivariate data collection from the system of data-loggers and at the meteorological station (Fig. 1); (2) quantification of stream discharge for the basin (Fig. 7), sub-basins and Tributary One; and, (3) regular habitat surveys in the habitat-types of all 4 reaches to measure physical habitat characteristics (e.g., slope, substrate particle size, bed roughness, channel shape). Specific details of the program and preliminary results for 1990-1992 data collection were outlined in Cunjak et al. (1993). This information and additional data collected from collaborative studies will be used to develop a comprehensive hydrological model for

Catamaran Brook. For example, rainfall interception by different forest stand-types in the basin affects the rate of evapotranspiration (Barry and Coté 1994) and the amount of water eventually reaching the stream.

The annual hydrograph for Catamaran Brook is characteristic of most Miramichi streams (Caissie and El-Jabi 1995). Highest flows are measured in the spring coincident with snowmelt with a secondary peak occurring in the autumn (Fig. 7). Lowest flows are measured in the summer but these are often punctuated by sudden storm storms. The few, and smaller, stream discharge peaks, particularly in the early summer (May/June) of 1992 compared with 1993 (Fig. 7) may partly explain the relatively low sedimentation measured in runs and pools in Catamaran Brook (Fig. 4). A continuous ice-cover over the stream (from early December until late March) and thick snow cover and cold temperatures result in a relatively stable, low flow over the winter (Fig. 7).

Groundwater can influence the thermal regime, discharge and chemistry of a stream (Newbury et al. 1969; Pinder and Jones 1969) as well as being extremely important to aquatic biota (Hynes 1983). Therefore, the ability to measure the relative contribution of groundwater in the hydrologic budget is important for assessing ecosystem productivity and how this may change after deforestation. As already noted, groundwater dynamics and chemistry are being monitored in the Catamaran Brook basin via a system of wells and piezometers (Jones and Bray 1994). More recently, the groundwater component of streamflow was estimated during several storm events using a hydrograph separation technique (Caissie et al. 1995). For a small event (rainfall = 19.6 mm), groundwater flow represented 91% of the total peak flow whereas for a more intense storm (rainfall = 77.0 mm), groundwater represented 55% of the total peak flow.

A significant aspect of the hydrologic budget of a forested catchment is loss of precipitation through canopy interception. The proportion of rainfall intercepted and evaporated directly from the leaves is greatly influenced by the type of forest stand. For rain events between June and September, 1993, throughfall and stemflow were measured (Barry and Coté 1994) within four different forest stands (Fig. 8). Stemflow was greatest in the mature tolerant hardwood stand whereas rainfall interception steadily increased as forest stand changed from hardwood to a spruce-fir hemlock stand (Fig. 8).

Biota

Many direct studies of the plants and animals in the catchment are also being carried out (see Cunjak et al. 1993). Probably the most labour intensive activity in the field is the operation of the fish-counting fence located near the mouth of Catamaran Brook (Fig. 1). A description of the design, protocol, and analysis of the fish movement data between 1990 and 1992 were provided by Cunjak et al. (1993). Briefly, all fish moving into and out of Catamaran Brook between May and mid-November are enumerated and measured prior to being released back into the stream. These data are important for assessing production of the system and for identifying population patterns which may change after the start of timber harvest.

The number of Atlantic salmon smolts emigrating from Catamaran Brook has varied five-fold in the 5 years prior to any forestry activity (Table 3) which indicates significant natural, inter-annual variability in those environmental factors affecting year-class survival and production. In contrast, the number of adult salmon returning to Catamaran Brook (Table 3) has varied relatively little in the first five years (2.4-fold difference, 1990-94). Smolt totals, when related to the total area of stream accessible to salmon indicate an average (1990-1994) annual production of approximately 1.3 smolts·100 m^2. This value is lower than the average of 4.6 smolts·100 m^2 estimated by Elson (1975) for the Miramichi between 1951 and 1971, but is within the range given by Symons (1979) for the Northwest Miramichi River. The annual estimates of smolt production in Catamaran Brook are based on the assumption that 18.5 river km are accessible to spawning salmon. In fact, this is rarely achieved as combinations of beaver dams and low autumn streamflow often limit spawning to less stream area thereby underestimating actual smolt production.

Salmon production from streams must also consider the emigration of salmon parr (> age 1) which outnumbered smolts in 4 of the 5 years in Catamaran Brook (Table 3). Although approximately 33% of the parr emigrants in the autumn are precocious (Cunjak et al 1993) and not likely to survive winter (Myers 1984), most are immature and would smoltify the following year(s). It should be noted that the annual numbers of emigrant parr are underestimates because the fish-counting fence is less efficient at catching smaller sizes of fish.

Electrofishing surveys conducted twice per year (July and November) provide estimates of fish density and growth rates in different habitat-types and different seasons for various age-classes and common stream fishes (e.g., salmon, brook trout, blacknose dace, lake chub, and slimy sculpin) throughout Catamaran Brook (see Cunjak et al. 1993). The results can also be used to estimate survival. For example, the survival of recently emerged salmon fry, in 1994, was estimated by comparing young-of-the-year (YOY) salmon density in July (approximately 3-4 weeks post-emergence) and egg deposition based on numbers and fecundity of mature female salmon entering the brook the previous autumn (1993) when 49 mature female salmon (12 grilse and 37 multi-sea-winter salmon) entered Catamaran Brook to spawn. Based on an average fecundity of 1,385 eggs·kg^{-1} (M. Hambrook, Miramichi Salmonid Enhancement Centre, pers. comm.), predicted egg deposition was 284,505 eggs which results in an average egg density of 2.92 eggs·m^{-2} as only the Middle and Lower sections of the brook were accessible to spawners in 1993 (Table 4). Relating the predicted egg deposition to the calculated YOY densities and abundance (Table 4) indicates that egg survival until 3-4 weeks post emergence in Catamaran Brook was 33% in 1993/94. For Atlantic salmon, wild egg to fry (mid-summer) survival has been estimated indirectly to be 8-20% for rivers in Maine (Meister 1962) and the

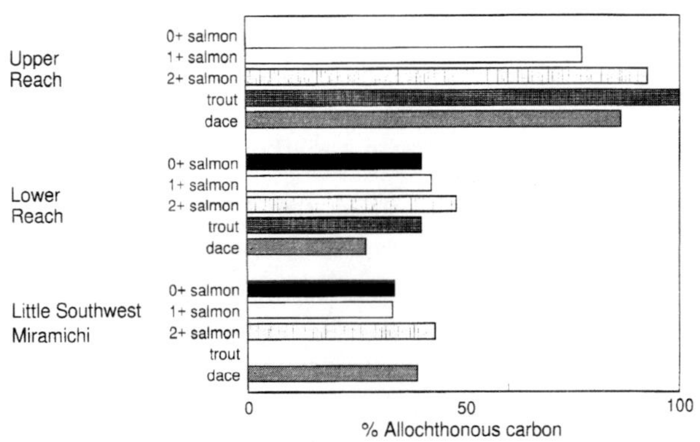

Fig. 9. Comparison among 3 locations in the Catamaran Brook - Little Southwest Miramichi River area of the relative importance of allochthonous (i.e., terrestrial) primary carbon sources to juvenile Atlantic salmon, brook trout, and blacknose dace collected during 1992-93. Percent contributions are mean values calculated by a 2-source mixing model using average leaf litter, and site-specific algal ^{13}C values (from Doucett 1994).

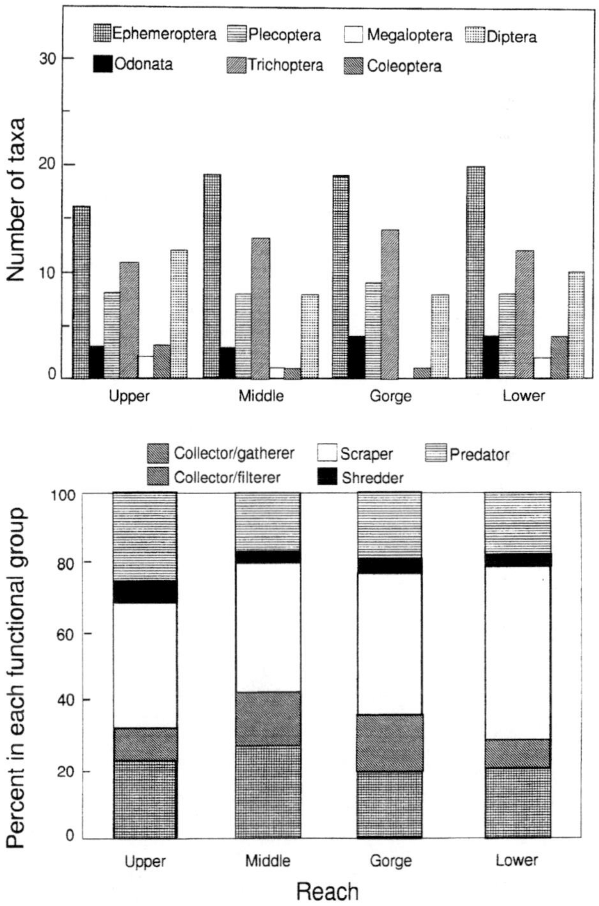

Fig. 10. Diversity and distribution of aquatic insects collected by kick-sampling from Catamaran Brook, May - October, 1990 (modified from Cunjak et al. 1993). Upper panel shows insect taxa richness (i.e., number of genera found in each order, except Chironomidae which were identified only to family) among the four stream reaches.

Miramichi (Elson 1957; Symons 1979).

In the past two years, electrofishing and mark-recapture experiments were carried out in the 8 sites of the Middle Reach to assess winter movement and survival of juvenile salmon. This study will permit us to establish (1) if in-stream movements are as common as suggested by Cunjak and Randall (1993) who found that winter site fidelity by juvenile salmon was < 30% in their study sites; and, (2) the winter mortality of salmon parr in such ice-covered streams. Preliminary results from eight sites in the Middle Reach of Catamaran Brook have found that winter mortality (measured as density change between November and April during 2 years) ranged from 35% – 65%, being greatest in the winter with least streamflow (Cunjak, unpubl. data).

Recently, Doucett (1994) used stable carbon isotope analysis to determine the relative importance of primary food sources leading to Atlantic salmon in Catamaran Brook and the Little Southwest Miramichi River. The results support the classical theory of stream ecology wherein food webs and juvenile Atlantic salmon in the headwaters of Catamaran Brook relied mostly on

Table 2. Age-representation (percent of sample population) of Atlantic salmon smolts from various rivers in New Brunswick and Nova Scotia. Age data from Catamaran Brook, Big Salmon River, and Pollet River are taken directly from smolts passing through counting fences; other ages are back-calculated from adult salmon returning to rivers.

River	Sample period	Age-2 Smolts (%)	Age-3 Smolts (%)
Restigouche River system[1]	1972-1980	27.9	66.4
Catamaran Brook	1990-1994	24.0	74.4
Miramichi River system[2,3]	1931	15.1	78.1
	1971-1992	38.9	59.0
Big Salmon River[4]	1966-1972	45.3	50.9
Pollet River[5]	1949-1953	88.3	11.6
Margaree River[6]	1987-1988	71.8	26.8

Sources: [1] Peppar and Pickard (1975), Pickard and Peppar (1979, 1983); [2] Blair (1935); [3] Courtenay et al. (1993); [4] Jessop (1975); [5] Elson (1962); [6] Claytor and Jones (1990).

Table 3. Numbers of emigrant parr, smolts, and adult Atlantic salmon enumerated at the counting-fence in Catamaran Brook, 1990-1994. Parr and smolt values after 1990 were corrected for capture efficiency at the fence. MSW refers to multi-sea-winter salmon.

	Downstream Movement		Upstream Movement	
Year	Migrant Parr[1]	Smolts	Grilse (<63 cm)	Adult Salmon MSW (≥63 cm)
1990	850[2]	760	83[2]	28 [2]
1991	2183	1515	79	48
1992	1434	2429	127	64
1993	1360	515	107	44
1994	2359	1002	56	24

[1] Total number of downstream migrant parr (ages > 1) , May-November.

[2] Incomplete count because of damage to counting-fence.

allochthonous carbon sources (i.e. from leaf litter falling in the stream from the closed canopy). In the Lower Reach of Catamaran Brook, as in the wide Little Southwest Miramichi River, there is a shift in dependence to autochthonous (in-situ) carbon sources such as algae (Fig. 9). These results are important as a reference for measuring future shifts in carbon pathways following perturbation (timber harvest) along several stream reaches.

Investigations on the macroinvertebrate community in Catamaran Brook are critical for trying to assess future impacts related to forest harvest in the basin because their response to changes in substrate composition, nutrient cycling, and environmental quality is often more immediate and more direct than for fishes. Benthic kick-samples taken in non-pool habitats between July and October, 1990, showed that Ephemeroptera were numerically dominant but that relative abundance of all major taxa was similar among the four stream reaches (Fig. 10). Subdivision of the 1990 invertebrate samples based on mode of feeding indicated several trends. Shredders (those chewing coarse particulate organic matter (CPOM) such as leaf litter) and predators were most abundant in the Upper Reach (Fig. 10). Between the Middle and Lower Reaches, collectors (which feed on fine particulate organic matter) decreased in relative abundance whereas scrapers (mainly algal feeders) increased (Fig. 10). Assuming that deforestation will result in an increase in water temperature and sedimentation, and a decrease in CPOM (Campbell and Doeg 1989, Garman and Moring 1991), it is likely that shredder abundance will decline and the proportions of scrapers and collectors will change depending on the amount of sunlight and fine sediments reaching the stream bottom (Cunjak et al. 1993).

The focus of current research is on comparative changes in community structure in the various habitat-types and experimental reaches between 1990-1994, including the effects of natural disturbance (e.g., floods). Recently, Garnett (1994) reported on the results of samples from 45 emergence traps set during the spring and summer in 1993; > 23,000 adult insects were counted, of which approximately 21,000 were Diptera (mainly chironomids and simuliids). That species diversity is high in Catamaran Brook is evidenced by the extensive list of taxa identified to date which includes 47 families of insects and approximately 100 species (D. Giberson, Univ. Prince Edward Island, pers. comm.).

In 1994, several field studies were initiated to assess the effectiveness of buffer-strips in protecting the biota which use this riparian zone. In the Catamaran Brook basin, a 30–60 m buffer strip will remain along any water course within a scheduled harvest block (Fig. 1). For those harvest blocks in the Gorge Reach and the Upper Reach, 3 studies have been designed to monitor amphibian populations and the diversity of understory plants. One of the amphibian studies has adopted the board cluster method (DeGraaf and Yamasaki 1992) for estimating relative abundance of salamanders (mainly red-backed salamanders, *Plethodon cinereus*) within riparian zones adjacent to clear-cuts and those where no harvest occurs nearby (controls). The second amphibian study is focussing on the two-lined salamander, *Eurycea*

Table 4. Mean density and abundance of young-of-the-year Atlantic salmon in different sections in Catamaran Brook, based on electrofishing surveys of 31 sites in July 1994.

Stream section	River km	Stream area (m^2)	Mean Density ($no.\cdot m^{-2}$)	Abundance per section
Upper	15.2-20.0	17,212	0	0
Middle	8.3-15.2	37,357	0.103	3,848
Lower	0-8.3	60,127	1.518	91,249
			Total =	95,097

bislineata, using field monitoring and laboratory experimentation to quantify distribution and habitat preferences in the stream and riparian zones adjacent to the proposed clear-cut area in the Upper Reach. The plant diversity study is being conducted along transects within riparian zones adjacent to a clear-cut to contrast with transects in a nearby control. All three projects will continue for 3-4 years.

Conclusion

As Phase 1 (the pre-logging phase) of the Catamaran Brook project nears completion and nearly five years of study have been carried out, it is useful to reflect on the status of the program. During the first 2 years, government and industry were the most important partners in getting the project going, financially and logistically. More recently, a diverse research program has been established through the involvement of private agencies and many universities and colleges (Table 5), specifically via graduate student projects. On their own, these studies may seem too specialized to be of practical significance for addressing forestry impacts. However, in providing solutions for the complex functioning of a stream ecosystem, they are contributing exactly the information necessary to assess land-use perturbation in the Miramichi, and elsewhere.

Was 5 years enough time to assess natural environmental variability in the system ? Between 1990 and 1994, many extreme events were experienced and quantified. Hurricane Bob, in August of 1991, brought the most intense rainfall measured to date (108 mm in 14 h). Severe ice-jamming in mid-April, 1994, in the Little Southwest Miramichi River, resulted in the highest-recorded flooding of the lower 700 m of Catamaran Brook. The highest (May 1991) and lowest (September 1994) stream discharge had estimated annual recurrence probabilities of 0.04 (i.e., 1 in every 25 yr) and 0.07 (1 in 15 yr), respectively. The extreme low water in the autumn of 1994 may be partly responsible for the relatively few salmon spawners entering the stream (Table 3), although the low smolt total in 1993 (Table 3) would also have contributed. Beaver dams in some years have limited the access to upstream spawning areas which may have important implications for density-dependent growth and relative survival of juvenile salmon in subsequent years. In 1994, a beaver dam was built approximately 1 km above the stream mouth whereas, in 1990, no main-channel beaver dams were present below river km 12. Therefore, much variability has been recorded but the simple answer is no, 5 years is not enough time to quantify all natural variability. It was the minimum practical period adopted in our study design because 5 years represents 1 complete generation for an Atlantic salmon (grilse) in Catamaran Brook.

How easily will impacts be detected and measured within the Catamaran Brook catchment ? Detecting changes at the scale of whole basin may be difficult with just 7% of the basin being clear-cut, and with 30m and 60m buffer strips being retained along the stream adjacent to harvest blocks . For example, 41% of the Carnation Creek basin was harvested over 6 years (Hartman and Scrivener 1990), and Bosch and Hewlett (1982) indicated that >20% of a basin had to be clear-cut in order to detect an impact. For the Catamaran Brook project, a primary objective of our research has been applicability to the typical timber harvest methods used in the province. The schedule of forestry activity in the basin was adopted *before* the study began and is based on considerations of suitable wood resources, accessibility, adjacency regulations, fish and wildlife conservation, and future wood supply projections. In this manner, the results should reflect the situation in other streams in the Miramichi system. Measurable changes in water quality parameters, hydrology, and biota are more likely to be found at a scale smaller than stream basin. In Catamaran Brook, therefore, environmental monitoring and regular sampling in specific sites (i.e., habitat-types) and those stream reaches immediately downstream of forestry activity are locations where change is likely to be detected.

Acknowledgments

Many people, including students, scientists,

Table 5. List of research projects initiated at Catamaran Brook during the pre-logging phase of the project, 1990-1994.

Research project(s) and responsibility	Agency / Institution	Address
Fish research, stream hydrology, physical habitat, project management	Department of Fisheries & Oceans, Science Branch	Moncton, NB
Water quality, chemistry Stream discharge monitoring	Environment Canada	Moncton, NB Fredericton, NB
Soil chemistry	Canadian Forest Service	Fredericton, NB
pH monitoring	N.B. Department of the Environment	Fredericton, NB
Forestry management, study design, historical data, logistics	N.B. Department of Natural Resources	Newcastle, NB
Timber harvest, road construction, forest spp. survey	Miramichi Pulp & Paper, Inc.	Newcastle, NB
Fish habitat modelling	G. Shooner, inc.	Québec, PQ
Hydrologic budget, meteorology, water temperature modelling Rainfall interception Fish behaviour	Université de Moncton – École de génie – Sciences forestières – Département de biologie	 Moncton, NB Edmundston, NB Moncton, NB
Fish ecology, aquatic fungi, salamander ecology	Mount Allison University – Biology Department	Sackville, NB
Fish behaviour	Concordia University – Biology Department	Montreal, PQ
 Fish biology, parasitology Groundwater	Univ. of New Brunswick Biology Department – Civil Engineering	Fredericton, NB
Aquatic invertebrate biology	University of Prince Edward Island – Biology	Charlottetown, PEI
Soil geochemistry, water chemistry	Univ. du Québec à Montréal – géographie	Montréal, PQ
Stream ecology, carbon cycling	Univ. of Waterloo – Biology	Waterloo, Ont.
Fluvial geomorphology	McGill University – Geography	Montreal, PQ
Hydrological modelling	Univ. du Québec INRS-EAU	Québec, PQ
Lake surveys	N.B. Community College	Chatham, NB

and conservation-minded individuals who respect the Miramichi have contributed greatly to the success which we have enjoyed at Catamaran Brook over the first five years. To all of them, I express my sincere appreciation. I also wish to thank D. Caissie, J. Conlon, R. Doucett, P. Hardie, and S. Komadina-Douthwright who contributed to the completion of this paper, and to C. Scrivener and R.J. Gibson who reviewed the original manuscript.

References

Bardonnet, A., P. Gaudin, and J.E. Thorpe. 1993. Diel rhythm of emergence and of first displacement downstream in trout (*Salmo trutta*), Atlantic salmon (*S. salar*) and grayling (*Thymallus thymallus*). J. Fish Biol. 43: 755-762.

Barry, R., and Coté, F. 1994. Interception de la précipitation par différents types de couverts forestiers dans la bassin du ruisseau Catamaran, N.-B. Rapport de Recherche, Université de Moncton (CUSLM), Moncton, N.B. 79p.

Bosch, J.M., and Hewlett, J.D. 1982. A review of catchment experiments to determine the effect of vegetation changes on water yield and evapotranspiration. J. Hydrol. 55: 3-23.

Blair, A.A. 1935. Ages at migration of Atlantic salmon (*Salmo salar*) in the Miramichi River. J. Biol. Board Can. 1: 159-169.

Bormann, F.H., and Likens, G.E. 1979. Pattern and process in a forested ecosystem. Springer-Verlag, New York, NY. 253 p.

Brown, G.W., and Krygier, J.T. 1970. Effects of clear-cutting on stream temperature. Water Resour. Res. 6: 1133-1139.

Caissie, D., and El-Jabi, N. 1995. Hydrology of the Miramichi River drainage basin, p. 83–93. *In* E.M.P. Chadwick [*editor*]. Water, science, and the public: the Miramichi ecosystem. Can. Spec. Publ. Fish. Aquat. Sci. 123.

Caissie, D., Pollock, T.L., and Cunjak, R.A. 1995. Variation in stream water chemistry and hydrograph separation in a small drainage basin. J. Hydrol. (in press).

Campbell, I.C., and Doeg, T.J. 1989. Impact of timber harvesting and production on streams: a review. Austr. J. Freshwater Res. 40: 519-539.

Cederholm, C.J., and Reid, L.M. 1987 Impact of forest management on coho salmon (*Oncorhynchus kisutch*) populations of the Clearwater River, Washington. A project summary, p. 373-398. *In* E.O Salo and T.W. Cundy [*editors*]. Streamside Management: Forestry and Fishery Interactions. Univ. Wash. Inst. For. Prod. Contrib. 57: 471p.

Chadwick, E.M.P. 1987. Causes of variable recruitment in a small Atlantic salmon stock. Amer. Fish. Soc. Symp. 1: 390-401.

Claytor, R.R., and Jones, R. 1990. Assessment of Atlantic salmon (*Salmo salar*) in the Margaree River, 1989. CAFSAC Res. Doc. 90/27. Dep. Fish. Oceans, Sci. Br., Moncton, N.B.

Courtenay, S.C., Moore, D.S., Pickard, R., and Nielsen, G. 1993. Status of Atlantic salmon in the Miramichi River in 1992. DFO Atlantic Fisheries Document 93/56. Department of Fisheries and Oceans, Science Branch, Moncton, N.B.

Cunjak, R.A. 1988. Behaviour and microhabitat of young Atlantic salmon (*Salmo salar*) during winter. Can. J. Fish. Aquat. Sci. 45: 2156-2160.

Cunjak, R.A., Caissie, D., and El-Jabi, N. 1990. The Catamaran Brook habitat research project: description and general design of study. Can. Tech. Rep. Fish. Aquat. Sci. 1751: 14p.

Cunjak, R.A., Caissie, D., El-Jabi, N., Hardie, P., Conlon, J.H., Pollock, T.L., Giberson, D.J., and Komadina-Douthwright, S. 1993. The Catamaran Brook (New Brunswick) habitat research project: biological, physical, and chemical conditions (1990-1992). Can. Tech. Rep. Fish. Aquat. Sci. 1914: 81p.

Cunjak, R.A., and Randall, R.G. 1993. In-stream movements of young Atlantic salmon (*Salmo salar*) during winter and early spring, p. 43-51. *In* R.J. Gibson and R.E. Cutting [*editors*]. Production of juvenile Atlantic salmon, *Salmo salar*, in natural waters. Can. Spec. Publ. Fish. Aquat. Sci. 118.

DeGraaf, R.M., and Yamasaki, M. 1992. A nondestructive technique to monitor the relative abundance of terrestrial salamanders. Wildl. Soc. Bull. 20: 260-264.

Dickison, R.B.B. 1988. Nashwaak experimental watershed project. Proceedings, Can. Hydrol. Symp., May 9-11, 1988. Banff, Alta: 59-66.

Doucett, R.R. 1994. Stable isotope analysis of food webs in the Miramichi River system, New Brunswick. M.Sc. thesis, Department of Biology, University of Waterloo, Waterloo, Ont.

Elson, P.F. 1957. Number of salmon needed to maintain stocks. Can. Fish Cult. 21: 19-23.

Elson, P.F. 1962. Predator-prey relationships between fish-eating birds and Atlantic salmon. Bull. Fish. Res. Board Can. 133: 87p.

Elson, P.F. 1967. Effects on wild young salmon of spraying DDT over New Brunswick forests. J. Fish. Res. Board Can. 24: 731-767.

Elson, P.F. 1974. Impact of recent economic growth and industrial development on the ecology of Northwest Miramichi Atlantic salmon (*Salmo salar*). J. Fish. Res. Board Can. 31: 521-544.

Elson, P.F. 1975. Atlantic salmon rivers, smolt production, and optimal spawning: an overview of natural production. Int. Atl. Salmon Found. Spec. Publ. Ser. 6: 96-119.

El-Kourdahi, G. 1993. Modélisations déterministe et stochastique de la température de l'eau en rivière. M.Sc.A. (Génie civil). Université de Moncton, Moncton, N.-B. 142p.

Everest, F.H., Beschta, R.L., Scrivener, J.C., Koski, K.V., Sedell, J.R., and Cederholm, C.J. 1987. Fine sediment and salmonid production: a paradox, p. 98-142. *In* E.O Salo and T.W. Cundy [*editors*]. Streamside Management: Forestry and Fishery Interactions. Univ. Wash. Inst. For. Prod. Contrib. 57: 471p.

Garman, G.C., and Moring, J.R. 1991. Initial effects of deforestation on physical characteristics of a boreal river. Hydrobiologia 209 : 29-37.

Garnett, H. 1994. Emergence of aquatic insects from Catamaran Brook (New Brunswick), with special reference to the Plecoptera. B.Sc. (Hon.) thesis, Department of Biology, University of Prince Edward Island, Charlottetown, P.E.I. 64p.

Grant, J.W.A., Englert, J., and Bietz, B.F. 1986. Application of a method for assessing the impact of watershed practices: effects of logging on salmonid standing crops. N. Amer. J. Fish. Manage. 6: 24-31.

Hartman, G.F., and Scrivener, J.C. 1990. Impacts of forestry practices on a coastal stream ecosystem, Carnation Creek, British Columbia. Can. Bull. Fish. Aquat. Sci. 223: 148p.

Hartman, G.F., Scrivener, J.C., Holtby, L.B., and Powell, L. 1987. Some effects of different streamside treatments on physical conditions and fish population processes in Carnation Creek, a coastal rainforest stream in British Columbia. *In* E.O. Salo and T.W. Cundy [*editors*]. Streamside Management: Forestry and Fishery Interactions. Univ. Wash. Inst. For. Prod. Contrib. 57: 471p.

Hall, J.D., Brown, G.W., and Lantz, R.L. 1987. The Alsea watershed study: a retrospective. *In* E.O. Salo and T.W. Cundy [*editors*]. Streamside Management: Forestry and Fishery Interactions. Univ. Wash. Inst. For. Prod. Contrib. 57: 471p.

Holtby, L.B. 1988. Effects of logging on stream temperatures in Carnation Creek, British Columbia, and associated impacts on the coho salmon (*Oncorhynchus kisutch*). Can. J. Fish. Aquat. Sci. 45: 502-515.

Hynes, H.B.N. 1983. Groundwater and stream ecology. Hydrobiologia 100: 93-99.

Jessop, B.M. 1975. Investigation of the salmon (*Salmo salar*) smolt migration of the Big Salmon River, New Brunswick, 1966-1972. Tech. Rep. Ser. MAR/T-75-1. Environ. Can., Fish. Mar. Serv. 57p.

Johnson, G.B. 1945. Miramichi woodsman. Press of Whitlet and Shepperson [?]. 102p.

Jones, A.R.M., and Bray, D.I. 1994. The Catamaran Brook groundwater study; summary of field activities and results of field data analysis for April 1993 to March 1994. Unpubl. Report, Univ. New Brunswick, Department of Civil Engineering, Fredericton, N.B. 86p.

Likens, G.E., Bormann, F.H., Johnson, N.M., Fisher, D.W., and Pierce, R.S. 1970. The effects of forest cutting and herbicide treatment on nutrient budgets in the Hubbard Brook watershed-ecosystem. Ecol. Monogr. 40: 23-47.

Likens, G.E., Bormann, F.H., Pierce, R.S., and Reiners, W.A. 1978. Recovery of a deforested ecosystem. Science 199: 492-496.

Meister, A.L. 1962. Atlantic salmon production in Cove Brook, Maine. Trans. Amer. Fish. Soc. 91: 208-212.

Meyer, J.L., Tate, C.M., Edwards, R.T., and Crocker, M.T. 1988. The trophic significance of dissolved organic carbon in streams, p. 269-278. *In* W.T. Swank and D.A. Crossley, Jr. [*editors*]. Forest hydrology and ecology at Coweeta. Springer-Verlag, New York, NY. 468p.

MREAC 1992. Summary-final report (1989-1992). Unpubl. report of the Miramichi River Environmental Assessment Committee, Newcastle, N.B. 157p.

Moring, J.R., and Lantz, R.L. 1975. The Alsea watershed study: effects of logging on the aquatic resources of three headwater streams of the Alsea River, Oregon. Fish. Res. Rep. 9, Oregon Dep. Fish Wildl., Corvallis, OR.

Myers, R.A. 1984. Demographic consequences of precocious maturation of Atlantic salmon (*Salmo salar*). Can. J. Fish. Aquat. Sci. 41: 1349-1353.

Peppar, J.L., and Pickard, R.R. 1975. Ages of migration of Atlantic salmon in the Restigouche River. Data Record Ser. MAR/D-75-8. Env. Can., Fish. Mar. Serv. 7p.

Peterman, R.M. 1981. Form of random variation in salmon smolt-to-adult relations and its influence on production estimates. Can J. Fish. Aquat. Sci. 38: 1113-1119.

Pickard, R.R., and Peppar, J.L. 1979. Ages of migration of Atlantic salmon in the Restigouche River, 1976-1978. Can. Data Rep. Fish. Aquat. Sci. 165: 5p.

Pickard, R.R., and Peppar, J.L. 1983. Ages of adult Atlantic salmon samples from the Dalhousie trap, Restigouche River system, 1972-1980. Can. Data Rep. Fish. Aquat. Sci. 393: 11p.

Powell, G.R. [*editor*]. 1983. Nashwaak experimental watershed project. Annual Report, 1981-1982. New Brunswick Forest Research Advisory Committee, 23p.

Randall, R.G. 1982. Emergence, population densities, and growth of salmon and trout fry in two New Brunswick streams. Can. J. Zool. 60: 2239-2244.

Randall, R.G., Healey, M.C., and Dempson, J.B. 1987. Variability in length of freshwater residence of salmon, trout, and char. Amer. Fish. Soc. Symp. 1: 27-41.

Rinne, J.N. 1990. The utility of stream habitat and biota for identifying potential conflicting forest land uses: Montane riparian areas. Forest Ecol. Manage. 33/34: 363-383.

Schindler, D.W. 1987. Detecting ecosystem responses to anthropogenic stress. Can. J. Fish. Aquat. Sci. 44(Suppl. 1): 6-25.

Swank, W.T., and Crossley, D.A., Jr. [*editors*] 1988. Forest hydrology and ecology at Coweeta. Springer-Verlag, New York, NY. 468p.

Swift, L.W., Jr. 1988. Forest access roads: design, maintenance, and soil loss, p. 313-324. *In* W.T. Swank and D.A. Crossley, Jr. [*editors*]. Forest hydrology and ecology at Coweeta. Springer-Verlag, New York, NY. 468p.

Symons, P.E.K. 1977. Dispersal and toxicology of the insecticide fenitrothion, predicting hazards of forest spraying.

Residue Rev. 68: 1-36.

Symons, P.E.K. 1979. Estimated escapement of Atlantic salmon (*Salmo salar*) for maximum smolt production in rivers of different productivity. J. Fish. Res. Board Can. 36: 132-140.

Titus, B., Kingston, D.G.O., and Mahendrappa, M.K. 1994. A lysimeter system for monitoring soil solution chemistry. Can. J. Soil Sci. (in press).

Webster, J.R., Gurtz, M.E., Hains, J.J., Meyer, J.L., Swank, W.T., Waide, J.B., and Wallace, J.B. 1983. Stability of stream ecosystems, p. 355-395. *In* J.R. Barnes and G.W. Minshall [*editors*]. Stream Ecology, Plenum Press, New York, NY.

Wesche, T.A., Reiser, D.W., Hasfurther, V.R., Hubert, W.A., and Skinner, Q.D. 1989. New techniques for measuring fine sediment in streams. N. Amer. J. Fish. Manage. 9: 234-238.

Wright, E.C. 1944. The Miramichi: a study of the New Brunswick river and of the people who settled along it. Tribune Press [?]: 79p.

CHAPTER 13

Atlantic Tomcod (*Microgadus tomcod*) and Smooth Flounder (*Pleuronectes putnami*) as Indicators of Organic Pollution in the Miramichi Estuary

Simon C. Courtenay

Environmental Studies Division, Science Branch, Department of Fisheries and Oceans, P.O. Box 5030, Moncton, New Brunswick, Canada E1C 9B6

P. Jim Williams

Department of Chemistry and Biochemistry, University of Moncton, Moncton, New Brunswick, Canada E1A 3E9

Catherine Vardy

Fireweed Consulting, 34 Johnson Avenue, Moncton, New Brunswick, Canada E1C 6X8

and Isaac Wirgin

Institute of Environmental Medicine, New York University Medical Center, Long Meadow Road, Tuxedo, NY 10987, USA

Courtenay, S.C., Williams, P.J., Vardy, C., and Wirgin, I. 1995. Atlantic tomcod (*Microgadus tomcod*) and smooth flounder (*Pleuronectes putnami*) as indicators of organic pollution in the Miramichi Estuary, p. 211-227. *In* E.M.P. Chadwick [*editor*]. Water, science, and the public: the Miramichi ecosystem. Can. Spec. Publ. Fish. Aquat. Sci. 123.

Abstract

Field and laboratory experiments were carried out to calibrate hepatic cytochrome P4501A (CYP1A) mRNA levels in Atlantic tomcod (*Microgadus tomcod*) and smooth flounder (*Pleuronectes putnami*) as a monitor of some of the most toxic organic contaminants found in estuaries of the southern Gulf of St. Lawrence. Many of these experiments were carried out in, or with fish from, the Miramichi Estuary. Results of these calibration studies are summarized in this paper. Intraperitoneal injection of two polycyclic aromatic hydrocarbons (PAHs) (ß-naphthoflavone, benzo-[*a*]-pyrene), 2,3,7,8 tetrachlorodibenzo-*p*-dioxin, and 3,3',4,4', tetrachlorobiphenyl resulted in elevated levels of CYP1A mRNA in tomcod. Thresholds and kinetics of responses differed between test substances. Smooth flounder showed a much smaller response than tomcod to intraperitoneal injection of ß-naphthoflavone and to environmental exposures. Tomcod sampled from three different sites in Miramichi Estuary showed similar CYP1A mRNA levels, but tomcod caged near the Miramichi Pulp and Paper Mill showed higher levels than fish caged upriver or downriver. CYP1A mRNA levels were higher in tomcod from the Miramichi Estuary than in those from two unindustrialized estuaries, but lower than in tomcod from the heavily contaminated Hudson River Estuary (New York). A low prevalence of hepatic DNA adducts and PAH metabolites in bile indicated that tomcod in the Miramichi Estuary had not been exposed to high concentrations of PAHs.

Résumé

Nous avons réalisé des expériences sur le terrain et en laboratoire pour mesurer les concentrations d'ARNm codant le cytochrome P4501A hépatique (CYP1A) chez le poulamon (*Microgadus tomcod*) et la plie lisse (*Pleuronectes putnami*) pour surveiller certains des polluants organiques les plus toxiques qui se retrouvent dans les estuaires du sud du golfe du Saint-Laurent. Bon nombre de ces expériences ont été réalisées dans l'estuaire de la Miramichi ou sur des poissons qui en provenaient. Les résultats de ces travaux d'étalonnage sont résumés dans notre étude. Après injection intrapéritonéale de deux hydrocarbures aromatiques polycycliques, ou HAP (ß-naphtoflavone et benzo-[α]-pyrène), de 2,3,7,8 tétrachlorodibenzo-*p*-dioxine et de 3,3,4,4, tétrachlorobiphényle, on a observé des concentrations élevées d'ARNm CYP1A chez le poulamon. Les seuils et la cinétique de réaction différaient d'une substance à l'autre. Chez la plie lisse, la réponse était beaucoup plus faible que chez le poulamon à l'injection intrapéritonéale de ß-naphtoflavone. Les poulamons échantillonnés à trois sites différents de l'estuaire de la Miramichi présentaient des concentrations similaires d'ARNm CYP1A, mais ceux qui étaient en cage près de l'usine de pâte et papier de la Miramichi présentaient des teneurs plus élevées que ceux qui se trouvaient en amont ou en aval. Ces teneurs étaient aussi plus élevées chez les poulamons de l'estuaire de la Miramichi que chez ceux provenant de deux estuaires non industrialisés, mais elles étaient plus basses que chez des poulamons provenant de l'estuaire de l'Hudson (New York), qui est très pollué. La faible prévalence d'adduits d'ADN et de métabolites d'HAP dans la bile indique que les poulamons de l'estuaire de la Miramichi n'avaient pas été exposés à de fortes concentrations d'HAP.

Introduction

The Miramichi Estuary receives bleached kraft mill effluent (BKME) from the largest pulp and paper mill in New Brunswick, Miramichi Pulp and Paper Incorporated (MPPI), located upriver from Newcastle (Fig. 1). MPPI is the only mill in New Brunswick to have discharged detectable levels of dioxins and furans in recent years (NB Department of the Environment 1991: 66-68; Table 1). Over 95% of organic wastes released into Miramichi Estuary originate from MPPI (N. Brodie, N.B. Dept. Environment, Fredericton, N.B., pers. comm.). Smaller contributers of organic contaminants to the estuary include an MPPI groundwood mill at Nelson-Miramichi, and several municipalities totalling approximately 50 000 people. Organic contaminants of concern include: dioxins and furans, which result from pulp and paper production and from combustion; polychlorinated biphenyls (PCBs), which were widely used as electrical insulators; polycyclic aromatic hydrocarbons (PAHs), originating from several sources in the Miramichi Estuary including a former wood preservation plant at Newcastle (closed in 1986), a former oil-burning electric power generating station near Chatham, and spills and runoff of petroleum products (NB Department of the Environment 1991); organochlorine pesticides including DDT and its degradation products (DDD and DDE); and a multitude of toxic compounds including chlorophenols that are also found in BKME.

Dioxins, furans, PCBs, and PAHs are contaminants of concern because of their toxicity and persistence in the environment (Safe 1991). Some of the dioxins are extremely toxic to both wildlife and humans. 2,3,7,8 -tetrachlorodibenzo-para-dioxin (2,3,7,8 TCDD) for example is among the most toxic synthetic compounds ever tested under laboratory conditions (Eisner 1986). Exposures to low doses of dioxins have been associated with carcinogenic, teratogenic, histopathologic, immunotoxic, and reproductive effects (Eisner 1986). Several of the PAHs are among the most potent carcinogens and mutagens known to exist, and their prevalence is probably a major contributor to the recent increase in human cancer rates reported for industrialized nations (Eisner 1987). Benzo *a* pyrene, for example, has been related to the presence of tumours in Beluga whales from the Gulf of St. Lawrence (Martineau et al. 1988). PCBs and organochlorine pesticides are acutely toxic to fish in ppm doses and produce subtle toxic responses at lower doses (Janz et al. 1992). The potential impact of these contaminants in Miramichi Estuary on humans comes not only from direct exposure to water, but also from fish consumed from the estuary. Miramichi Estuary supports extensive commercial fisheries for gaspereau (*Alosa pseudoharengus* and *A. aestivalis*), American eels (*Anguilla rostrata*) and rainbow smelt (*Osmerus mordax*) as well as an important recreational fishery for Atlantic salmon (*Salmo salar*) (Chaput 1995). Effects on fish may be pronounced because a number of species utlilize the estuary for spawning, early life history, and overwintering (Locke and Courtenay 1995a, b; Bradford et al. 1995; Hanson and Courtenay 1995).

In 1990 research was begun to monitor the impact of organic pollution on fish populations living in the Miramichi and other estuaries of the southern Gulf of St. Lawrence. This research included direct analyses of organic contaminants such as dioxins, furans, pesticides, PCBs, and

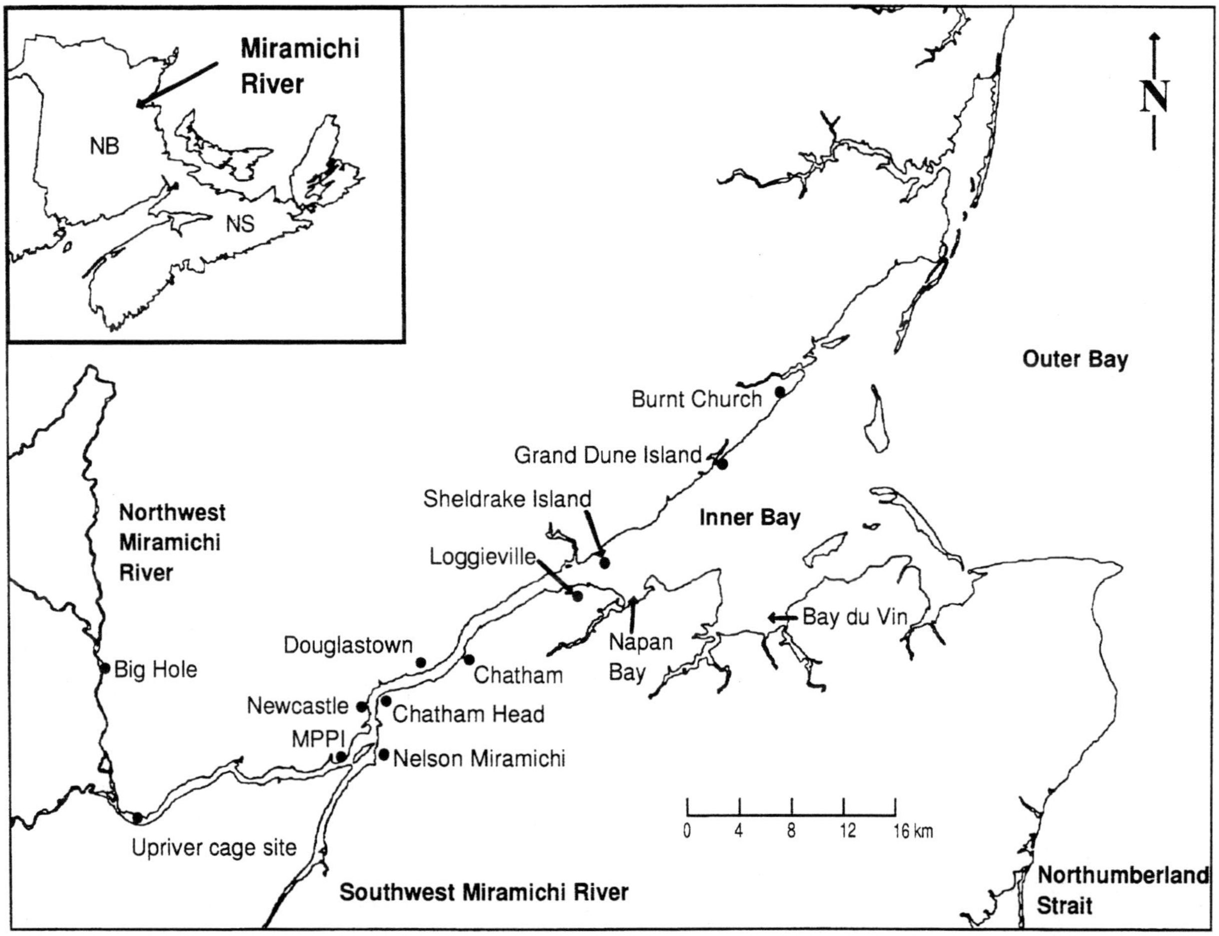

Fig. 1. A map of the Miramichi Estuary showing locales referred to in text.

PAHs, and inorganic contaminants such as heavy metals in fish and aquatic invertebrates. Also sampled was hepatic cytochrome P4501A (CYP1A) as a bioindicator of exposure to organic contaminants. Since 1993, fish have been sampled in spring and fall from three estuaries which receive effluents from pulp and paper mills as well as other industries and municipalities (Miramichi, Restigouche, Pictou Harbour) and two estuaries without industry or municipal effluents (Kouchibouguac, Margaree) (Fig. 2). A number of laboratory and field experiments were carried out to develop protocols for the monitoring program. Many of these experiments were conducted with Miramichi fish or in the Miramichi Estuary. This paper summarizes what has been learned to date about the bioindicator CYP1A and organic contamination in the Miramichi Estuary.

Background

In rivers, contaminants would typically be distributed in a decreasing gradient downstream of the source (Vannote et al. 1980). However, distribution of contaminants in an estuary can be more complex. In addition to bidirectional movement of water by tides, the interaction of fresh water and salt water causes changes in pH and ionic strength that promote flocculation, the formation of small masses of organic and inorganic material (MacKnight 1990). Dissolved organic compounds attach to these organic flocs and fall to the sediments where they can be trapped, moved upstream by flood tides, or flushed out of the estuary during

Table 1. Concentrations of 2,3,7,8 tetrachlorodibenzo-*p*-dioxin (TCDD) and 2,3,7,8 tetrachlorodibenzo-p-furan (TCDF) in final effluent of Miramichi Pulp and Paper Mill, Newcastle, N.B. (courtesy of J. F. MacDougall and O. Kristiansen, Miramichi Pulp and Paper Incorporated, Newcastle, N.B., and J. Desmarais, Environment Canada, Fredericton, N.B.). ppq: parts per quadrillion (10^{-15}), 24 h composite sample. Analyses performed by Seakem Labs and AXYS Analytical Services, Sydney B.C.

DATE	TCDD (ppq)	TCDF (ppq)
Dec. 1988	100	950
13 Feb.1989	93	970
Spring, 1989*	210	1300
24 Sept. 1990	27	
July 1991	<15	79
9-13 June 1992	37	110
20 Oct. 1992	38	150
10/11 Jan. 1993	33	68
27/28 April 1993	30	58
12/13 July 1993	21	43
2/3 Nov. 1993	6	19
18/19 Nov. 1993	8	20
23/24 Jan. 1994	15	29
31 Jan. 1994	11	32
22/23 Feb. 1994	13	36
13/14 Mar. 1994	14	33
26/27 April 1994	10	34
8/9 May 1994	5	12
Federal Limit	15	50
Method Detection Limit	2	2

* NB Department of the Environment 1991: 68.

re-suspension of sediments by strong currents or other disturbances.

In the Miramichi Estuary, the zone of maximal flocculation (i.e., maximum turbidity zone) occurs between Loggieville and Sheldrake Island (Fig. 1) during the spring freshet, and moves upstream of Newcastle during the low flows in summer. The majority of suspended solids enter Miramichi Inner Bay during the spring freshet (Buckley and Winters 1983). Therefore, contaminants originating upriver are dispersed in sediments between Newcastle and Sheldrake Island rather than occurring in a decreasing gradient down river. This section of the Estuary has been referred to as a "contaminant sink" (MacKnight 1990). Concentrations of organic carbon, most of which originate from MPPI, are highest in the deeper parts of the Estuary and around the Loggieville to Grand Dune Island areas (Philpott 1978) (Fig. 1). Organic contaminants are believed to follow this distribution (Philpott 1978).

Measurements of PAHs and PCBs in the sediments of the Miramichi Estuary were reported by MREAC (1992), and measurements of of dioxins and furans taken from sediments in 1988 near MPPI were reported by Trudel (1991). PCB concentrations were low, varying from 5 to 400 ppb with some high levels found between Douglastown and Chatham. PAH concentrations were moderate, from 141 to 5 190 ppb with the highest levels just below Chatham. Concentrations of over 1 000 ppb occurred in patches from Newcastle to Sheldrake Island. One sample taken upriver of Newcastle showed a PAH level of 61 489 ppb which is above the highest level recorded for any North American estuary: Boston Harbour at 57 778 ppb (National Status and Trends 1988). The most toxic of the dioxins, TCDD, was not detected in sediments

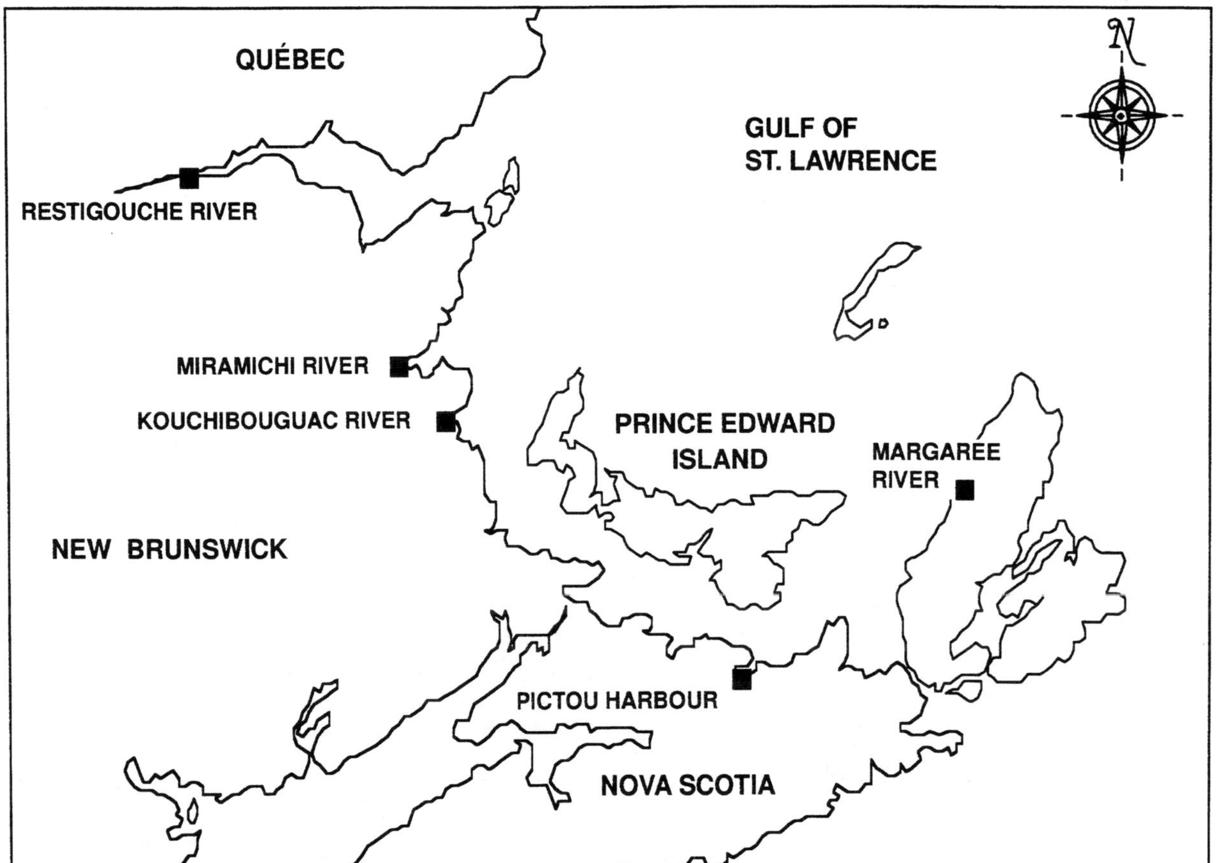

Fig. 2. Estuaries in the southern Gulf of St. Lawrence from which tomcod and smooth flounder were sampled seasonally. The Restigouche, Miramichi, and East River (Pictou) estuaries receive effluents of pulp and paper mills, other industries, and municipalities, whereas the Kouchibouguac and Margaree estuaries do not.

(i.e., <3-10 pptr dry weight). However, hepta- and octachlorodibenzo-dioxins were detected at concentrations of 250 and 960 pptr respectively at a site 1.5 km upstream of the mill, and 71 and 490 pptr respectively 0.5 km downstream of the mill. Furans, including 2,3,7,8 tetrachlorodibenzofuran, were detected at levels between 4 and 150 pptr. Dioxin and furan levels were low relative to those detected in sediments near certain mills in B.C., Ontario, and Quebec (Trudel 1991). More recent sampling around and downstream from the MPPI effluent discharge site failed to detect unusually high concentrations of PAHs, PCBs, or extractable organic halogens (EOX), which include dioxins (R. Parker, Environment Canada, Halifax, pers. comm.). Grab samples of sediments were taken by Environment Canada in June 1993 from eight sites between Chatham and the railway bridge upstream of MPPI. Maximum levels were 1 890 ppb for PAHs and 170 ppb for PCBs, and non-detectable for EOX (R. Parker, Environment Canada, pers. comm.).

MREAC (1992) also presented historic data collected on PCB, PAH, and dioxin levels in the biota of Miramichi Estuary. TCDD was detected at levels of 3-16 pptr (wet-weight) in tissues from finfish including Atlantic tomcod (*Microgadus tomcod*) and 16-22 pptr in the tomalley (i.e., hepatopancreas) of lobsters (*Homarus americanus*). Samples of lobsters collected from four estuaries in Maine that receive BKME had similar levels. Based on those data, the State of Maine issued an advisory to pregnant women and children to avoid eating lobster tomalley (Bangor Daily News, February 3, 1994). TCDD concentrated by the biota of Miramichi Inner Bay almost certainly originated from MPPI (MREAC 1992). Various PCB congeners were detected in finfish tissues at low levels (e.g., 3-31 ppb in the liver of a tomcod from Napan Bay). For comparison, levels of 20-16 100 ppb were detected in Hudson River tomcod (Wirgin et al. 1992). Concentrations of PAHs were generally low in invertebrates examined, and were not examined in finfish which metabolize PAHs rapidly (Varanasi and Stein 1991).

Methods and Materials

The Measure

The levels of cytochrome P4501A gene products in fish can serve as effective indicators of exposure and metabolic response to certain organic contaminants (Addison 1984; Payne et al. 1987; Stegeman and Kloepper-Sams 1987; Haasch et al. 1989; Hodson et al. 1991). Cytochrome P450 enzymes function in the metabolism and excretion of endogenous substances such as steroid hormones, as well as exogenous organic pollutants including dioxins, coplanar PCBs, and PAHs (Andersson and Forlin 1992; Goksoyr and Forlin 1992; Haasch et al. 1992; Guengerich 1993). CYP1A production is inducible by exposure to these contaminants and can be readily measured. Induction of the CYP1A system has been measured at three points; at the pretranslational level by measurement of levels of CYP1A mRNA, at the translational level by measurement of the CYP1A protein, and at the enzymatic level by measurement of the catalytic activity of the enzyme on various substrates. Historically, the measurement of the catalytic activity of the enzyme has been the most commonly used indicator of induction. This method is relatively inexpensive to perform, and has been shown to be very sensitive, with induction occurring at near undetectable levels of inducing contaminant (Haasch et al. 1993). There are, however, some drawbacks to measuring induction at this level. Some contaminants can competitively inhibit enzyme activity, particularly at high contaminant levels (Goksoyr and Forlin 1992; Haasch et al. 1993). Successful measurement of enzyme activity also requires stringent sample collection protocol and rigid sample storage conditions (Goksoyr and Forlin 1992). Measurement at the intermediate point in the induction process, translation of the CYP1A protein, is not susceptable to competitive inhibition, and also requires less stringent sample storage than does measurement of enzyme activity. However, some heavy metals,which may be present with organic containants, have been shown to reduce immunodetectable CYP1A protein levels (Goksoyr and Forlin 1992). One laboratory trial found that induction, as indicated by both catalytic enzyme activity and detection of the protein, was inversely related to the concentration of the inducing agent (Haasch et al. 1992). If the detection of induction is to be used as an early warning indicator of potential contaminant effects, there is a strong case to be made for measuring induction at the earliest step, by measuring CYP1A mRNA (Haasch et al. 1993). Detection of induction at this transcriptional level is relatively inexpensive, has been shown to be sensitive with respect to low levels of contaminant (Haasch et al. 1993), and is not subject to competitive inhibition. Upon exposure to an inducing agent, the levels of CYP1A mRNA have been shown to increase before increases in the protein itself are detectable (Andersson and Forlin 1992). Because it cannot be assumed that translation of CYP1A mRNA to the protein and subsequent catalytically active enzyme will occur, measurement of CYP1A mRNA is the most reliable indicator of exposure to an inducing contaminant.

There are several reasons why monitoring cytochrome P4501A mRNA is a useful complement to chemical analyses of organic contaminants in fish. First, CYP1A monitoring can be a cost-effective way of detecting the presence of certain organic contaminants such as TCDD, which are too costly to measure directly with any frequency. CYP1A is a relatively inexpensive indicator of whether more detailed chemical analyses are warranted. Second, CYP1A levels indicate ongoing or recent exposure to contaminants whereas chemical analyses may not. Because certain contaminants are stored for long periods in biota, elevated levels may reflect environmental insult long past. Third, CYP1A can be induced by toxic chemicals that have not yet been identified (as in the case of BKME effluents) and thus are not yet detected by routine chemical analyses. Fourth, CYP1A measurement indicates not only that the animal has been exposed to a contaminant, but that the contaminant has interacted with and altered the animal's physiology. Fifth, because CYP1A is induced by low levels of contaminant, elevated levels of CYP1A mRNA will be seen long before more serious effects appear in the fish or their consumers. At present, CYP1A induction is used as an indicator of exposure to certain organic contaminants but it has also been suggested that the affected detoxification system may elicit secondary effects on growth, liver structure, diseases including cancers (Lehtinen 1990; Janz et al. 1992) and reproduction (Addison 1995).

The Fish

Atlantic tomcod and smooth flounder (*Pleuronectes putnami*) were selected as potential sentinel species because: they are abundant in most Gulf of St. Lawrence estuaries and easily and inexpensively obtained as bycatch from commercial eel, gaspereau, and smelt fisheries; they spend most of their lives within the same estuary (e.g., Dew 1991); they migrate within the estuary and therefore likely present an integrated measure of the estuary's contaminants; they are bottom-dwelling and consume organisms likely to accumulate benthic contaminants like amphipods, sandshrimp, and bivalves (Grabe 1980; McLaren et al. 1988); they have large easily sampled livers; and they are relatively small, hardy, and survive well in captivity - an important consideration for calibration experiments such as those described below. In addition, these species are found on the east coast of North America from southern Labrador to North Carolina (tomcod), and Ungava Bay to Rhode Island (smooth flounder) (Scott and Scott 1988) and although neither species has been used by other investigators in studies of CYP1A response to environmental pollution, closely related species have been used. CYP1A data have been collected from burbot (*Lota lota*), a relative of tomcod, (D. Muir, DFO-Freshwater Institute, Winnipeg Manitoba, pers. comm.), and from the winter flounder (*Pleuronectes americanus*) (Stegeman et al. 1987; Monosson and Stegeman 1994) and plaice (*Pleuronectes platessa*) (Boon et al. 1992), both relatives of smooth flounder. We selected smooth flounder because it remains in Miramichi Estuary year-round (Hanson and Courtenay 1995). Smooth flounder spawn in December/January and young-of-the-year are found in large numbers in the area of MPPI during the summer (Hanson and Courtenay, unpubl. data). Larger fish are available at all times of year, and do not appear to leave the estuary in summer as do tomcod and winter flounder (Hanson and Courtenay 1995)

Tomcod spawn in fresh water near Big Hole (Fig. 1) between December and February (R. Cunjak, Gulf Fisheries Centre, Moncton, NB, pers. comm.). Young-of-the-year remain in the Estuary all year (Locke and Courtenay 1995a, b; Courtenay and Hanson, unpubl. data). Larger fish are absent from the estuary during July and August, otherwise, they can be sampled both upstream and downstream from the mill and can be sampled in all seasons.

Our experiments can be divided into four parts: time and dose experiments on the test species; experiments that examined variation within the estuary; experiments that tested the role of MPPI; and finally, a survey which compared Miramichi with other estuaries along the North American east coast.

1. Time and Dose Experiments

Our first step was to demonstrate that CYP1A gene induction in tomcod and flounder responded to contaminants of interest. Dose-response experiments were used to determine the response-threshold and to show whether the magnitude of response was dose-dependent. Time experiments examined the dynamic of response to a particular dose of contaminant (i.e., time to first response, maximal response, and end of response).

Tomcod were exposed by intraperitoneal injection to two PAHs (ß-naphthoflavone - BNF, and benzo-*a*-pyrene - B[*a*]P), a dioxin (TCDD), and a PCB (3,3',4,4', tetrachlorobiphenyl - TCB). Smooth flounder were injected with BNF only. Controls were uninjected or injected with corn oil. Fish had been obtained from the Miramichi or Hudson Estuary at least 3 weeks before exposure to allow CYP1A mRNA to return to baseline levels (Kreamer et al. 1991; Courtenay et al. 1993). For details of experimental exposure and mRNA analysis, see Wirgin et al. (1992) and Courtenay et al. (1993).

2. Variation within Miramichi Estuary

To compare different sites within Miramichi Estuary, surveys of CYP1A mRNA levels in tomcod and smooth flounder in Miramichi Estuary were made during the autumns of 1991 and 1993, and spring of 1992. Fish were obtained from three sites: Newcastle, Chatham, and Loggieville which are 5, 12, and 26 km downstream of MPPI respectively (Fig. 1). This study was prompted by a similar survey carried out by Addison et al. (1991) who recorded low and similar levels of CYP1A enzyme activity (EROD and CN-ECOD) in flounders caught at Newcastle, Chatham, Loggieville, and a fourth site further downstream (Burnt Church).

3. Role of MPPI

Effluents from bleached kraft mills contain CYP1A inducers for some species of fish (e.g., Andersson et al. 1988; Munkittrick et al. 1991, 1992a, 1992b; Haasch et al. 1992; Gagne and Blaise 1993; Ahokas et al. 1994; Martel et al. 1994) and we were interested to determine whether MPPI effluent might be responsible for the moderately high CYP1A mRNA levels observed in Miramichi tomcod. In one experiment we compared CYP1A mRNA levels in tomcod caged for 9-10 d at the mill site to levels in tomcod held in clean water in the laboratory and tomcod caged downriver at Newcastle and at Chatham (Courtenay et al. 1993). A second experiment examined the rate of CYP1A mRNA clearance as a first step towards identifying the inducer. Tomcod were caged for 10-14d at the mill site and then transferred to clean water and sacrificed over a 10 d period. For details of experimental treatment and CYP1A mRNA analysis, see Courtenay et al. (1993).

4. Comparison of Miramichi to other east coast estuaries.

Tomcod from the Miramichi, St. Lawrence (Quebec), Margaree (Nova Scotia), Saco and Royal (Maine), and Hudson (New York) Estuaries were sampled between 1987 and 1993 (Fig. 3). The Hudson Estuary is one of the most contaminated estuaries in the USA, the St. Lawrence and Miramichi Estuaries are moderately polluted, and the Saco, Royal and Margaree Estuaries are relatively pristine (Wirgin et al. 1994). This study also addressed the question of the identity of the CYP1A inducer in the Hudson and Miramichi Estuaries. Two indicators of exposure to PAHs, fluorescent aromatic hydrocarbons in bile and hepatic DNA adducts, were measured in tomcod from all estuaries except the Saco and Royal. Details of sampling and analyses are given in Wirgin et al. 1994.

Results

1. Time and Dose Experiments

Some of these results have been published (Courtenay et al. 1993) and others are being prepared for publication elsewhere. Examples of time and dose responses are shown in Fig. 4 and 5. The conclusions of experiments with tomcod are as follows:

1. All three classes of contaminants caused significant CYP1A induction
2. Thresholds of response varied for the different

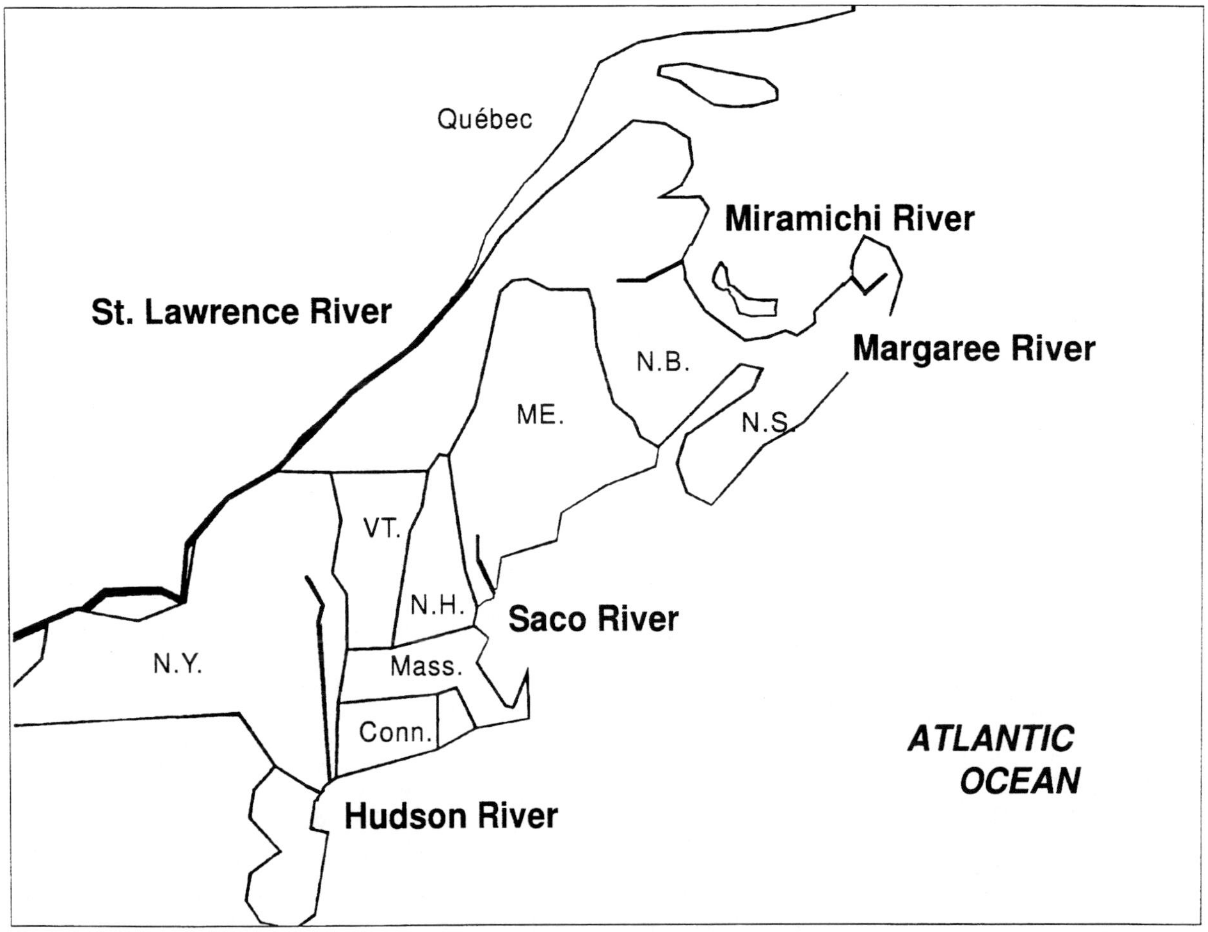

Fig. 3. Estuaries from which tomcod were sampled for levels of CYP1A mRNA, hepatic DNA adducts, and PAH metabolites in bile. The figure is adapted from Wirgin et al. (1994).

contaminants between 100 pptr to 10 ppm. Sensitivity was greatest to TCDD, followed by B[a]P, TCB, and BNF. Because of small sample sizes and large individual variability, thresholds of response could be considerably lower than indicated by these data.
3. Magnitudes of response were dose-dependent.
4. Responses to the two PAHs were seen within a day and were maintained for at least a week. Responses to TCDD and TCB were first seen 5-7 d after exposure and remained elevated at the end of the experiments 25 and 70 d later, respectively.

The fact that exposure to a dioxin, a coplanar PCB and PAHs resulted in elevated CYP1A mRNA levels in tomcod lends support to the use of this system as an indicator. These results are preliminary in the sense that thresholds for response are probably affected by gender and reproductive state (Courtenay et al. 1994) and other variables such as age that were not stratified in these experiments. In addition, the dynamics of responses may differ at higher or lower dosages than the single dose used in our experiments (Beebe et al. 1990). Of greater concern was the indication that responses were stock-specific. Intraperitoneal injection of TCB elevated CYP1A mRNA in tomcod from the Miramichi Estuary but not the Hudson Estuary (Wirgin et al. 1992).

Few experiments have been conducted with smooth flounder, but early indications are that the CYP1A system of this species may be less responsive than that of tomcod. Intraperitoneal injection of 10 ppm BNF resulted in induction only 14% that observed in tomcod.

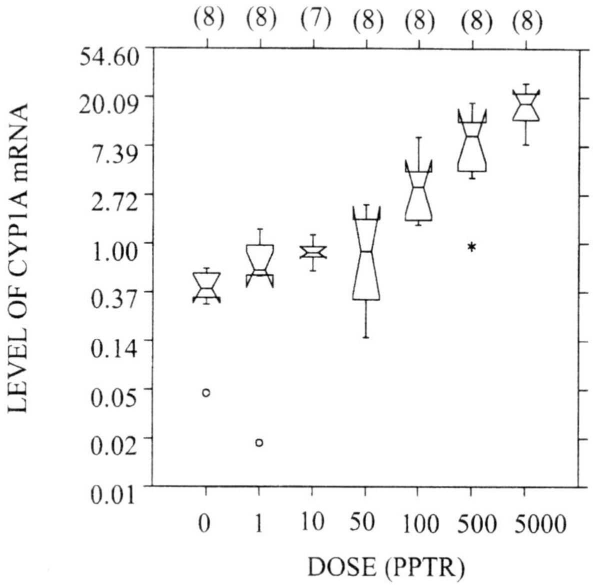

Fig. 4. Levels of CYP1A mRNA in tomcod treated with a single intraperitoneal injection 10 days earlier of TCDD of dose 1 to 5000 parts per trillion (pptr). Levels of CYP1A mRNA are shown as optical density (OD) units. The numbers in parentheses above the figure indicate the number of samples analyzed. Boxes represent the median (central horizontal line) bordered below and above by the 25 and 75% quartiles. Boxes are notched at the median and return to full width at the approximate limits of the 95% confidence interval around the median. Whiskers show the range of values falling within 1.5 interquartile-ranges of the quartiles. Asterisks represent outside values (>1.5 interquartile-ranges from the quartiles), and open cicle represent far outside values (>3 interquartile-ranges from the quartiles) (Systat for Windows: Graphics, Version 5 Edition. Evanston, IL: SYSTAT, Inc., 1992. 636 p.).

2. *Variation within Miramichi Estuary*

We did not find a gradient in tomcod CYP1A

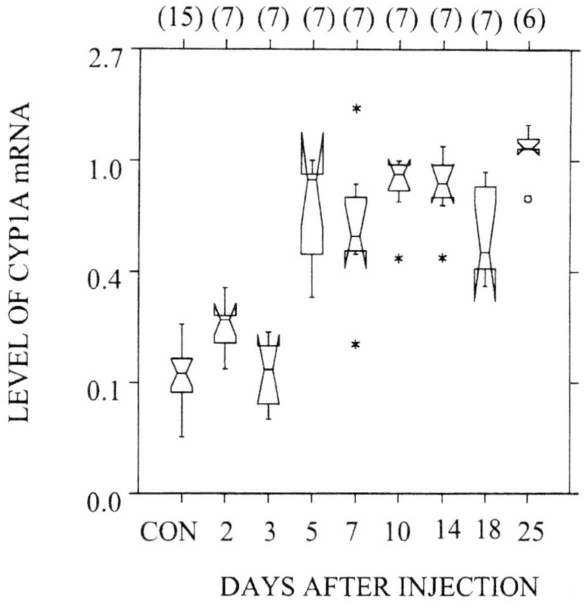

Fig. 5. Levels of CYP1A mRNA in tomcod treated with a single intraperitoneal injection of 500 pptr TCDD and sacrificed 2-25 d later. CON represents a sham-injected control group. Figure is adapted from Courtenay et al. (1993). The rest of the legend is as in Fig. 4.

mRNA levels with increasing distance downstream of MPPI (Fig. 6). CYP1A mRNA levels were an order of magnitude higher in spring than fall, and males showed higher levels of CYP1A mRNA than females in fall. There is evidence in other fishes that sex, reproductive state, and age can affect both basal CYP1A activity and the extent to which CYP1A can be induced (Van Der Kraak et al. 1992; Monosson and Stegemann 1994; Stegeman and Hahn 1994; Addison 1995).

Smooth flounder showed CYP1A mRNA levels much lower than those of tomcod at all three sites sampled, and again, there was no obvious gradient among sites. These data were consistent with those reported by Addison et al. (1991). The difference in CYP1A levels between tomcod and smooth flounder sampled from the same environment is unlikely to be related to reproductive state, as both species spawn between December and February. Diet could be a factor as the two species differ in diet (Hanson and Courtenay, unpubl. data) and it has been shown that hepatic CYP1A response can be induced by contaminants in the diet (e.g., Boon et al. 1992; Parrot 1993) as well as by water-borne contaminants (e.g., Martel et al. 1994). But even if distributions and diets were identical for the two species, different species may show different inducibility of the CYP1A gene. For example, we have observed that in the contaminated Hudson Estuary CYP1A mRNA levels were high in tomcod but low in hogchokers (*Trinectes maculatus*) and striped bass (*Morone saxatilis*)(Wirgin et al. 1995).

3. Influence of MPPI

Cageing tomcod for 9 d in the effluent of MPPI resulted in 11 fold CYP1A induction over laboratory controls (Fig. 7) (Courtenay et al. 1993). Tomcod caged downriver at Newcastle and Chatham showed lower levels of induction. In a second experiment, CYP1A mRNA levels in tomcod caged at the mill were significantly greater than those in fish caged upriver, beyond the influence of mill effluent (Fig. 1). The high levels of CYP1A mRNA in effluent-exposed fish returned to baseline within 10d in fish moved to clean water in the lab. This experiment should be repeated with larger samples sizes, but the implication was that whatever the inducing agents were in MPPI effluent, they were cleared relatively quickly. The transient nature of induction caused by BKME has

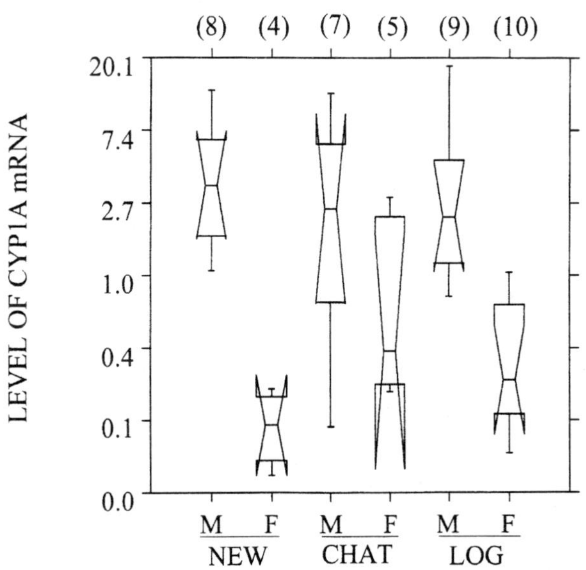

Fig. 6. Levels of CYP1A mRNA in tomcod sampled from three sites in the Miramichi Estuary in October 1993. NEW: Newcastle; CHAT: Chatham; LOG: Loggieville; M: male; F: female. The rest of the legend is as in Fig. 4.

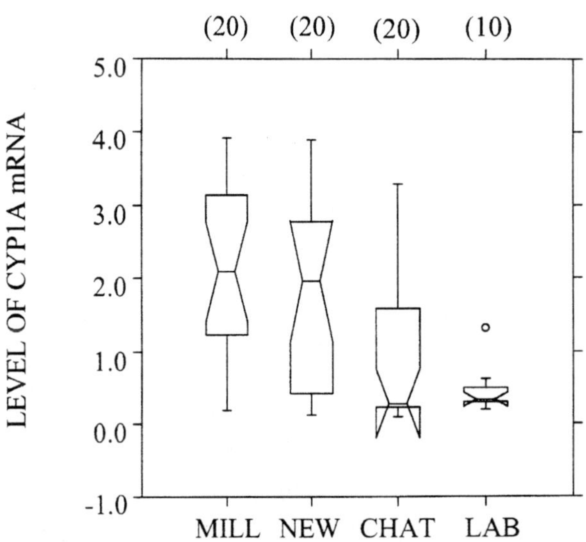

Fig. 7. Levels of CYP1A mRNA in tomcod caged in the Miramichi Estuary at: Miramichi Pulp and Paper Mill (MILL), Newcastle (NEW), Chatham (CHAT), or not exposed to the river (LAB). The figure is adapted from Courtenay et al. (1993). The rest of the legend is as in Fig. 4.

been reported by other workers as well (Munkittrick et al. 1992a). The dynamic of CYP1A mRNA clearance might seem to rule out dioxins such as TCDD, which in our intraperitoneal injection experiments resulted in elevated CYP1A mRNA levels for at least 25d. However, this conclusion is premature. Clearance times have been shown in other species to be dose-dependent (Beebe et al. 1990) and our injection experiment utilized a single, fairly high dose (0.5 ppb) of TCDD.

4. Comparison of Miramichi to other East Coast estuaries

Miramichi tomcod showed significantly higher CYP1A mRNA levels than tomcod from the St. Lawrence, Saco-Royal, and Margaree estuaries, but lower levels than tomcod from the highly contaminated Hudson Estuary (Fig. 8). This pattern reflects the relative degree of chemical contamination in these estuaries with the exception of the St. Lawrence. This latter result may be explained by the fact that St. Lawrence fish were sampled in two relatively pristine areas, unrepresentative of the St. Lawrence as a whole, and at the time of spawning when inducibility of the CYP1A gene is reduced (Courtenay et al. 1994).

Low levels of DNA adducts and PAH bile-metabolites were found in Miramichi tomcod. These indicators of exposure to PAHs were comparable to levels found in Margaree tomcod and significantly lower than levels recorded in tomcod from the Hudson and St. Lawrence Estuaries. CYP1A induction in Miramichi tomcod would therefore appear to be a response to something other than PAHs, and in this regard the Miramichi differs from the Hudson Estuary. These results are consistent with reported levels of PAHs in sediments of these Estuaries. Some of the highest levels in USA estuaries are found in the Hudson (7 100 - 34 000 ppb dry weight; Wirgin et al. 1994) compared to moderate levels in the Miramichi Estuary (<5 200 ppb except at the MPPI site; see above) and St. Lawrence Estuary (200 - 3 900 ppb; Pham et al. 1993).

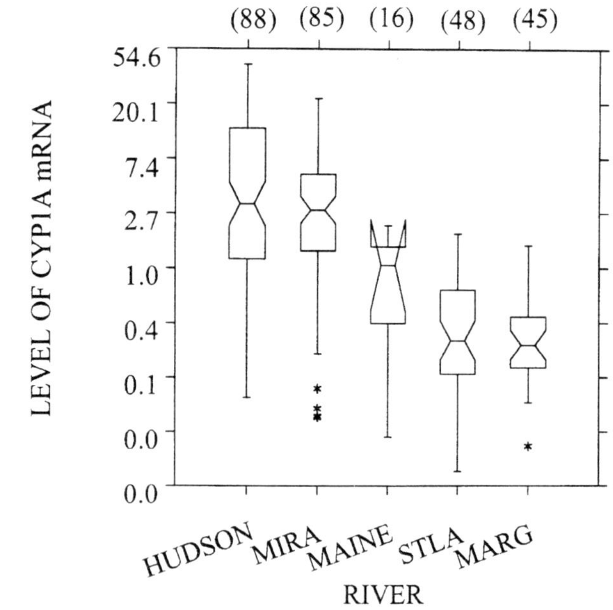

Fig. 8. Levels of CYP1A mRNA in tomcod collected from five estuaries along the Atlantic coast of North America: the Hudson, Miramichi (MIRA), Saco and Royal (MAINE), St. Lawrence (STLA), and Margaree (MARG). The figure is adapted from Wirgin et al. (1994). The rest of the legend is as in Fig. 4.

Discussion

Time and dose response experiments confirmed that CYP1A induction in Atlantic tomcod is a sensitive indicator of exposure to certain PAHs, PCBs, and dioxins, and CYP1A mRNA levels in tomcod sampled from different rivers generally correlate with known levels of organic contaminants in those rivers. By contrast, the CYP1A system of smooth flounder appears much less sensitive, suggesting that in this species, CYP1A levels may not be a useful bioindicator. Each pulp and paper mill in Canada is now required to monitor the impact of its effluent on two species of endemic fish. In the case of MPPI, both tomcod and smooth flounder are being considered as sentinel species. If CYP1A measurements are to be included, a species other than smooth flounder might provide a better indicator. The mummichog (*Fundulus heteroclitus*) would be a good candidate.

A number of researchers have confirmed CYP1A induction in mummichogs (e.g., Elskus and Stegeman 1989; Kloepper-Sams and Stegeman 1992; Van Veld et al. 1992; C. Couillard, Fisheries and Oceans Canada, Institut Maurice Lamontagne, Mont Joli, PQ, pers. comm.). Mummichogs tend to be sedentary and useful for comparisons of different sites within the same estuary.

The absence of a gradient in CYP1A mRNA levels in tomcod sampled between Newcastle and Loggieville may reflect the distribution of contaminants, or the movement of tomcod, along this 20 km stretch of Miramichi Estuary. Two pieces of information suggest that CYP1A inducers do follow a declining gradient downriver of MPPI. First, a gradient in CYP1A mRNA was observed in tomcod caged at MPPI, Newcastle, and Chatham. Second, an ongoing investigation of CYP1A enzymatic activity in mummichogs (*Fundulus heteroclitus*) in Miramichi Estuary by C. Couillard (Fisheries and Oceans Canada, Institut Maurice Lamontagne, Mont Joli, PQ, pers. comm.) and J. Bureau (Environment Canada, St. Lawrence Centre, Montreal, PQ) has found higher EROD activity at Strawberry Marsh (between MPPI and Newcastle) than at sites downriver near Loggieville and Bay du Vin. The finding of a gradient in CYP1A activity in mummichogs but not tomcod downriver of MPPI may be explained by the fact that mummichogs were sampled nearer MPPI than were tomcod and onshore versus in the channel, but more likely, by the fact that mummichogs probably migrate less extensively along the estuary than tomcod.

Experiments with caged tomcod indicated that MPPI was probably a source of CYP1A inducer(s), but this needs to be confirmed with exposures to MPPI effluent in the laboratory. Further experiments are planned, to determine whether recent changes in MPPI effluent (e.g., Table 1) have reduced CYP1A inducers in the upper estuary. Laboratory experiments are also being conducted to determine whether Miramichi tomcod are being exposed to CYP1A inducers through their diet, the water, and/or contaminated substrates.

The identity of the inducer in the Miramichi Estuary remains unknown, but the lack of DNA adducts and PAH metabolites in the bile of tomcod suggests that it is not a PAH. Nor is the inducer likely to be a PCB because high concentrations of PCBs are not found in Miramichi Estuary. Dioxins cannot be ruled out at this stage, but the CYP1A inducer could also be some other class of compound. Considerable effort is being expended by other researchers in the identification of CYP1A inducers in effluents of Canadian pulp and paper mills (K. Munkittrick, DFO Burlington, pers. comm.). It is already clear that effluents from mills that do not bleach with chlorine, and therefore do not emit dioxins, still contain CYP1A inducers for fish (e.g., Martel et al. 1994).

While CYP1A induction is a useful indicator of exposure to certain organic contaminants, the more important question is whether elevated levels of CYP1A mRNA correlate with detrimental effects on survival, growth, and reproduction of fish, or with body burdens of contaminants which have such effects on fish or their consumers. A careful examination of the health of the Miramichi population of tomcod has not been conducted, but to date, the only disease reported has been an atypical strain of furunculosis caused by the bacterium Aeromonas salmonicida (Olivier 1992). Population characteristics such as the truncated age distribution and high incidence of hepatic tumours noted in the Hudson River tomcod population (McLaren et al. 1988; Dey et al. 1993) have not been observed in the Miramichi population. The health of tomcod, and other estuarine residents, may well be compromised by exposure to contaminants in Miramichi Estuary. The area into which MPPI discharges its effluent is an important nursery for several species of fish including tomcod (Locke and Courtenay 1994; 1995). Effects of BKME have been observed on the growth and survival of juveniles of other fish species (e.g., Shenker and Cherr 1990; Vuorinen and Vuorinen 1991). Juveniles and older Miramichi tomcod may well show some of the histological, vertebral, blood composition, metabolic or growth effects of BKME exposure documented for other species (e.g., Hardig et al. 1988; Santos et al. 1990; McMaster et al. 1991).

A first step in addressing the question of whether higher level biological effects are associated with CYP1A induction in tomcod and smooth flounder of the southern Gulf of St. Lawrence will be made through our ongoing biomonitoring program. In addition to CYP1A levels, the biomonitoring program collects data on body condition (weight at length), reproductive preparedness (gonadosomatic index), and hepatosomatic index. The last measure may be an indicator of exposure to contaminants and activity in detoxifying them (Addison 1984; Munkittrick et al. 1991). A subsample of fish from each river are also sampled for parasites of the blood, gills, and guts. Pulp and paper mill effluents have been shown to influence the prevalence, intensity, and pathogenicity of fish parasites (Lehtinen et al. 1984; Thulin et al. 1988; Axelsson and Norrgren 1991; Khan et al. 1992; Barker et al. 1994). In addition, fish from the Miramichi and Kouchibouguac Rivers are sampled for size and number of eggs, and sex steroids in blood (11 keto testosterone, 17 ß-estradiol). Egg number and size provide information on reproductive potential and quality of eggs. Changes in steroid hormone levels have been noted in response to pulp and paper mill effluents (McMaster et al. 1991) and have been discussed as a long-term indicator of environmental stress (Donaldson 1990). Data on steroid levels are also useful in interpreting CYP1A levels in the same fish, as high titres of 17 ß-estradiol have been associated with depressed CYP1A levels and induceability (e.g., Snowberger-Gray et al. 1991), and CYP1A induction may also affect metabolism of steroid hormones (reviewed by Addison 1995). Data on body burdens of organic contaminants in tomcod from the Miramichi and other Gulf estuaries are being collected by other researchers, and will be examined for correlation with CYP1A mRNA levels.

The biomonitoring program will provide a temporal record of the presence of some of the most toxic organic contaminants in Gulf estuaries. With increasing environmental awareness, significant changes will be made to the way in which wastes are discharged to Miramichi and other estuaries. The largest contributer of organic matter to the Miramichi estuary - MPPI - has recently reduced its output of dioxins to a fraction of their concentration 5 years ago (Table 1). By 1996, the groundwood mill at Nelson-Miramichi will be piping its toxic effluent across the river to the MPPI effluent lagoon for treatment and release with the kraft mill effluent. And a third source of organic contamination will also be modifying its effluent. There are plans to build a central sewage treatment facility at Chatham Head to provide secondary treatment for the effluents of all of the municipalities on both sides of the Estuary between Douglastown and Newcastle/Chatham Head (i.e., the new Miramichi City area) (H. Collins, MREAC, Chatham, N.B., pers. comm.). Completion of this project is scheduled for 1997. These changes, while well-intentioned, may or may not reduce the impact of toxic contaminants on the environment. CYP1A monitoring will be an effective tool in making this determination.

Acknowledgements

We thank Sophie Bastien-Daigle and Albert Vromans for preparation of figures, Jasmin Desmarais (Environment Canada, Fredericton, NB), John MacDougall and Ole Kristiansen (Miramichi Pulp and Paper Incorporated, Newcastle, NB) for providing data on MPPI effluent, and Mark Hanson, Catherine Couillard, Richard Addison and Mike Chadwick for helpful comments on earlier drafts of the manuscript. This work was supported by the Toxic Chemicals Program of Canada's Green Plan for the Environment (S. Courtenay) and U.S. National Institute of Environmental Health Sciences grants, numbers ES05541 and ES05003 (I. Wirgin).

References

Addison, R.F. 1984. Hepatic mixed function odixase (MFO) induction in fish as a possible biological monitoring system. *In* V.W. Cairns, P.V. Hodson and J.O. Nriagu [*editors*]. Contaminant effects on fisheries. Adv. Envir. Sci. Technol. 16: 51-60.

Addison, R.F. 1995. Relating biochemical responses to "higher order effects". *In* Molecular aspects of oxidative drug metabolizing enzymes: their significance in environmental toxicology, chemical carcinogenesis and health. NATO Advanced Study Institute. (in press)

Addison, R.F., Hansen, P.-D., and Wright, E.C. 1991. Hepatic mono-oxygenase activities in American plaice

(*Hippoglossoides platessoides*) from the Miramichi Estuary, N.B. Can. Tech. Rept. Fish. Aquat. Sci. 1800: 18 p.

Ahokas, J.T., Holdway, D.A., Brennan, S.E., Goudey, R.W., and Bibrowska, H.B. 1994. MFO activity in carp (*Cyprinus carpio*) exposed to treated pulp and paper mill effluent in Lake Coleman, Victoria, Australia, in relation to AOX, EOX, and muscle PCDD/PCDF. Environ. Toxicol. Chem. 13: 41-50.

Andersson, T., and Forlin, L. 1992. Regulation of the cytochrome P450 enzyme system in fish. Aquat. Toxicol. 24:1-20.

Andersson, T., Bengtsson, B.-E., Forlin, L., Hardig, J., and Larsson, A. 1988. Physiological disturbances in fish living in coastal water polluted with kraft pulp mill effluents. Can. J. Fish. Aquat. Sci. 45: 1525-1536.

Axelsson, B., and Norrgren, L. 1991. Parasite frequency and liver anomalies in three-spined stickleback, *Gasterosteus aculeatus* (L.), after long-term exposure to pulp mill effluents in marine mesocosms. Environ. Contam. Toxicol. 1991: 505-513.

Barker, D.E., Kahn, R.A., Lee, E.M., Hooper, R.G., and Ryan, K. 1994. Anomalies in sculpins (*Myoxcephalus spp.*) sampled near a pulp and paper mill. Arch. Environ. Contam. and Toxicol. 26: 491-496.

Beebe, L., Parke, S.S., and Anderson, L.M. 1990. Differential enzyme induction of mouse liver and lung following a single low or high dose of 2,3,7,8 tetrachlorodibenzo-*p*-dioxin (TCDD). J. Biochem. Toxicol. 5: 211-219.

Boon, J.P., Everaarts, J.M., Hillebrand, M.T.J., Eggens, M.L., Pijnenburg, J., and Goksoyr, A. 1992. Changes in levels of hepatic biotransformation enzymes and haemoglobin levels in female plaice (*Pleuronectes platessa*) after oral administration of a technical polychlorinated biphenyl mixture (Clophen A40). The Science of the Total Environment, 114: 113-133.

Bradford, R.G., Robichaud, K.A., and Courtenay, S.C. 1995. By-catch in commercial fisheries as an indicator and regulator of striped bass (*Morone saxatilis*) abundance in the Miramichi River Estuary, p. 249–259. *In* E.M.P. Chadwick [*editor*] Water, science and the public: the Miramichi ecosystem. Can. Spec. Publ. Fish. Aquat. Sci. 123.

Buckley, D.E., and Winters, G.V. 1983. Geochemical transport through the Miramichi Estuary. Can. J. Fish. Aquat. Sci. 40. (Suppl. 2): 162-182.

Chaput, G.J. 1995. Temporal distribution, spatial distribution, and abundance of diadromous fish in the Miramichi River watershed, p. 121-139. *In* E.M.P. Chadwick [*editor*]. Water, science and the public: the Miramichi ecosystem. Can. Spec. Publ. Fish. Aquat. Sci. 123.

Courtenay, S., Grunwald, C., Kreamer, G.-L., Alexander, R., and Wirgin, I. 1993. Induction and clearance of cytochrome P4501A mRNA in Atlantic tomcod caged in bleached kraft mill effluent in the Miramichi River. Aquat. Toxicol. 27: 225-244.

Courtenay, S., Williams, P.J., Grunwald, C., Konkle, B., Ong, T.-L., and Wirgin, I.I. in press. Assessment of within group variation in CYP1A mRNA inducibility in chemically treated and environmentally exposed Atlantic tomcod. Environ. Health Perspec.

Dew, C.B. 1991. Early life history and population dynamics of Atlantic tomcod (*Microgadus tomcod*) in the Hudson River Estuary, New York. Ph.D. Thesis, City University of New York, NY.

Dey, W.P., Peck, T.H., Smith, C.E., and Kreamer, G.-L. 1993. Epizoology of hepatic neoplasia in Atlantic tomcod (*Microgadus tomcod*) from the Hudson River Estuary. Can. J. Fish. Aquat. Sci. 50: 1897-1907.

Donaldson, E.M. 1990. Reproductive indices as measures of the effects of environmental stressors in fish. Am. Fish. Soc. Symp. 8: 109-122.

Eisner, R. 1986. Dioxin hazards to fish, wildlife, and invertebrates: a synoptic review. U.S. Fish Wildl. Serv. Biol. Rep. 85(1.8): 37 pp.

Eisner, R. 1987. Polycyclic aromatic hydrocarbon hazards to fish, wildlife, and invertebrates: a synoptic review. U.S. Fish Wildl. Serv. Biol. Rep. 85(1.11). 81 pp.

Elskus, A.A., and Stegeman, J.J. 1989. Induced cytochrome P450 in *Fundulus heteroclitus* associated with environmental contamination by polychlorinated biphenyls and polynuclear aromatic hydrocarbons. Mar. Environ. Res. 27: 31-50.

Gagné, F., and Blaise, C. 1993. Hepatic metallothioinein level and mixed function oxidase activity in fingerling rainbow trout (*Oncorhynchus mykiss*) after acute exposure to pulp and paper mill effluents. Wat. Res. 27(11): 1669-1682.

Goksoyr, A., and Forlin, L. 1992. The cytochrome P-450 system in fish, aquatic toxicology and environmental monitoring. Aquat. Toxicol. 22: 287-312.

Grabe, S. 1980. Food of age 1 and 2 Atlantic tomcod, *Microgadus tomcod*, from Haverstraw Bay, Hudson River, New York. Fish. Bull. 77(4): 1003-1006.

Guengerich 1993. Cytochrome P450 enzymes. American Scientist 81: 440-447.

Haasch, M.L., Wejksnora, P.J., Stegeman, J.J., and Lech, J.J. 1989. Cloned rainbow trout liver P(1)450 complementary DNA as a potential environmental monitor. Toxicol. Appl. Pharmacol., 98(2): 362-368.

Haasch, M.L., Quardokus, E.M., Sutherland, L.A., Goodrich, M.S., Prince, R., Cooper, K.R., and Lech, J.J. 1992. CYP1A1 protein and mRNA in teleosts as an environmental bioindicator: laboratory and environmental studies. Mar. Environ. Res. 34: 139-145.

Haasch, M.L., Prince, R., Wejksnora, P.J., Cooper, K.R., and Lech, J.J. 1993. Caged and wild fish: induction of hepatic cytochrome P450 (CYP1A) as an environmental biomonitor. Environ. Toxicol. Chem. 12: 885-895.

Hanson, J.M., and Courtenay, S.C. 1995. Seasonal abundance and distribution of fishes in the Miramichi Estuary, p. 141-160. *In* E.M.P. Chadwick [*editor*] Water, science and the public: the Miramichi ecosystem. Can. Spec. Publ. Fish. Aquat. Sci. 123.

Hardig, J., Andersson, T., Bengtsson, B.-E., Forlin, L. and Larsson, V. 1988. Long-term effects of bleached kraft mill effluents on red and white blood cell status, ion balance, and vertebral structure in fish. Ecotoxicology and Environmental Safety 15: 96-106.

Hodson, P.V., Kloepper-Sams, P.J., Munkittrick, K.R., Lockhart, W.L., Metner, D.A., Luxon, P.L., Smith, I.R., Gagman, M.M., Servos, M., and. Pain, J.F. 1991. Protocol for measuring mixed function oxygenases of fish liver. Can. Tech. Rep. Fish. Aquat. Sci. 1829: 1-5.

Janz, D.M., Metcalfe, T.L., Metcalfe, C.D., and Haffner, G.D. 1992. Relative concentrations of cytochrome P450-active organochlorine compounds in liver and muscle of rainbow trout from Lake Ontario. J. Great Lakes Res. 18(4): 759-765.

Khan, R.A., Barker, D., Hooper, R., and Lee, E.M. 1992. Effect of pulp and paper effluents on a marine fish, *Pseudopleuronectes americanus*. Bull. Environ. Contam. Toxicol. 48: 449-456.

Kloepper-Sams, P.J. and Stegeman, J.J. 1992. Effects of temperature acclimation on the expression of hepatic cytochrome P4501A mRNA and protein in the fish *Fundulus heteroclitus*. Arch. Biochem. Biophys. 299: 38-46.

Kreamer, G.-L., Squibb, K., Gioeli, D., Garte, S.J., and Wirgin, I. 1991. Cytochrome P4501A mRNA expression in feral Hudson River tomcod. Environ. Res. 55: 64-78.

Lehtinen, K.-J. 1990. Mixed-function oxygenase enzyme responses and physiological disorders in fish exposed to kraft pulp-mill effluents: a hypothetical model. Ambio 19(5): 259-265.

Lehtinen, K.-J., Notini, M., and Landner, L. 1984. Tissue damage and parasite frequency in flounders, *Platichtys flesus* (L.) chronically exposed to bleached kraft pulp mill effluents. Ann. Zool. Fennici 21: 23-28.

Locke, A., and Courtenay, S.C. 1995a. Effects of environmental factors on ichthyoplankton communities in the Miramichi estuary, Gulf of St. Lawrence. J. Plankton Research 17(2): 333–339.

Locke, A., and Courtenay, S. C. 1995b. Ichthyoplankton and invertebrate zooplankton of the Miramichi Estuary: 1918–1993, p. 97–120. *In* E.M.P. Chadwick [*editor*] Water, science and the public: the Miramichi ecosystem. Can. Spec. Publ. Fish. Aquat. Sci. 123.

MacKnight, S. 1990. Contaminants in the Miramichi estuary system. *In*: Review of Impacts of channel dredging on biological resources of the Miramichi system. Acadia Centre for Estuarine Research, Acadia University, Wolfville NS.

Martel, P.H., Kovacs, T.G., O'Connor, B.I., and Voss, R.H. 1994. A survey of pulp and paper mill effluents for their potential to induce mixed function oxidase enzyme activity in fish. Wat. Res. 28(8): 1835-1844.

Martineau, D., Lagacé, A., Béland, P., Higgins, R., Armstrong, D., and Shugart, L.R. 1988. Pathology of stranded Beluga whales (*Delphinapterus leucas*) from St. Lawrence estuary, Quebec, Canada. J. Comp. Path. 98: 287-311.

McLaren, J.B., Peck, T.H., Dey, W.P., and Gardinier, M. 1988. Biology of Atlantic tomcod in the Hudson River Estuary. Am. Fish. Soc. Monograph 4: 102-112.

McMaster, M.E., Van Der Kraak, G.J., Portt, C.B.,Munkittrick, K.R., Sibley, P.K., Smith, I.R., and Dixon, D.G. 1991. Changes in hepatic mixed-function oxygenase (MFO) activity, plasma steroid levels and age at maturity of a white sucker (*Catostomus commersoni*) population exposed to bleached kraft mill effluent. Aquat. Toxicol. 21: 199-218.

MREAC. 1992. Summary Final Report. Miramichi River Environmental Assessment Committee. Chatham, N.B. 157p.

Monosson, E., and Stegeman, J.J. 1994. Induced cytochrome P4501A in winter flounder, *Pleuronectes americanus*, from offshore and coastal sites. Can. J. Fish. Aquat. Sci. 51: 933-941.

Munkittrick, K.R., Portt, C.B., Van Der Kraak, G.J., Smith, I.R., and Rokosh, D.A. 1991. Impact of bleached kraft mill effluent on population characteristics, liver MFO activity, and serum steroid levels of a Lake Superior white sucker (*Catostomus commersoni*) population. Can. J. Fish. Aquat. Sci. 48: 1371-1380.

Munkittrick, K.R., Van Der Kraak, G.J., McMaster, M.E., and Portt, C.B. 1992a. Response of hepatic MFO activity and plasma sex steroids to secondary treatment of bleached kraft pulp mill effluent and mill shutdown. Environ. Toxicol. Chem. 11: 1427-1439.

Munkittrick, K.R., McMaster, M.E., Portt, C.B., Van Der Kraak, G.J., Smith, I.R., and Dixon, D.G. 1992b. Changes in maturity, plasma sex steroid levels, hepatic mixed-function oxygenase activity, and the presence of external lesions in lake whitefish (*Coregonus clupeaformis*) exposed to bleached kraft mill effluent. Can. J. Fish. Aquat. Sci. 49: 1560-1569.

National Status and Trends Program for Marine Environmental Quality. 1988. Progress report: a summary of selected data on chemical contaminants in sediments collected during 1984, 1985, 1986, and 1987. NOAA Technical Memorandum NOS OMA 44, Rockville, MD.

New Brunswick Department of the Environment. 1991. The Miramichi River industrial point source report. submitted as part of the Miramichi Environmental Assessment Committee Study. 229p.

Olivier, G. 1992. Furunculosis in the Atlantic provinces: an overview. Bull. Aquacult. Assoc. Can. 92(1): 4-10.

Parrot, J.L. 1993. Relative potency of polychlorinated dibenzo-p-dioxins and dibenzofurans for inducing mixed-function oxygenase activity in rainbow trout. Ph.D. thesis, Dept. of Biology, University of Waterloo, Waterloo Ontario.

Payne, J.F., Fancey, L.L., Rahimtula, A.D., and Porter, E.L. 1987. Review and perspective on the use of mixed-function oxygenase enzymes in biological monitoring.

Comp. Biochem. Physiol. 86C: 233-245.

Pham, T., Lum, K., and Lemieux, C. 1993. Sources of PAHs in the St. Lawrence River (Canada) and their relative importance. Chemosphere 27(7): 1137-1149.

Philpott, K.L. 1978. Miramichi channel study: a Canada New Brunswick Project. Public Works Canada, Design and Construction Marine Directorate, Cat. No. W31-33/1978: 284 p.

Safe, S. 1991. Polychlorinated dibenzo-*p*-dioxins and related compounds: sources, environmental distribution and risk assessment. Environ. Carcinog. Ecotoxicol. Rev. C9(2): 261-302.

Santos, M.A., Pires, F., and Hall, A. 1990. Metabolic effects of kraft mill effluents on the eel *Anguilla anguilla* L. Ecotoxicol. Environ. Saf. 20: 10-19.

Scott, W.B., and Scott, M.G. 1988. Atlantic fishes of Canada. Can. Bull. Fish. Aquat. Sci. 219: 731 p.

Shenker, J.M., and Cherr, G.N. 1990. Toxicity of zinc and bleached kraft mill effluent to larval English sole (*Parophrys vetulus*) and topsmelt (*Atherinops affinis*). Arch. Environ. Contam. Toxicol. 19: 680-685.

Snowberger-Gray, E., Woodin, B.R., and Stegeman, J.J. 1991. Sex differences in hepatic monooxygenases in winter flounder (*Pseudopleuronectes americanus*) and scup (*Stenotomus chrysops*) and regulation of P450 forms by estradiol. J. Exp. Zool. 259: 330-342.

Stegeman, J.J., and Hahn, M.E. 1994. Biochemistry and molecular biology of monooxygenases: Current perspectives on forms, functions, and regulation of cytochrome P450 in aquatic species. pp. 87-206 *In* Malins, D.C., and G.K. Ostrander [*editors*]. Aquatic toxicology. Molecular, biochemical, and cellular perspectives. Lewis Publishers, Ann Arbor, MI.

Stegeman, J.J., and Kloepper-Sams, P.J. 1987. Cytochrome P-450 isozymes and monooxygenase activity in aquatic animals. Environ. Health Perspect. 71: 87-95.

Stegeman, J.J., Teng, F.Y., and Snowberger, E.A. 1987. Induced cytochrome P450 in winter flounder (*Pseudopleuronectes americanus*) from coastal Massachusetts evaluated by calalytic assay and monoclonal antibody probes. Can. J. Fish. Aquat. Sci. 44: 1270-1277.

Thulin, J., Hoglund, J., and Lindesjoo, E. 1988. Diseases and parasites of fish in a bleached kraft mill effluent. Water Sci. Technol. 20(2): 179-180.

Trudel, L. 1991. Dioxins and furans in bottom sediments near the 47 Canadian pulp and paper mills using chlorine bleaching. Water Quality Branch, Inland Water Directorate, Environment Canada, Ottawa, Ont.

Van Der Kraak, G.J., Munkittrick, K.R., McMaster, M.E., Portt, C.B., and Chang, J.P. 1992. Exposure to bleached kraft mill effluent disrupts the pituitary-gonadal axis of white sucker at multiple sites. Toxicol. Appl. Pharmacol. 115: 224-233.

Van Veld, P.A., Vogelbein, W.K., Smolowitz, R., Woodin, B.R., and Stegeman, J.J. 1992. Cytochrome P4501A1 in hepatic lesions of a teleost fish (*Fundulus heteroclitus*) collected from a polycyclic aromatic hydrocarbon contaminated environment. Aquat. Toxicol. 17: 117-132.

Vannote, R.L., Minshall, G.W., Cummins, K.W., Sedell, J.R., and Cushing, C.E. 1980. The river continuun concept. Can. J. Fish. Aquat. Sci. 37: 130-137.

Varanasi, U., and Stein, J.E. 1991. Disposition of xenobiotic chemicals and metabolites in marine organisms. Environ. Health Perspect. 90: 93-100.

Vuorinen, M., and Vuorinen, P.J. 1991. Mortality and growth of rainbow trout (*Oncorhynchus mykiss*) early life phases in exposure to bleached kraft mill effluent (BKME). Finn. Fish. Res. 12: 111-123.

Wirgin, I.I., Konkle, B., Pederson, M., Grunwald, C., Williams, J., and Courtenay, S. 1995. A comparison of cytochrome P4501A (CYP1A) mRNA inducibility in four species of Atlantic coast anadromous fishes. Estuaries. (In press).

Wirgin, I.I., Grunwald, C., Courtenay, S., Kreamer, G.-L., Reichert, W.L., and Stein, J. E. 1994. A biomarker approach to assessing xenobiotic exposure in Atlantic tomcod from the North American Atlantic Coast. Environ. Health Perspect. 102(9): 764-770.

Wirgin, I.I., Kreamer, G.-L., Grunwald, C., Squibb, K., Garte, S.J., and Courtenay, S. 1992. Effects of prior exposure history on cytochrome P4501A mRNA induction by PCB congener 77 in Atlantic tomcod. Mar. Environ. Res. 34: 103-108.

Striped bass is a popular sport fish whose abundance fluctuates cyclically in Miramichi River. In recent years, we have learned that Miramichi is an important spawning and rearing area for this species.

CHAPTER 14

The Effect of Fisheries on the Biological Characteristics and Survival of Mature Atlantic Salmon (*Salmo salar*) from the Miramichi River

D.S. Moore, G.J Chaput, and P.R. Pickard

Department of Fisheries and Oceans, Science Branch, P.O. Box 5030, Moncton, New Brunswick, Canada E1C 9B6

Moore, D.S., Chaput, G.J., and Pickard, P.R. 1995. The effect of fisheries on the biological characteristics and survival of mature Atlantic salmon (*Salmo salar*) from the Miramichi River, p. 229-247. *In* E.M.P. Chadwick [*editor*]. Water, science, and the public: the Miramichi ecosystem. Can. Spec. Publ. Fish. Aquat. Sci. 123.

Abstract

The analysis of the 22-year time series of biological characteristics of Atlantic salmon from the Miramichi River has verified the predictions of the effects of fisheries on the biological characteristics. When the size-selective commercial fisheries are closed, there is an observed increase in the size-at-age and in the proportion of previous spawner 1SW salmon. When commercial and recreational fisheries which harvested 2SW salmon are closed, there is an increase in the abundance of previous spawner 2SW salmon. Salmon returning to the river in June and July were characterized by shorter fork length but higher condition factor than those sampled in September and October. Previous spawners of maiden 1SW origin were more abundant in the June and July samples whereas previous spawners of maiden 2SW origin were more abundant in the September and October samples. Survival to a second spawning migration increased to more than 30% for the 1989 and 1990 maiden 2SW spawning groups from less than 10% between 1976 and 1983. Changes in management of Atlantic salmon fisheries have had the greatest impact on the survival rates of salmon during their maiden and in subsequent spawning migrations. The high proportions of previous spawners in the returns of large salmon to the Miramichi are mainly the result of increased survival of salmon spawners because of reduced fisheries exploitation. The importance of previous spawners to the Miramichi River population cannot be understated as these have contributed more than 40% of the total egg depositions to the river in recent years.

Résumé

L'analyse d'une série chronologique de 22 ans portant sur les caractéristiques biologiques du saumon atlantique de la Miramichi nous a permis de vérifier les prévisions des effets des pêches sur ces caractéristiques. Lorsque les pêches commerciales sélectives par la taille sont fermées, on observe une augmentation de la taille en fonction de l'âge et de la proportion des saumons unibermarins à ponte multiple. Quand les pêches commerciales et sportives qui capturaient des saumons dibermarins sont fermées, il y a une hausse de l'abondance des saumons dibermarins à porte multiple. Les saumons qui reviennent à la Miramichi en juin et juillet se caractérisent par une plus faible longueur à la fourche, mais par un facteur de condition plus élevé que ceux qui sont capturés en septembre et octobre. Les saumons à ponte multiple ayant eu leur première ponte après un an en mer étaient plus abondants dans les échantillons de juin et juillet, tandis que ceux ayant eu leur première ponte après deux ans en mer étaient plus abondants dans les échantillons de septembre et d'octobre. Le taux de survie à une deuxième migration reproductrice montait à plus de 30 % pour les groupes de géniteurs vierges après deux ans en mer en 1989 et 1990, tandis qu'il n'était que de 10 % entre 1976 et 1983. Les changements apportés à la gestion des pêches du saumon atlantique ont eu une incidence extrêmement forte sur les taux de survie des saumons pendant leur première migration reproductrice et les migrations suivantes. La forte proportion de saumons à ponte multiple dans les retours de gros saumons vers la Miramichi est causée principalement par la hausse de la survie des géniteurs du fait de la réduction de l'exploitation par les pêches. Il ne faut pas sous-estimer l'importance des saumons à ponte multiple dans la population de la Miramichi, car ces dernières années ils ont contribué pour plus de 40 % à la ponte totale.

Introduction

The true biological characteristics of salmon stocks are difficult to define when these stocks are subjected to selective fishery exploitation. Fisheries can selectively harvest different sizes, ages, or even stocks of fish. The indigenous characteristics of a stock would only be expressed in the absence of such fisheries. For example, Atlantic salmon can spawn several times yet the proportion of previous spawners in the adult returns to rivers is normally low, from 3-6% (Mills 1989). Schaffer and Elson (1975) suggested that the true age composition of the spawning run of the Miramichi River had been distorted by commercial fisheries which eliminated the older age classes of maiden fish.

Low returns of previous spawners are in large part associated with recreational and commercial exploitation. The proportion of previous spawners in the Miramichi River averaged less than 5% in 1929 (Kerr 1961), 5% in 1931 (Blair 1935), and 4% of all sea age groups from 1954 to 1971 (Ruggles and Turner 1973). The combined exploitation of commercial (distant water + homewater) and angling fisheries accounted for an average of 97% of large salmon (>= 63 cm fork length) and 56% of small salmon (<63 cm fork length) returning to the Miramichi River during the 1960s (Saunders 1969). Salmon from Big Salmon River were assumed to be subjected to low exploitation in sea fisheries and the proportion of previous spawners in the returns averaged 46% over a seven year period (Jessop 1986). Since 1984 most maritime fisheries for large salmon have been closed and the proportion of multiple spawners in the Miramichi River has increased to nearly 40% of the large salmon returns (Chaput et al. 1994).

The fishing gear used in the commercial fisheries of Atlantic salmon is size-selective for the larger individuals from the maiden one-sea-winter (1SW) component (Chadwick and Claytor 1990, Reddin 1984). Large older salmon are frequently selectively fished out (Chadwick 1988) and in New Brunswick from at least 1931 until 1981, regulations prohibited the retention of salmon less than 2.3 kg (Blair 1932; Marshall 1988). In addition, because size-at-age is also affected by the season of return, variations in abundance by season may mask any effects of size-selective fisheries.

The primary objective of the changes in the management of salmon fisheries since 1971 was to increase the spawning escapements of Atlantic salmon in Canada. The management of the fisheries can be categorized into three time periods. Although the commercial salmon fishery in New Brunswick was closed from 1971-1980, the bycatch of salmon in nonsalmon commercial gears could be retained and sold. Salmon from the Miramichi River were also intercepted in other fisheries from the Gulf of St. Lawrence (Claytor et al. 1987). The commercial fishery reopened between 1981 and 1983 under quota restrictions. Since 1984, the commercial fishery has been closed and the retention of salmon from non-salmon gear (by-catch) has been prohibited. In addition, anglers have been required to release all large salmon.

An analysis of a time series of age and length data collected at the Millbank trapnet, provides a description of the biological characteristics of Atlantic salmon from the Miramichi River and allows us to test the following predictions regarding the effects of fisheries on the biological characteristics of the stock:

1) Size-at-age of salmon increased after size-selective commercial fisheries were closed.
2) Proportions of previous-spawner 1SW salmon increased after commercial fisheries were closed.
3) Abundance of previous-spawner 2SW salmon increased when commercial and recreational fisheries for this size group were closed.

Because fish sampled at Millbank had mostly passed through the commercial sea fisheries but had not yet been exploited in the recreational fisheries it is possible to examine selectivity of commercial fisheries and the effects of recreational fisheries on survival of spawners.

Materials and Methods

Biological characteristics of Atlantic salmon returning to the Miramichi River were collected annually between 1971 and 1992 at the Millbank estuarine trapnet (Fig. 1). The entire run of Atlantic salmon was monitored; the trapnet fished in most years from mid-May until the end of October and sometimes into early November. Samples of fork length and scales were collected systematically and additional samples were available from mortalities at the trapnet. Whole weight and internal determination of sex were recorded from sacrificed fish and incidental mortalities. Large salmon were not sacrificed after 1984. Scales were collected from the standard zone and ages were determined according to ageing protocols defined by International Council for Exploration of the Sea (Shearer 1989).

Spawning histories were designated as follows:

- 1 maiden 1SW salmon having spent one winter at sea and spawning for the first time
- 2 maiden 2SW salmon having spent two winters at sea and spawning for the first time
- 3 maiden 3SW salmon having spent three winters at sea and spawning for the first time
- 1C previous spawner, originally mature as 1SW and returning for a second spawning in the year immediately following the first spawning (consecutive spawner)
- 1A previous spawner, originally mature as 1SW and returning for a second spawning in the alternate year after the first spawning

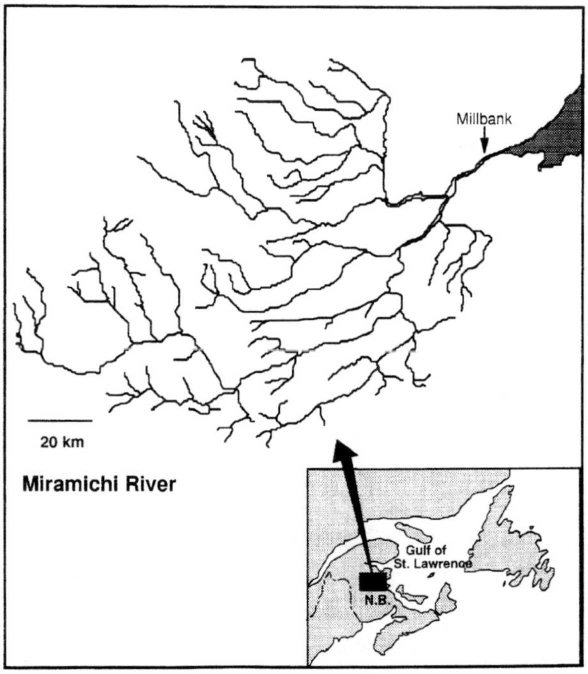

Fig. 1. Miramichi River indicating location of Millbank trapping facility.

2C previous spawner, originally mature as 2SW and returning for a second spawning in the year immediately following the first spawning (consecutive spawner)

2A previous spawner, originally mature as 2SW and returning for a second spawning in the alternate year after the first spawning

1CR previous consecutive spawner, originally mature as 1SW and returning for a third or greater spawning

1AR previous alternate spawner, originally mature as 1SW and returning for a third or greater spawning

2CR previous consecutive spawner, originally mature as 2SW and returning for a third or greater spawning

2AR previous alternate spawner, originally mature as 2SW and returning for a third or greater spawning

3R previous spawner, originally mature as 3SW regardless of previous spawning history.

Logarithmic transformations of weight and length data were used for calculating the weight-length relationship and covariance analysis was used to test for differences in slope and intercept between sexes. Sex ratios by spawning history and sea-age groups were compared using Chi-square statistics.

In a first set of analyses we examine the biological characteristics of Atlantic salmon for the time period when homewater commercial fisheries were closed in order to describe the inherent variability in an unexploited population. We describe variations in length, weight and condition factor by age of Atlantic salmon for the years 1986 to 1992. Analysis of variance (ANOVA) methods were used for determining the relative contribution of sex, month and year of sampling to the observed variability. Condition factors were calculated using analysis of covariance under the assumptions of multiplicative errors (Patterson 1992). Only length was examined for 2SW maiden salmon because no fish of this age group were sacrificed at Millbank after 1984. All ANOVA analyses were performed using SAS (1990).

In the second set of analyses, we compare the size-at-age among three management time periods (1971 to 1980 commercial closed with legal bycatch; 1981 to 1983 commercial open; 1984 to 1992 commercial closed, hook and release of large salmon in recreational fishery). The analyses were divided into two seasons: summer - June and July; and fall - September and October. Fish sampled in August were excluded to permit a sharper distinction between summer and fall seasons. We also analyzed the differences in the proportions of each age between the two seasons as well as in each season during the three management time periods. Because no scale samples of large salmon were obtained in the fall of 1983, this year was omitted from the subsequent analyses. These comparisons were performed using ANOVA but with the significance determined using randomization procedures (Edgington 1987). A total of 2000 replicates were used to calculate the *P*-value of the null hypothesis. Unless otherwise stated, statistical significance is determined at a probability value (α) = 0.05.

Estimates of survival between the first and second spawning migrations were obtained from the estimated total returns to Millbank. Returns by age group were stratified by month as determined from scale samples collected at the trapnet and the proportion of the total run intercepted monthly. Total returns for the year were obtained from published assessments (Chaput et al. 1994). Survival to second spawning for each year-class was calcu-

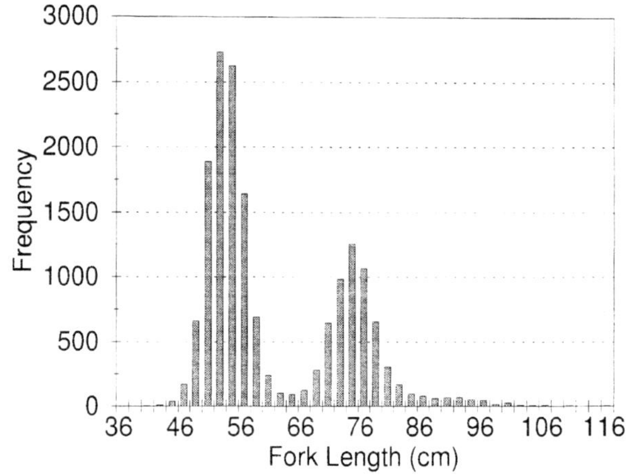

Fig. 2. Fork length (cm) frequency distribution of all Atlantic salmon sampled from the Millbank trapnet, 1971 to 1992.

lated as the ratio of the sum of the returns of consecutive and alternate spawners relative to the returns of maiden spawners. Differences in survival rates between time periods were determined from a t-test with the significance of the test statistic evaluated by randomization procedures.

Results

The fork lengths of Atlantic salmon sampled at the Millbank trapnet between 1971 and 1992 varied from 35 to 115 cm fork length (Fig. 2). Salmon ranging in length between 49 and 57 cm fork length comprised 52% of the total. Whole weights, obtained from 13,924 salmon, varied from 1kg to 17 kg (maximum weight obtained from weight to length relationship in Table 1). Female salmon were significantly heavier at length than males (Table 1).

Maiden salmon were the most abundant spawning history group representing 95% of the 16,112 age determinations. The oldest salmon sampled were 8 sea-years old (n=10). These fish were on their fourth spawning migration and all had originally matured as 2SW salmon. One previous spawner 1SW fish was on its fifth spawning migration. In the 1SW maturing component, alternate spawning was the most commonly observed previous spawner category, representing 89% of the multiple spawners (Table 2). For the 2SW maturing component, alternate spawning was also more common but consecutive spawning was recorded in 25% of these samples (Table 3).

Males were the dominant component in the 1SW maturation category at 74% in the maiden component but decreasing to 66% in the returns of alternate spawners (Table 2). This difference in the sex ratios was significant (χ^2 = 5.17; df = 1; $0.01<\alpha<0.05$). The 2SW maiden salmon were mostly female and previous spawners of this maturation category had a similar sex ratio to the maiden spawner group (χ^2 = 1.45; df = 1; $\alpha>0.10$) (Table 3).

Biological characteristics (1986 to 1992)

In general, length increased with the number of years spent at sea. 1SW maiden salmon had a median fork length of 55.0 cm, varying between 49.0 and 61.0 (Fig. 3, Table 2). Previous spawner 1SW salmon had median lengths of 65.0 cm as consecutive spawners (1C) and 80.0 cm as alternate spawners (1A). 2SW maiden salmon, at about 72.0 cm, were shorter than alternate 1SW spawners but the 2SW consecutives (2C) were of similar size to the latter. The 2SW alternates (2A) were generally greater than 90.0 cm fork length (Fig. 3). All salmon greater than 92 cm fork length were previous spawners.

Table 1. Analysis of variance of the weight-to-length relationship of Atlantic salmon from the Miramichi River. Proportion of the total variance (total and model) explained by a main effect is calculated for each effect individually and with the interaction term(s) excluded.

		Proportion of Variance		
Model Effects	*P*-value	Total	Model	df
Overall	0.001	0.94		8198
Fork length	0.001	0.93	0.99	1
Sex	0.001	0.23	0.24	1
Fork length*sex	0.001	<0.01		1
Whole weight (kg) = α * Fork length (cm)$^{\beta}$				
	α	β		
Male	0.000015	2.903		
Female	0.000008	3.069		1

Table 2. Fork length and weight, by spawning history, of one-sea-winter Atlantic salmon from the Miramichi River based on samples collected at Millbank trapnet between 1971 and 1992.

		Fork length (cm)			Whole weight (kg)			
Spawning history	Sex	Mean	SD	*N*	Mean	SD	*N*	Sex Ratio
1	Male	53.8	3.1	4399	1.59	0.28	4013	74%
	Female	52.1	2.8	1437	1.50	0.27	1396	26%
	Both	52.5	3.2	10840	1.58	0.29	9410	
1C	Male	64.8	4.5	20	3.24	0.72	16	94%
	Female	73.0	-	1	4.50	-	1	6%
	Both	65.1	5.1	47				
1A	Male	81.1	5.5	167	5.49	0.98	101	66%
	Female	76.9	4.0	58	5.21	0.73	52	34%
	Both	79.1	5.2	393	5.32	0.93	218	
1CR	Male	74.9	10.1	10	4.27	1.90	4	100%
	Female	–	–	–	–	–	–	–
	Both	73.7	8.7	16	3.90	1.68	6	
1AR	Male	91.7	10.5	13	7.77	0.66	5	100%
	Female	83.3	7.6	7	–	–	–	–
	Both	88.6	10.3	34	6.81	1.47	11	

Table 3. Fork length and weight, by spawning history, of multi-sea-winter Atlantic salmon from the Miramichi River based on samples collected at Millbank trapnet between 1971 and 1992. Spawning history designation 3R refers to all previous spawner salmon which first matured after 3 years at sea.

		Fork length (cm)			Whole weight (kg)			
Spawning history	Sex	Mean	SD	*N*	Mean	SD	*N*	Sex ratio
2	Male	75.2	4.5	402	4.43	0.91	348	14%
	Female	73.8	3.4	2782	4.43	0.72	2188	86%
	Both	73.4	3.7	5183	4.43	0.75	3942	
2CMale		80.3	8.3	5	6.15	0.69	3	14%
	Female	81.2	4.8	50	5.15	1.15	18	86%
	Both	80.8	4.9	71	5.39	1.15	29	
2A	Male	89.9	7.1	17	8.80	3.80	11	19%
	Female	90.2	6.2	104	8.19	1.79	46	81%
	Both	90.4	6.0	193	8.35	2.84	74	
2CR	Male	78.5	2.9	2	4.05	–	1	–
	Both	89.7	6.2	38	6.55	1.80	5	
2AR	Male	92.8	1.8	2	6.10	-	1	13%
	Female	95.8	5.6	44	11.22	6.59	7	87%
	Both	97.1	6.0	81	10621	5.24	14	
3	Male	90.5	–	1	6300	–	1	7%
	Female	87.7	4.8	15	7576	2.06	14	93%
	Both	86.9	4.7	24	7270	1.92	22	
3R	Female	95.0	6.6	5	12250	0.64	2	–

Fork length of 1SW maiden salmon varied relative to the sex, month, and year of sampling. The main effects (sex, month and year) accounted for 31% of the total variation in fork length (Table 4). Monthly variations in fork length were more important than year effects which were more important than variations due to the sex of the fish. The fork length of both sexes increased from June to October (Fig. 4). June and July sampled fish were 4 cm shorter and August fish were 3 cm shorter than fall-run fish. Relative to 1992, 1SW maiden salmon were significantly shorter in 4 of 6 years (except 1986 and 1990) by as much as 2 cm (Fig. 5). Male 1SW maiden salmon were significantly longer than female 1SW maiden salmon in all months, averaging 0.9 cm longer (Fig. 4).

Maiden 2SW salmon fork lengths also showed significant differences among months and years, with month of sampling contributing almost three times the variability than year (Table 5). Salmon sampled in June were on average 3 cm shorter than those in October while salmon from August and September were about 1 cm shorter (Fig. 5). There were significant differences among years; salmon sampled in 1987 to 1989 were larger, by up to 2 cm, than those from 1992 (Fig. 5). Annual differences in length-at-age were not predictable between different maturation schedules of the same cohort. Large 1SW maiden salmon in one year were followed by either large or small 2SW salmon the following year. These limited data (6 years) indicate that there is not a strong environmental signal determining growth rates.

Year of sampling was the most important variance component of whole weight of maiden 1SW salmon (Table 4). Salmon from 1986 and 1992 had significantly higher mean whole weights than other years; they were about 0.2 kg heavier. Male 1SW salmon were heavier than females; this is consistent with males being longer than females (Fig. 4). Salmon tended to be heavier in August. The condition factor varied more consistently. Whole weight was strongly correlated with fork length and month was the most important effect contributing to the variation in condition (Table 4; Fig. 4). Year and sex were of minimal importance. The average condition decreased from June to October but most so during the last 2 months.

Effects of fisheries

Year and season effects accounted for 25% of the total variation in length of 1SW maiden salmon with year being slightly more important than sea-

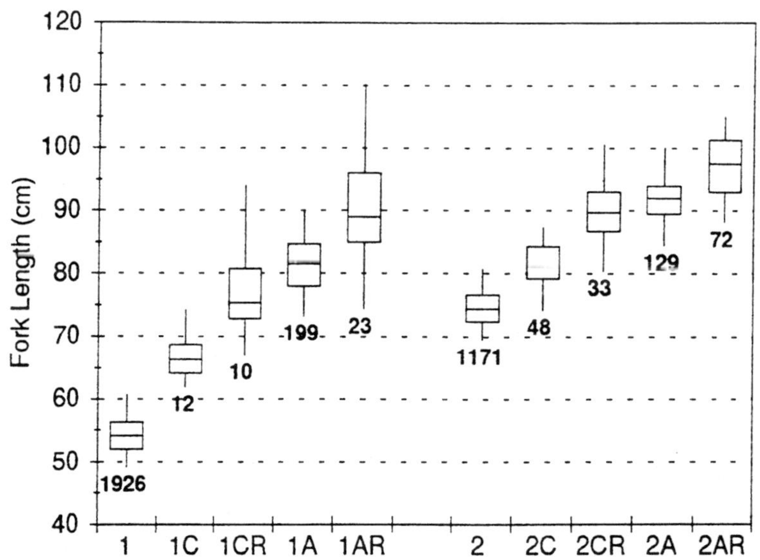

Fig. 3. Fork length (cm) at sea age distributions for the spawning histories of Atlantic salmon from the Miramichi River. Vertical bars represent the interquartile range, vertical line are the 5th to 95th percentiles, horizontal line is the median, and number below each plot is the sample size.

son (Table 6). Although the season by year interaction term was significant, it accounted for only 1% of the total variation (Table 6). Generally, fork lengths of maiden 1SW salmon were similar between 1971 and 1985 but were significantly smaller than fish from 1986 to 1992 (Fig. 6). Salmon in 1986, 1990 and 1992 were significantly larger than all other years (Fig. 7). The change in length over that time period was significant for both summer-run and fall-run fish.

There were significant annual and seasonal variations in fork length of maiden 2SW salmon. Year and season accounted for 18% of the total variation in length of 2SW maiden fish; the interaction term was significant but small (1% of the

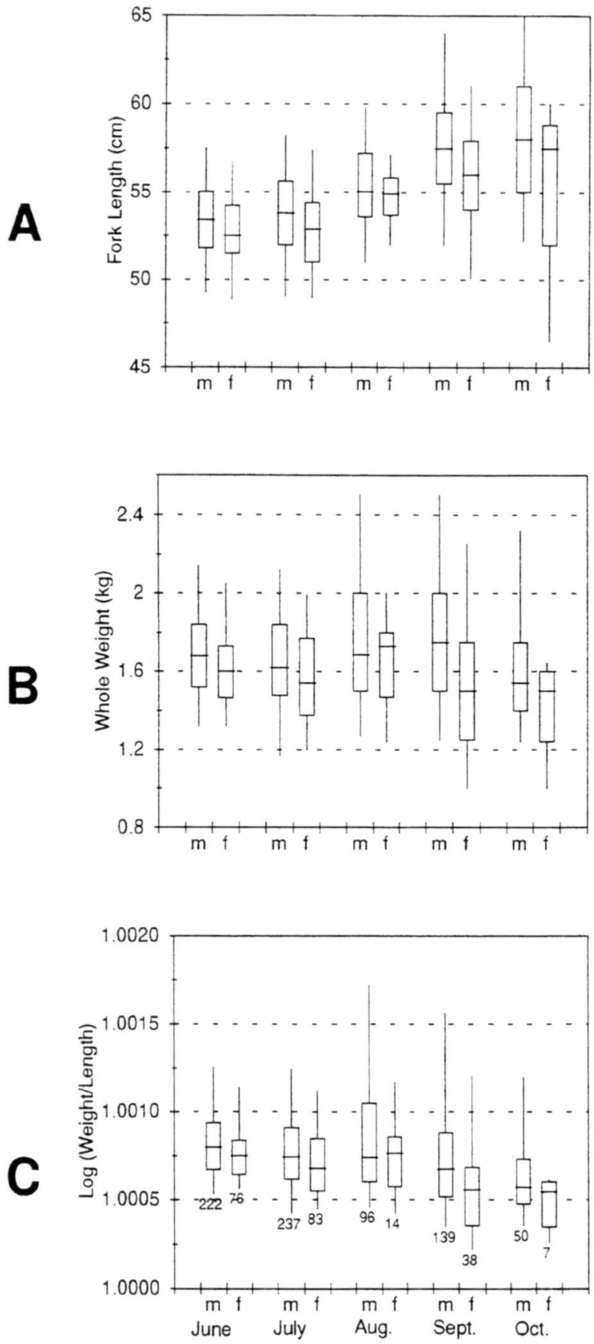

Fig. 4. Fork length (A), whole weight (B) and condition factor (C) distributions by sex and month of 1SW maiden Atlantic salmon from the Miramichi River, 1986 to 1992. Bars and lines as in Fig. 3. "m" = male and "f" = female.

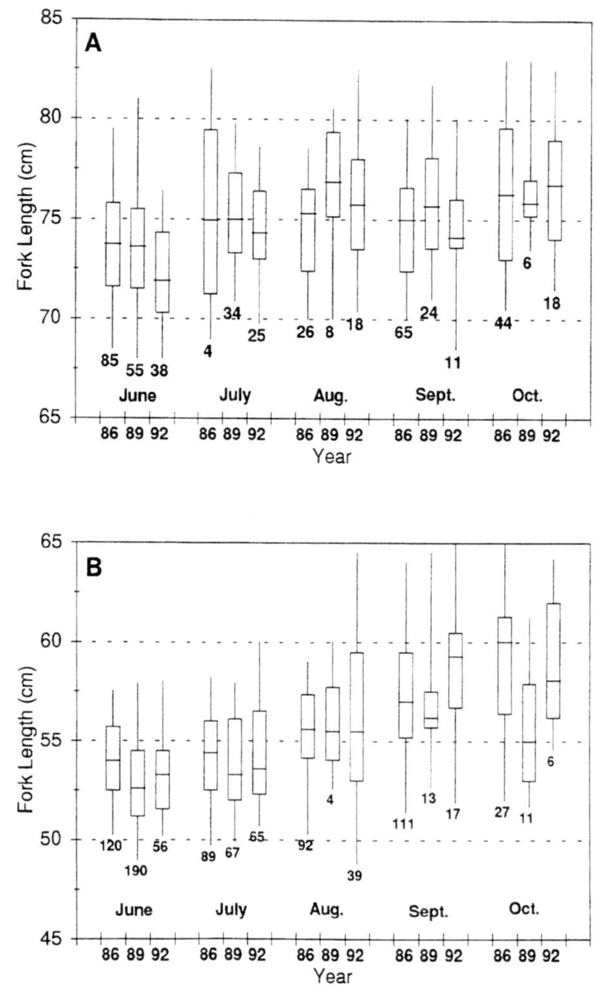

Fig. 5. Annual and monthly variations in the observed fork length of 2SW (A) and 1SW (B) maiden Atlantic salmon from the Miramichi River, 1986 to 1992. Bars and lines as in Fig. 3.

Table 4. Analysis of variance of sex, month and year effects on the observed biological characteristics of 1SW maiden Atlantic salmon from the Miramichi River. Proportion of variance (total and model) explained by a main effect is calculated for each effect individually and with the interaction term(s) excluded.

Model Effects	*P*-value	Proportion of Variance		df
		Total	Model	
Fork length				
Overall	0.001	0.35		961
Sex	0.001	0.03	0.09	1
Month	0.001	0.25	0.78	4
Year	0.001	0.13	0.41	6
Sex*Month	0.533	0.01		4
Sex*Year	0.832	<0.01		6
Month*Year	0.253	0.02		22
Sex*Month*Year	0.876	0.01		14
Whole weight				
Overall	0.001	0.17		961
Sex	0.001	0.02	0.21	1
Month	0.008	0.01	0.12	4
Year	0.001	0.09	0.78	6
Sex*Month	0.604	<0.01		4
Sex*Year	0.300	0.01		6
Month*Year	0.344	0.02		22
Sex*Month*Year	0.892	0.01		14
Condition factor				
Overall	0.001	0.50		961
Length	0.001	0.34	0.69	1
Sex	0.034	<0.01	0.01	1
Month	0.001	0.12	0.25	4
Year	0.040	0.01	0.02	6
Sex*Month	0.796	0.01		4
Sex*Year	0.292	<0.01		6
Month*Year	0.128	0.02		22
Sex*Month*Year	0.915	<0.01		14

Table 5. Analysis of variance of month and year effects on the fork length 2SW maiden Atlantic salmon from the Miramichi River. Proportion of variance (total and model) explained by a main effect is calculated for each effect individually and with the interaction term(s) excluded.

Model Effects	*P*-value	Proportion of Variance		df
		Total	Model	
Overall	0.001	0.14		8198
Month	0.001	0.08	0.68	1
Year	0.001	0.03	0.24	1
Month*Year	0.103	0.03		1

explained variance) (Table 6). Season was a more important effect than year; fall-run salmon were always larger than those sampled in the summer (Fig. 6). Predicted fork lengths of 2SW salmon were much more variable than those of 1SW salmon; the largest fish were observed in 1976 and in 1987 to 1990 (Fig. 7).

Size differences in the fork lengths of 1SW maiden salmon were noted among the three management regimes. About 22% of the total variation in fork length was explained by the effects of management regime and season with the largest effect attributed to changes in management (Table 7). Lengths of 1SW maiden salmon during the 1984 to

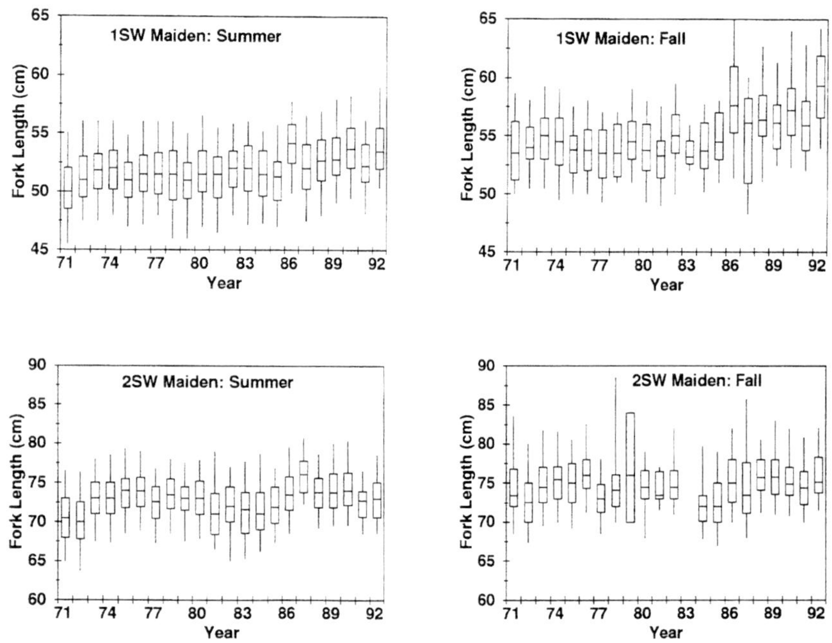

Fig. 6. Fork length distributions of 1SW and 2SW maiden Atlantic salmon by summer (June and July) and fall (September and October) runs from 1971 to 1992. Bars and lines as in Fig. 3.

Table 6. Analysis of variance of season and year effects on the observed fork lengths of 1SW and 2SW maiden Atlantic salmon from the Miramichi River, 1971 to 1992. Reference categories are "summer" for the season effect and "1992" for the year effect. Proportion of variance (total and model) explained by a main effect is calculated for each effect individually and with the interaction term(s) excluded.

		Proportion of Variance		
Model Effects	*P*-value	Total	Model	df
1SW maiden salmon				
Overall	0.001	0.26	0.25	9645
Season	0.001	0.13	0.53	1
Year	0.001	0.17	0.69	21
Season*Year	0.001	0.01		21
2SW maiden salmon				
Overall	0.001	0.18	0.18	4674
Season	0.001	0.13	0.76	1
Year	0.001	0.08	0.43	21
Season*Year	0.002	0.01		21

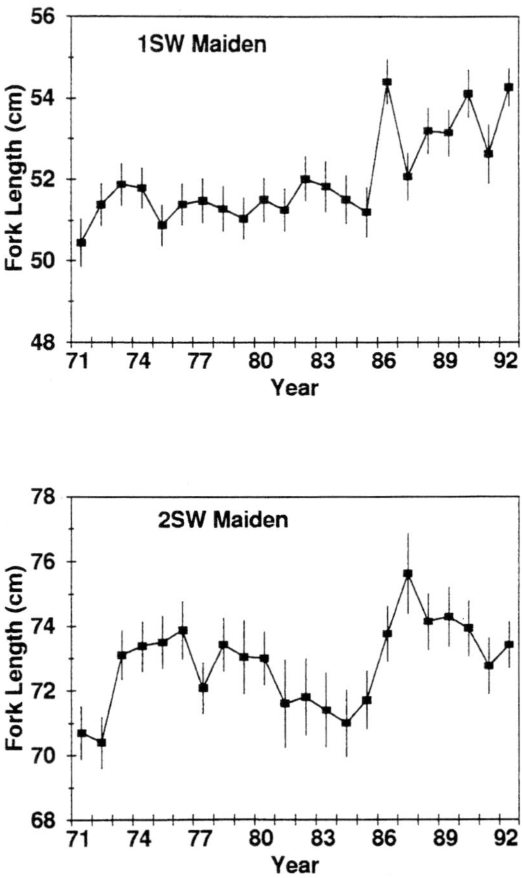

Fig. 7. Predicted fork length of 1SW and 2SW maiden Atlantic salmon from the analysis of variance of length by year and season (summer and fall). Predicted lengths are based on summer and 1992 as reference levels. Squares are predicted value, vertical bars are ± 2 standard errors.

1992 commercial closure were significantly longer by about 1.5 cm than those during 1971 to 1980 and 1981 to 1983 periods. Lengths of salmon during 1981 to 1983 were also significantly longer, by 0.3 cm, than those of 1971 to 1979. The fact that lengths of fish in both the summer and fall seasons were significantly longer after the closure of 1984 indicates that larger individuals from both the early and late runs were being selectively exploited in the sea fisheries.

There were also significant size differences in the fork lengths of 2SW maiden salmon directly attributable to the three management regimes. Only 9% of the total variation in fork length was explained by the effects of management regime and season but the former was a more important effect (Table 7). 2SW salmon sampled at Millbank in the 1984 to 1992 period were significantly longer than those in the other management time periods (Fig. 7). Exploitation during the 1981 to 1983 period was highly selective on the larger individuals of the 2SW salmon component; fork lengths of these fish during 1970 to 1980 were significantly larger, by 1 cm, than from 1981 to 1983.

Sea age of salmon was also effected by the commercial fisheries. Since 1971, small salmon were consistently 97 to 100% maiden 1SW salmon and this group was not examined further. The age composition of large salmon, however, differed

Table 7. Analysis of variance of management regime and season on the observed fork lengths of 1SW and 2SW maiden Atlantic salmon from the Miramichi River, 1971 to 1992. Reference categories are "summer" for the season effect and "1981 to 1983" for the management regime effect. Proportion of variance (total and model) explained by a main effect is calculated for each effect individually and with the interaction term(s) excluded.

Model Effects	*P*-value	Proportion of Variance		df
		Total	Model	
Maiden 1SW salmon				
Overall	0.001	0.22	0.22	9645
Management regime	0.001	0.05	0.53	2
Season	0.001	0.17	0.69	1
Management*Season	0.001	<0.01		2
Maiden 2SW salmon				
Overall	0.001	0.09	0.09	4674
Management Regime	0.001	0.08	0.87	2
Season	0.001	0.02	0.27	1
Management*Season	0.196	<0.01		2

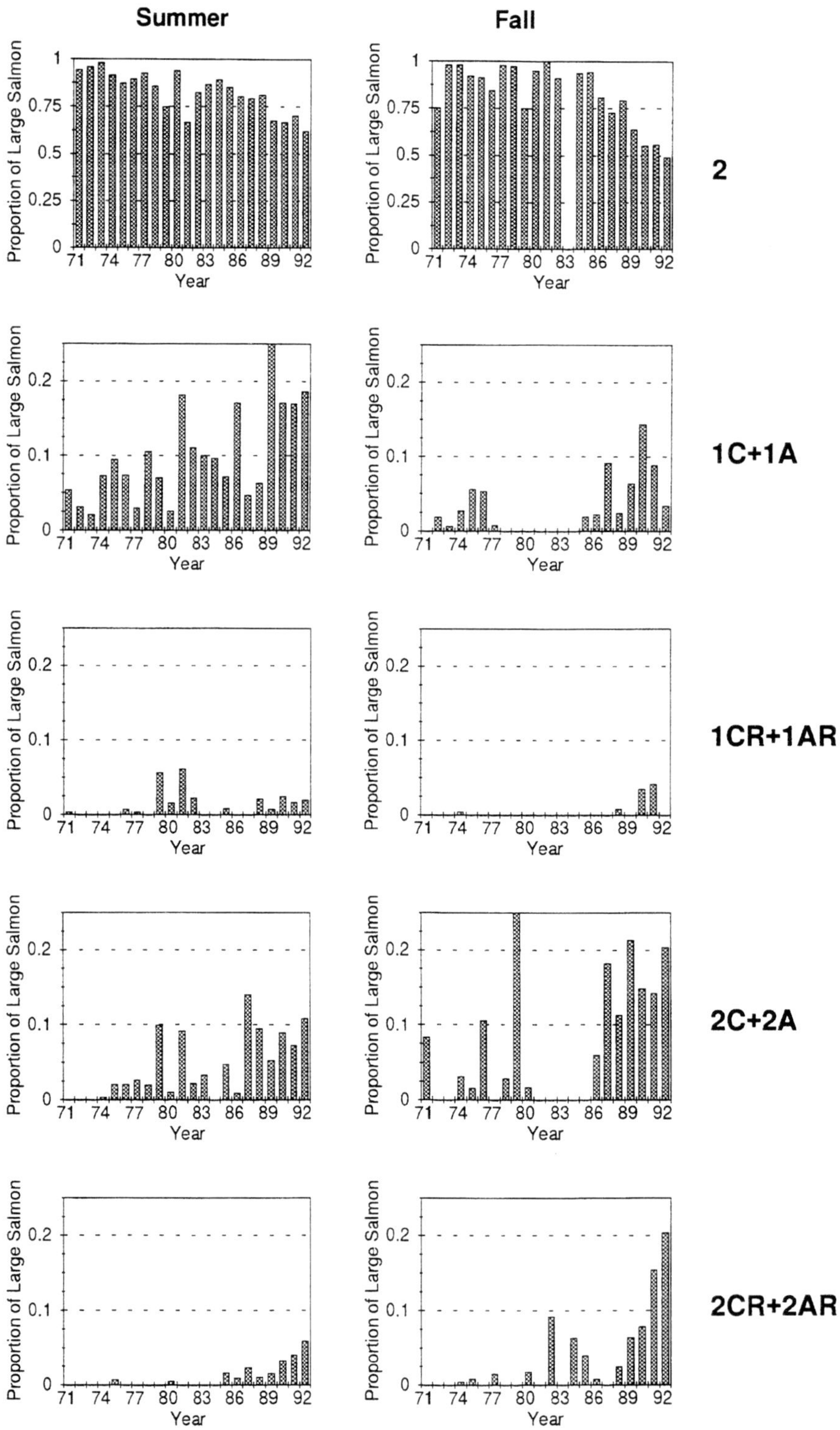

Fig. 8. Proportion by spawning history in the large salmon group (>= 63 cm fork length) for summer (June and July) and fall (September and October) runs of Atlantic salmon to the Miramichi River, 1971 to 1992. Spawning history designations are defined in text.

between years and between summer and fall seasons (Fig. 8). Maiden 2SW salmon typically accounted for over 80% of all large salmon for both summer and fall runs during the 1970's. The proportions of maiden 2SW salmon in the large salmon component were not significantly different between summer and fall for either the 1971 to 1982 period or the 1984 to 1992 period (Table 8). Previous spawner 2SW salmon have been more abundant in the fall since 1984 (Table 8; Fig. 8). By contrast, previous spawner 1SW salmon were significantly more abundant in the summer for both the 1971 to 1982 and the 1984 to 1992 periods (Table 8).

The proportions of previous spawners changed significantly between the three manage-

Table 8. Relative proportions of previous spawner 1SW and 2SW Atlantic salmon by season for different management regimes, 1971 to 1992.

		Relative proportions			
		Median	Min.	Max.	*P*-value of comparisons
Previous 1SW spawners					
1971 to 1982	Summer	0.077	0.021	0.242	0.01
	Fall	0.003	0.0	0.056	
1984 to 1992	Summer	0.171	0.047	0.258	0.03
	Fall	0.034	0.0	0.178	
Previous 2SW spawners					
1971 to 1982	Summer	0.020	0.0	0.099	0.35
	Fall	0.031	0.0	0.250	
1984 to 1992	Summer	0.105	0.0	0.167	0.04
	Fall	0.182	0.039	0.407	

Table 9. Relative proportions of previous spawner 1SW and 2SW Atlantic salmon among the different management regimes for summer and fall seasons. Similar lower case letters refer to group of management regimes being compared under the null hypothesis that the proportions during the regimes are similar.

		Relative proportions			*P*-value of comparisons			
		Median	Min.	Max.				
Previous 1SW spawners								
Summer	1971-80	0.065	0.021	0.127	a			
	1981-83	0.133	0.100	0.242	a	b		
	1984-92	0.171	0.047	0.258	a	0.01	b	0.83
Fall	1971-82	0.171	0.047	0.258	c			
	1984-92	0.034	0.0	0.178	c	0.01		
Previous 2SW spawners								
Summer	1971-80	0.018	0.0	0.0	99d			
	1981-83	0.033	0.022	0.091	d		e	
	1984-92	0.105	0.0	0.167	d	0.01	e	0.10
Fall	1971-82	0.105	0.0	0.167	f			
	1984-92	0.182	0.039	0.407	f	0.01		

ment periods. Previous spawners of both 1SW and 2SW groups were significantly more abundant in the periods after 1980 (Table 9; Fig.8). There were no significant differences between the 1981 to 1983 period and the 1984 to 1992 period for the summer run, but in the fall runs, 1SW and 2SW previous spawners were significantly more abundant after 1983 (Table 9; Fig. 8).

Because the recruitment of maiden 2SW salmon to the Miramichi River has not been constant we examined its effect on the proportions at age. Between 1976 and 1992, the estimated returns of maiden 2SW salmon to Millbank varied between 5,300 and 43,000 fish, with a median return of 17,100, but have been increasing since 1989 (Fig. 9). Summer returns have generally been stronger than fall returns but the relative sizes of

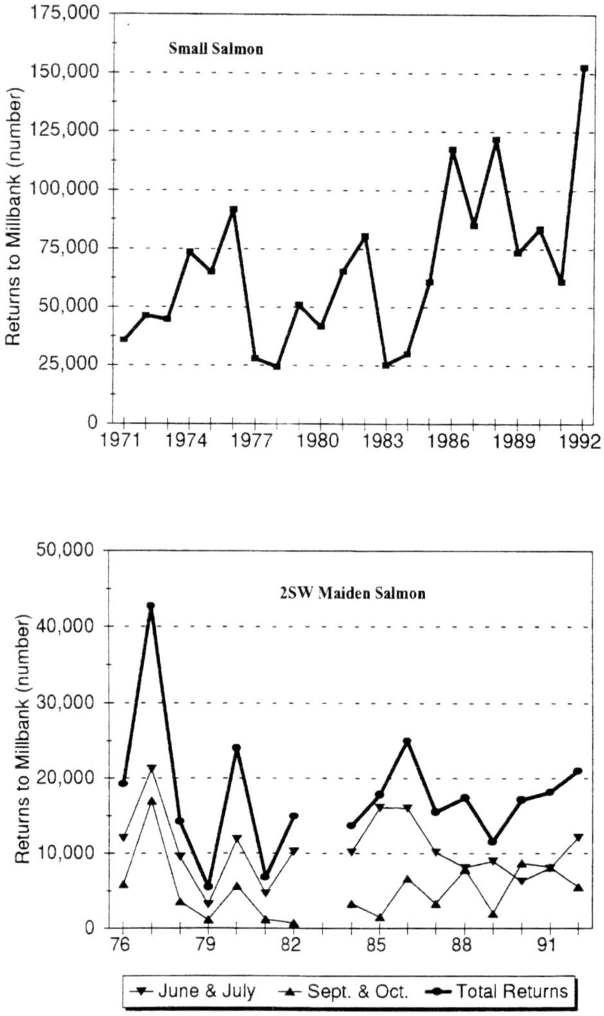

Fig. 9. Estimated returns (number) of small and large Atlantic salmon to Millbank, 1976 to 1992.

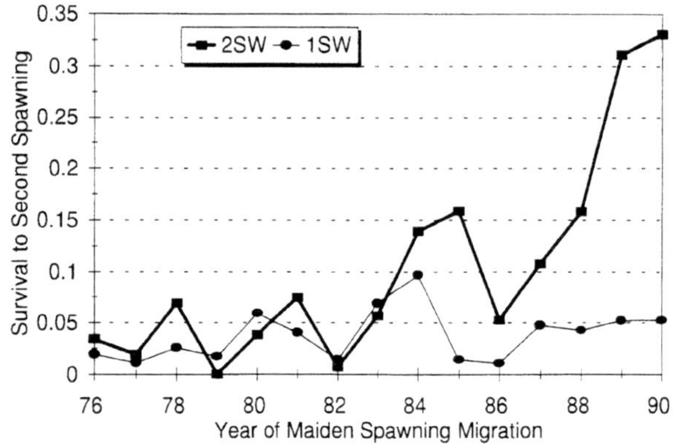

Fig. 10. Estimated survival rates to a second spawning migration of 1SW and 2SW Atlantic salmon from the Miramichi River.

Table 10. Estimated survival rates to a second spawning migration of 1SW and 2SW maiden Atlantic salmon from the Miramichi River relative to two periods of different management regime. Years refer to year of maiden spawning migration.

		Survival to second spawning			*P*-value of comparison
		Median	Min.	Max.	
1SW	1976 to 1983	0.022	0.011	0.069	0.39
	1984 to 1990	0.048	0.011	0.097	
2SW	1976 to 1983	0.036	0.0	0.074	0.01
	1984 to 1990	0.158	0.052	0.331	

each run component have become more variable in recent years. Increased proportions of previous spawners in the total run can not be attributed to a decline in the maiden 2SW component. For the summer returns, the number of maiden 2SW salmon has been increasing in conjunction with the proportion of previous spawners in the large salmon component (compare Figs. 8 and 9). For both the summer and fall runs, the larger increase in previous spawners can be mostly attributed to an increase in survival.

The proportion of 2SW maiden salmon which survive the fisheries and return to Millbank to spawn a second time, has increased significantly since 1984 (Table 10; Fig. 10). By contrast, survival of 1SW salmon to a second spawning return has not changed in spite of the closure of the commercial fisheries and the reduction in the seasonal bag limits in the angling fisheries (Fig. 10). The greatest change in post-spawner survival of large salmon can be ascribed to reductions in homewater fisheries exploitation. Prior to 1985, large salmon could be retained by anglers. Since 1984, the total exploitation rate on large salmon has declined to 7% (median 1984-1992) compared to 54% before 1984 (Fig. 11). Exploitation rates on small salmon have also declined since 1984 but much less than for large salmon. The median was 31% between 1984 and 1992 compared to a median of 42% prior to 1984 (Fig. 11). Slight reductions in exploitation and increasing returns of maiden 1SW salmon have resulted in a relatively greater proportion of 1SW previous spawners in the large salmon returns to the Miramichi River since 1984.

While changes in homewater fisheries appear to have affected survival at sea, changes in distant commercial fisheries off Newfoundland, Labrador, and Greenland have not had a detectable impact. Previous spawners tagged at Millbank have been intercepted every year in the coastal fisheries of Newfoundland and Labrador, Greenland and Quebec (Table 11). The proportion of available tags to tags returned from previously spawned 2SW salmon caught in marine fisheries did not decline significantly, between the 1971-1975 and 1985-1990 time periods (Table 11). The exploitation rate of tagged previously spawned 1SW salmon also did not change between the two time periods (Table 11).

Discussion

Our analysis has verified the prediction that fisheries affect the biological characteristics of Atlantic salmon. When the size-selective commercial fisheries were closed, there was an increase in the size-at-age, in the proportion of previous spawner 1SW salmon, and in the abundance of previous spawner 2SW salmon in the returns.

There is strong evidence from this analysis and from other studies that the commercial fisheries which operated in the Maritime provinces

Table 11. Recaptures in the sea fisheries of large and small salmon tagged at the Millbank trapnet, 1971 to 1975 and 1985 to 1990. The null hypothesis is that the proportions of available tags returned during the two time periods are similar.

			Annual proportion returned			
Time period	Available tags	High seas recaptures	Median	Min.	Max.	*P*-value of comparison
Large salmon						
1971-1975	1536	30	0.016	0.008	0.043	
1985-1990	1609	13	0.007	0.0	0.018	0.08
Small salmon						
1971-1975	2611	21	0.007	0.007	0.011	
1985-1990	6035	44	0.006	0.002	0.015	0.84

were size-selective. These commercial salmon fisheries used fishing gear with a minimum stretched mesh size of 127 mm, for both gillnets and trapnets (Dunfield 1974) and as such the fishery was restricted to capturing salmon weighing more than 2.3 kg (Kerswill 1971). The selectivity of such gear for the larger fish of the maiden 1SW component and on the larger previous spawners and 2SW salmon has been documented for the Miramichi (Reddin 1984) as well as for other stocks (Chadwick 1988; Chadwick and Claytor 1990; Doubleday and Reddin 1981). Other studies

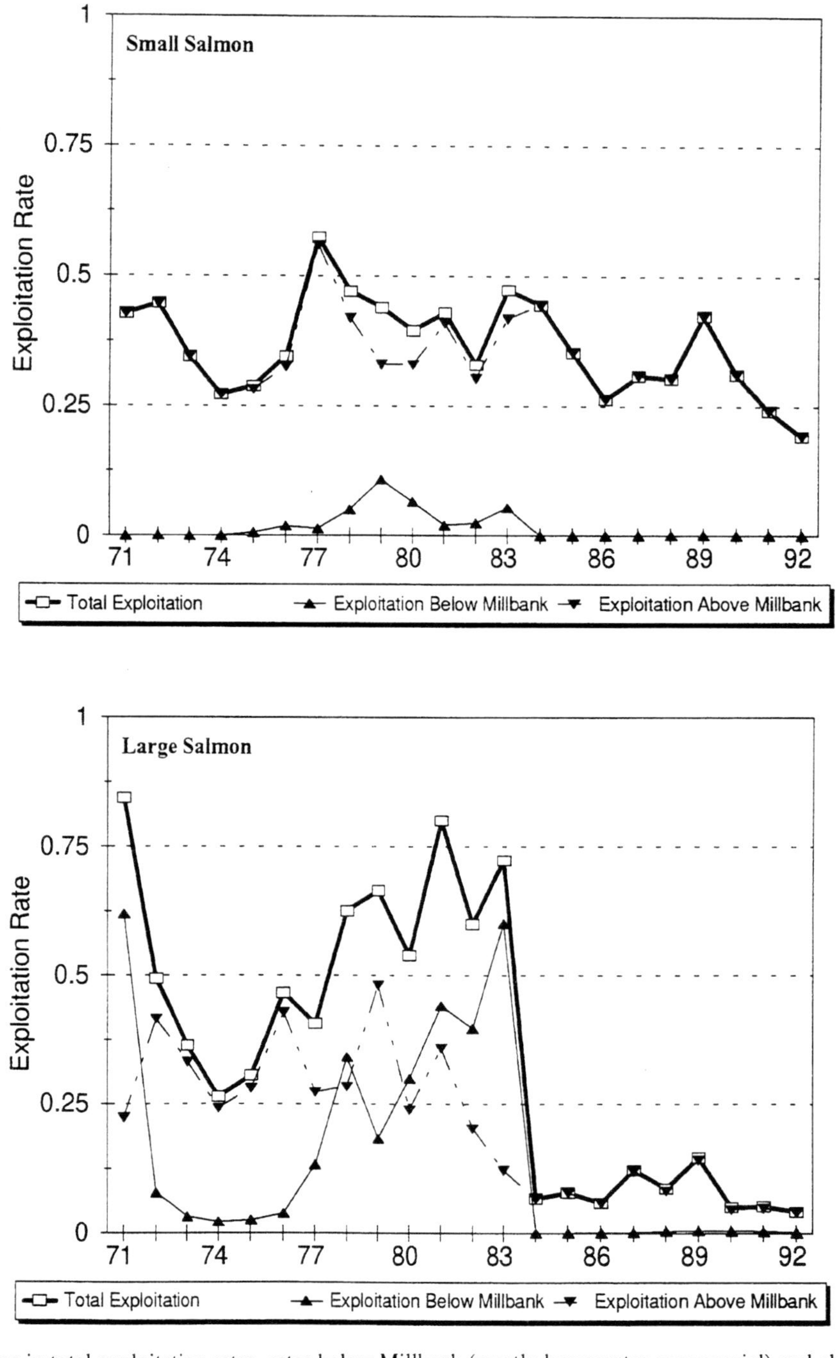

Fig. 11. Changes in total exploitation rates, rates below Millbank (mostly homewater commercial) and above Millbank (commercial, recreational, and food fisheries) on small and large Atlantic salmon of the Miramichi River, 1971 to 1992.

have described changes in size-at-age which have been subsequently attributed to commercial fisheries (D'Amours 1993). When such size-selective fisheries were closed, the size-at-age increased. This phenomenon was evident in our study, in the analysis of the size-at-age in Gaspé rivers (D'Amours 1993), and in west coast stocks of Iceland (Scarnecchia et al. 1989).

The increased proportion of previous spawners in both the 1SW and 2SW components was also consistent with the elimination of a fishery which was size-selective rather than age selective. The exploitation rate on large salmon in the commercial fishery was higher than on small salmon (Kerswill 1971) and the combined exploitation rates from both commercial and recreational fisheries reached 97% of the returns as compared to 56% on small salmon (Saunders 1969). Between 1971 and 1983, the combined exploitation rate on large salmon varied between 26% and 78% while on small salmon, the rate varied between 26% and 55% (this study). It was evident that the Maritime fisheries were exploiting both early and late run components of the Miramichi River because size-at-age of salmon increased in both seasons when the commercial fisheries were closed.

The dramatic reduction in the fishing mortality of the 2SW maiden salmon has resulted in sea survival rates to a second spawning that are six times higher than survival to a second spawning of maiden 1SW salmon. Jonsson et al. (1991b) reported that the post-spawning survival of males was lower than females because of the energetic costs of reproduction; males and females had similar total energetic expenditures but male losses were somatic whereas female losses were mostly gonadic material. In the Miramichi, small salmon are about 75% male and our analysis suggests that males may indeed have lower post-spawner survival than females (change in sex ratio between 1SW maiden salmon and 1SW alternate spawners). This would in turn suggest that small salmon survival rates should be lower than large salmon in the Miramichi in the absence of fisheries exploitation. The survival rates of 1SW maiden salmon to a second spawning of 2% to 4% are slightly lower than the 3.1% for 1SW salmon and 8.8% for 2SW salmon reported for Saint John River (Marshall and MacPhail 1987). Differences may be due to different methods of calculating these rates. In the Saint John River study they were calculated after in stream fisheries, which includes catch-and-release of large salmon. Jonsson et al. (1991a) reported that sea survival of large salmon from Norwegian rivers was lower than for small salmon and attributed this to the differential costs of reproduction. Our study would suggest that lower apparent sea survival to a second spawning of the Norwegian large salmon is probably related to fisheries exploitation rather than natural survival. Following changes to fisheries in 1984, survival rates of maiden 2SW spawners increased to a median value of 16%. When Newfoundland and Labrador fisheries were restricted by quota in 1990 and 1991 and closed completely for insular Newfoundland in 1992 survival rates increased to 30%.

Natural mortality of spent Atlantic salmon varies from river to river. Radio tracking studies of post-spawning male and female Atlantic salmon in France found that post-spawning mortality was high (Baglinière et al. 1990, 1991). Post-spawning natural mortality in the River Conon, U.K., averaged 26% over 6 years (Mills 1989) whereas at Western Arm Brook, Newfoundland, overwinter survival rates were as high as 88% (Chadwick et al. 1978). High survival rates to a second spawning in the absence of exploitation may be directly related to overwintering habitat and the abundance of food available to the kelts in the spring. Kelts are known to actively feed on smelt and this food item has been used to initiate feeding of kelts in captivity (Crim et al. 1992). There is a large smelt population in the Miramichi, upwards of several 100 million fish (McKenzie 1964; Chaput 1995) and these are readily available to kelts from March through June. Rivers without an abundance of prey during spring may not favour survival. We hypothesize that the high proportions of previous spawners in the returns of large salmon to the Miramichi are the result of increased survival due to reduced fisheries exploitation and perhaps due to the extensive suitable overwintering habitat and abundant feed for the kelts in the spring. Overwintering habitat may be one of the critical components to survival. Western Arm Brook has an extensive pond system with suitable habitat for overwintering survival rates have varied between 29% and 88% (Chadwick et al. 1978).

The importance of previous spawners cannot be understated as they can contribute significantly to egg depositions. Previous spawners contributed more than 40% of the egg depositions for the Miramichi in 1992 and 1993 (Chaput et al. MS1994) and from 3% to 26% of the egg depositions for the Saint John River (Marshall and MacPhail 1987). In addition previous spawners mitigate the effects of variable sea survival of maiden salmon (Chadwick 1988; Schaffer and Elson 1975).

In a summary of the age composition from 24 Canadian rivers, Schaffer and Elson (1975) indicated that the proportion of previous spawners was less than 5% in 17 rivers; it was between 6% and 12% in 6 rivers; and it was 61% on one river, Big Salmon River. The proportion of multiple spawners in the overall annual returns to a river will depend upon the size of the maiden return in previous years as well as post-spawning survival rates. Low proportions of previous spawners in the 1930s to 1970s in the Miramichi River were most likely the result of fisheries exploitation. Further closures of high seas fisheries in recent years should result in increased numbers of previous spawners; proportions of 50% or more previous spawners should not be unexpected in some rivers.

References

Baglinière, J.-L., Maisse, G., and Nihouarn, A. 1990. Migratory and reproductive behavior of female adult Atlantic salmon, *Salmo salar* L., in a spawning stream. J. Fish. Biol. 36: 511-520.

Baglinière, J.-L., Maisse, G., and Nihouarn, A. 1991. Radio tracking of male adult Atlantic salmon, *Salmo salar* L., during the last phase spawning migration in a spawning stream (Brittany, France). Aquat. Living Resour. 4: 161-167.

Blair, A.A. 1932. Salmon (*Salmo salar*) of the Miramichi River System. M.A. Thesis, U. of Toronto. 70 p.

Blair, A.A. 1935. Ages at migration of Atlantic salmon (*Salmo salar*) in Miramichi River. J. Biol. Bd. Can. 1: 159-169.

Chadwick, E.M.P. 1988. Relationship between Atlantic salmon smolts and adults in Canadian rivers. p. 301-324. *In* D. Mills and D. Piggins [*editors*]. Atlantic salmon: Planning for the Future. Proc. 3rd International Atlantic Salmon Symposium. Croom Helm, London.

Chadwick, E.M.P., and Claytor, R.R. 1990. Predictability in a small commercial Atlantic salmon fishery in western Newfoundland. Fish. Res. 10: 15-28.

Chadwick, E.M.P., Porter, T.R., and Downton, P. 1978. Analysis of growth of Atlantic salmon (*Salmo salar*) in a small Newfoundland river. J. Fish. Res. Board Can. 35:60-68.

Chaput, G.J. 1995. Temporal distribution, spatial distribution, and abundance of diadromous fish in the Miramichi River watershed, p.121-139. *In* E.M.P. Chadwick [*editor*]. Water, science, and the public: the Miramichi ecosystem. Can. Spec. Publ. Fish. Aquat. Sci. 123.

Chaput, G., Moore, D.S., Biron, M., and Claytor, R. 1994. Abundance of Atlantic salmon (*Salmo salar*) in the Miramichi River in 1993. Can. Atl. Fish. Res. Doc. 94/20. 80p.

Claytor, R.R., Léger, C.E., and. Gray, R.W. 1987. Stock composition of Northumberland Strait, Nova Scotia Atlantic salmon (*Salmo salar*) commercial fisheries. Can. Tech. Rep. Fish. Aquat. Sci. 1563: viii + 19p.

Crim, L.W., Wilson, C.E., So, Y.P., Idler, D.R., and Johnston, C.E. 1992. Feeding, reconditioning, and rematuration of captive Atlantic salmon (*Salmo salar*) kelt. Can. J. Fish. Aquat. Sci. 49: 1835-1842.

D'Amours, P. 1993. Variations annuelles du poids moyen des captures sportives de Saumon atlantique *Salmo salar* (Linné) gaspésien. M.Sc. Thesis. Université de Moncton. 115 p.

Doubleday, W.G., and Reddin, D. G. 1981. An analysis of the implications of alternative mesh sizes for gillnets and opening dates for the commercial salmon fishing season at West Greenland. Can. Atl. Fish. Sci. Adv. Comm. Res. Doc. 81/3: 56p.

Dunfield, R.W. 1974. Types of commercial salmon fishing gear in the Maritime provinces - 1971. Resources Development Branch Maritimes Region Information Publication No. MAR/N-74-1. 43p.

Edgington, E.S. 1987. Randomization Tests 2nd Edition.Marcel Dekker Inc., New York, NY. 341p.

Jessop, B.M. 1986. Atlantic salmon (*Salmo salar*) of the Big Salmon River, New Brunswick. Can. Tech. Rep. Fish. Aquat. Sci. No. 1415: xii+50p.

Jonsson, N., Hansen, L.P., and Jonsson, B. 1991a. Variation in age, size and repeat spawning of adult Atlantic salmon in relation to water discharge. J. Animal Ecol. 60: 937-947.

Jonsson, N., Jonsson, B., and Hansen, L.P. 1991b. Energetic cost of spawning in male and female Atlantic salmon (*Salmo salar* L.). J. Fish Biol. 39: 739-744.

Kerr, R.B. 1961. Scale to length ratio, age and growth of Atlantic salmon in the Miramichi fisheries. J. Fish. Res. Board Can. 18: 117-124.

Kerswill, C.J. 1971. Relative rates of utilization by commercial and sport fisheries of Atlantic salmon (*Salmo salar*) from the Miramichi River, New Brunswick. J. Fish. Res. Board Can. 28:351-363.

Marshall, T.L. 1988. Harvest and recent management of Atlantic salmon in Canada. p. 117-142. *In* D. Mills and D. Piggins [*editors*]. Atlantic salmon: Planning for the Future. Proc. 3rd International Atlantic Salmon Symposium. Croom Helm, London. 117-142.

Marshall, T.L., and MacPhail, D.K. 1987. Black salmon fishery and repeat spawning salmon of the Saint John River, N.B. 1987. Can. Atl. Fish. Sci. Adv. Comm. Res. Doc. 87/100. 14p.

McKenzie, R.A. 1964. Smelt life history and fishery in the Miramichi River, New Brunswick. Fish. Res. Board Can. Bull. No. 144: 77p.

Mills, D. 1989. Ecology and management of Atlantic salmon. Chapman and Hall. New York, NY. 351 p.

Patterson, K.R. 1992. An improved method for studying the condition of fish, with an example using Pacific sardine *Sardinops sagax* (Jenyns). J. Fish Biol. 40: 821-831.

Reddin, D.G. 1984. Results of Atlantic salmon tagging study in Miramichi Bay, New Brunswick. Can. Tech. Rep. Fish. Aquat. Sci. No. 1301: iv+22p.

Ruggles, C.P., and Turner, G.E. 1973. Recent changes in the stock composition of Atlantic salmon (*Salmo salar*) in the Miramichi River, New Brunswick. J. Fish. Res. Board Can. 30: 779-786.

SAS. 1990. SAS User's Guide: Statistics. SAS Institute Inc. Cary, North Carolina.

Saunders, R.L. 1969. Contributions of salmon from the Northwest Miramichi River, New Brunswick, to various fisheries. J. Fish. Res. Board Can. 28: 269-278.

Scarnecchia, D.L., Isaksson, A., and White, S.E. 1989. Effects of oceanic variations and the West Greenland fishery on age at maturity of Icelandic west coast stocks of Atlantic salmon. Can. J. Fish. Aquat. Sci. 46: 16-27.

Schaffer, W.M., and Elson, P.F. 1975. The adaptive significance of variations in life history among local populations of Atlantic salmon in North America. Ecology 56: 577-590.

Shearer, W.M. 1989. Report of the second Atlantic salmon scale reading workshop – Aberdeen, Scotland, 12-14 October 1988. ICES CM 1989\M:7.

CHAPTER 15

By-Catch in Commercial Fisheries as an Indicator and Regulator of Striped Bass (*Morone saxatilis*) Abundance in the Miramichi River Estuary

Rodney G. Bradford

New Brunswick Wildlife Federation, P.O. Box 20211,
Fredericton, New Brunswick, Canada E3B 7A2

Kimberly A. Robichaud

Department of Biology, University of New Brunswick, P.O. Box 4400,
Fredericton, New Brunswick, Canada E3B 5A3

and Simon C. Courtenay

Department of Fisheries and Oceans, Science Branch, 343 Archibald St.,
Moncton, New Brunswick, Canada E1C 9B6

Bradford, R.G., Robichaud, K.A., and Courtenay, S.C. 1995. By-catch in commercial fisheries as an indicator and regulator of striped bass (*Morone saxatilis*) abundance in the Miramichi River Estuary, p. 249-259. *In* E.M.P. Chadwick [*editor*]. Water, science, and the public: the Miramichi ecosystem. Can. Spec. Publ. Fish. Aquat. Sci. 123.

Abstract

By-catch of striped bass (*Morone saxatilis*) in the commercial fisheries of the Miramichi River Estuary occurs at all life-history stages beyond the larval stage. Gaspereau traps (May-June) catch both spawning adult and immature (age 2^+) striped bass with a potential exploitation rate for adult fish of about 25%. Fyke nets set for American eel (July to October) catch predominantly immature striped bass <2 years of age. During autumn (October to December), smelt nets catch both immature and adult striped bass, but 0^+ fish dominate by-catch. Few (<<1 per net per 24 h) striped bass are captured in the winter smelt fishery (January-March). Total catch of striped bass for all fisheries are in the mid-tens of thousands of age 0^+ fish and in the thousands for both older juveniles and adults. Mark recapture experiments suggested that adult striped bass remained near the spawning grounds during spawning and that they migrate along the coast of the southern Gulf of St. Lawrence during summer. Some adult fish may return to the Miramichi River Estuary to overwinter. Both immature and adult striped bass marked elsewhere in the southern Gulf of St. Lawrence were recaptured in the Miramichi River Estuary as either migrant, or overwintering or spawning fish.

Résumé

On observe des prises accessoires de bar rayé (*Morone saxatilis*) dans les pêches commerciales de l'estuaire de la Miramichi à tous les stades biologiques, passé le stade larvaire. Les trappes à gaspareau (mai-juin) retiennent à la fois des adultes reproducteurs et des immatures (âge 2^+) de bar rayé, avec un potentiel d'exploitation des poissons adultes d'environ 25 %. Les verveux installés pour capturer l'anguille (de juillet à octobre) retiennent principalement des bars rayés immatures de moins de deux ans. Pendant l'automne (d'octobre à décembre), les filets à éperlan capturent à la fois des immatures et des adultes de bar rayé, mais les poissons d'âge 0^+ dominent dans les prises accessoires. Un très petit nombre (moins de un par filet par 24h) de bars rayés est capturé à la pêche hivernale de l'éperlan (de janvier à mars). Au total, les captures de bar rayé dans toutes les pêches sont de l'ordre de 5 000 poissons d'âge 0^+ et de quelques milliers pour les juvéniles plus âgés et les adultes. Les expériences de marquage-recapture indiquent que les bars rayés adultes restent près des frayères pendant la ponte, puis migrent le long de la côte sud du golfe du Saint-Laurent pendant l'été. Certains adultes peuvent revenir hiverner dans l'estuaire de la Miramichi. On a repris dans l'estuaire des bars rayés immatures et adultes marqués ailleurs dans le sud du golfe du Saint-Laurent; ces poissons étaient soit en migration, soit en hivernage, soit en période de fraye.

Introduction

Greater than 90% of all reported landings of striped bass in the Maritime Provinces occur in the Gulf of St. Lawrence as unregulated by-catch in the estuarine commercial fisheries of New Brunswick (Chaput and Randall 1990; Melvin 1991). Knowledge of the impact of by-catch on the population dynamics of striped bass is necessary if recreational fisheries planned for this species (Loftus et al. 1993) are to develop. Angling for striped bass along the eastern seaboard of the United States generates over $100 million annually (Norton et al. 1983). Neither the size composition nor the life-history stages of striped bass captured in the New Brunswick fisheries have been examined (Hooper 1991) even though by-catch has been cited, since before the turn of the century, as a contributing factor to the cyclic nature of striped bass abundance in eastern Canada (Venning 1885). Landings from recreational and Aboriginal food fisheries which occur in most estuaries of the southern Gulf of St. Lawrence have not been systematically reported until very recently.

We examined the extent to which estuarine commercial fisheries in New Brunswick could influence current striped bass abundance, which is at a low level relative to the long term average (Chaput and Randall 1990; Chaput 1995). We also examined the usefulness of using by-catch to estimate adult and juvenile abundance. By-catch of striped bass was restricted in 1993 to fish < 38 cm total length, or < 35.5 cm fork length (FL). By-catch reduction was part of a striped bass management plan that was designed to arrest declines in abundance, to rebuild the resource, and to maximize economic benefits.

Miramichi River estuary (Fig. 1) is a known spawning site for striped bass (Scott and Scott 1988) and the location of well-developed fixed-gear commercial fisheries (Chaput 1995). Commercial fishing has been an important component of the local economy for over 100 years (Perley 1852) with total landings of about 2000 - 3000 t yr^{-1} (LeBlanc and Chaput 1991). Systematic sampling of the Miramichi fisheries, and their by-catch has been initiated only recently (Hanson and Courtenay 1995).

Description of Fixed-Gear Fisheries

Three fixed-gear commercial fisheries occur seasonally in the Miramichi River Estuary. Gaspereau (alewife (*Alosa pseudoharengus*) and blueback herring (*A. aestivalis*)) are fished with box traps during May and June. Ichthyoplankton surveys have shown that the gaspereau fishery occurs at the same time as spawning of striped bass, and partially at the same location (Robichaud, unpublished data; Fig. 1a). American eels (*Anguilla rostrata*) are fished with fyke nets during June to October. Rainbow smelt (*Osmerus mordax*) and Atlantic tomcod (*Microgadus tomcod*) are fished with either box or bag nets from October to March. Tomcod is a by-catch in the smelt fishery. Beach-seining surveys indicated that eel and smelt fishing occurs in the same locations as the summer-autumn distributions of young-of-the-year striped bass (Robichaud, unpublished data; Fig. 1b) and older striped bass (Chaput and Randall 1990). Seasons, locations, mesh size of fishing gear and distribution of effort for each fishery are given in Table 1.

Methods

During May and June, 1993, gaspereau traps were sampled in the Northwest (NW) Miramichi River Estuary (Fig. 1a). This area was known for striped bass spawning (Robichaud, unpublished data; Fig. 1a) and for the largest recorded by-catch (Chaput and Randall 1991). The fishery exploits the predictable annual spawning run of gaspereau (median peak catches, 1978-1987, was June 4; Chadwick and Claytor 1989). Peak spawning activity for striped bass is mid-May to mid-July (Robichaud unpublished data; Bradford unpublished data). One to six traps were visited daily. Traps near Chatham and Loggieville (Fig. 1b) were visited between 4 June - 15 June; only one to two traps were sampled per day. Estimates of by-catch were based on 73 of 338 trap records from the NW Miramichi section from which 920 striped bass were sampled.

In 1993, fishers were permitted to keep striped bass smaller than 35.5 cm fork length (FL), the approximate length at first maturity for northern populations (Hogans and Melvin 1984). Therefore, live striped bass were measured to fork length (FL; 0.1 cm) and samples were separated into 2 size groups: ≤ 35.5 cm and > 35.5 cm. Age structure of the legal catch was obtained from length-stratified sampling (minimum of 5 fish per 0.5 cm interval). Scales were removed from the region between the lateral line and the anterior dorsal fin (Chaput and Robichaud 1995). Fish were assigned an arbitrary birth date of January 1.

Striped bass ≥35.5 cm were tagged and released to estimate potential exploitation rates in the by-catch fisheries. Individually numbered, fluorescent-orange Dennison anchor tags (length 8.9 cm) were inserted between the first two spines of the anterior dorsal fin (Dunning et al. 1987). A $5.00 reward was offered for each tag returned with information on date and location of capture. Each tag carried the address of the Gulf Fisheries

Table 1. Seasons, location, and gear number, type and mesh size (cm) for commercial fisheries since 1993.

Fishery		Duration	Location	Gear	Nets	Mesh Size
Gaspereau		15 May – 20 June	NW Miram.	Box	13	3.1
			Chatham		11	
			Loggieville		6	
			Napan R.		6	
Eel		1 July – 5 October	NW Miram.	Fyke	44	2.4
			Loggieville		15	
			Napan R.		24	
Smelt						
	Autumn	15 October– 6 December	Chatham to Napan R.	Bag	21	2.4
	Winter	5 January– 5 March	Miramichi Bay	Bag Box	>300	

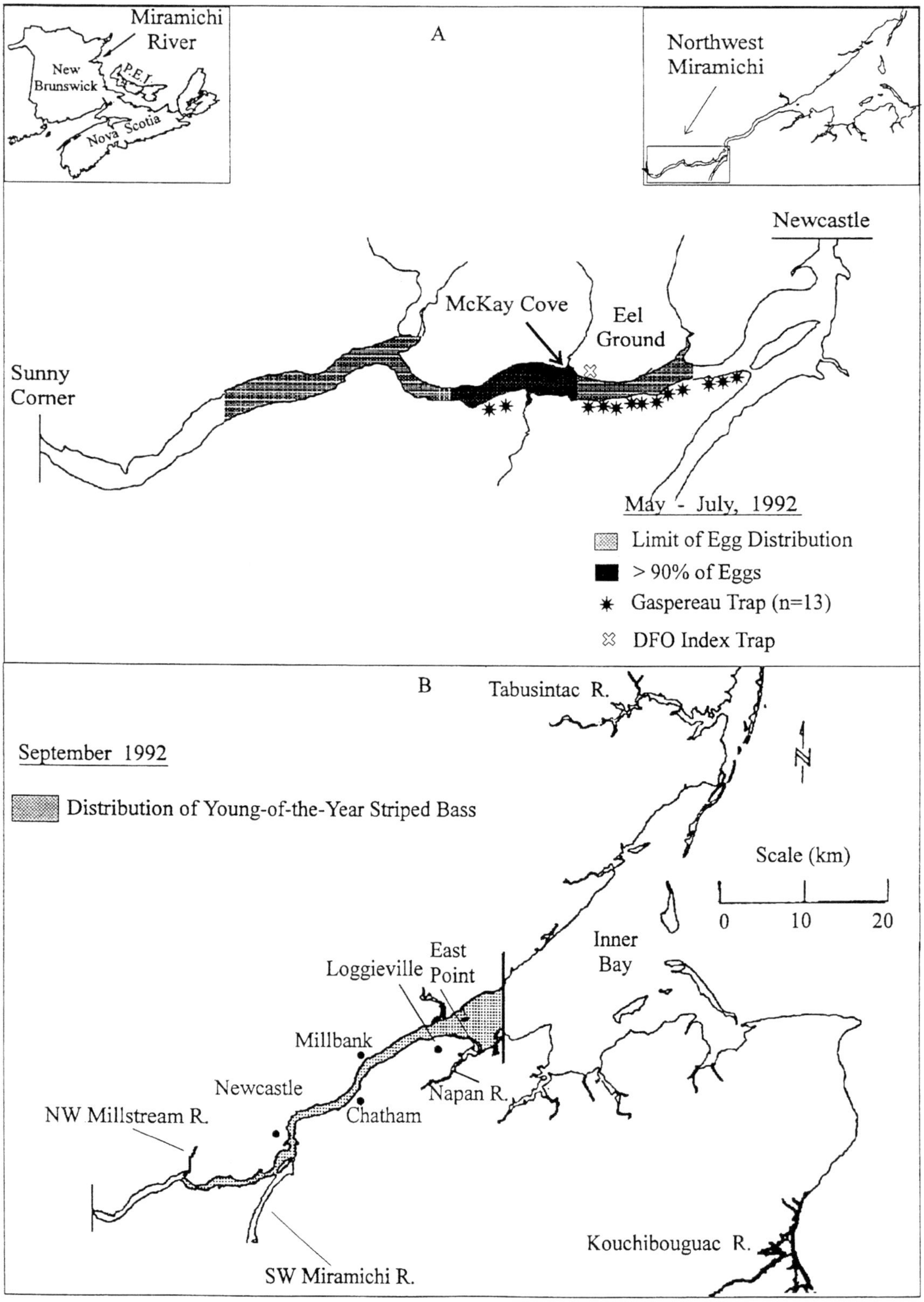

Fig. 1. Map of study area with place names used in the text; (a) distribution of striped bass eggs (shaded) and gaspereau traps in Northwest Miramichi River Estuary, (b) distribution of young-of-the-year striped bass during September. The limits of sampling are denoted by solid vertical lines.

Centre in Moncton, but many were recovered directly from fishers on the same day of recapture. Marked fish that were recaptured when the sampler was present were returned to the water after tag number and date of capture were recorded. Recaptures were weighted by fishing effort and the potential exploitation rate for adult striped bass captured in gaspereau traps was estimated as μ=recaptures ÷ marks (Ricker 1975). Bass caught in traps nets operated by the Department of Fisheries and Oceans at Eel Ground (Fig. 1a) were examined and marked as described above.

Seasonal and annual movements of striped bass were examined from fish marked during May-June and from a study in Kouchibouguac River estuary (Years 1983-present; Hogans and Melvin 1984, Eric Tremblay unpublished data).

Logbooks were completed by 2 fishers who fished about 70 % of the fyke nets set within the Miramichi River estuary. The number of traps fished, hours fished, and the total catch of striped bass per day were recorded. Daily catch of juveniles (Ages 0^+ and 1^+; < 15 cm) was reported separately from adolescents and adults (fish that would spawn the following spring; > 15 cm). By-catch of bass was recorded for individual fyke nets fished in the NW branch of the Miramichi River on four occasions between 2 August and 1 September.

There were two parts to the smelt fishery. The first part took place from October 15 to ice-up, December 8, 1993. The second part started when the ice was solid, January 4, 1994 and continued to March 5, 1994. Fishers were visited weekly during autumn and bi-weekly during winter. Effort was recorded as number of nets and hours fished. Striped bass could not be completely sorted from the catch until brought to shore, where they were counted and measured.

By-catch of striped bass was standardized to catch-per-unit-effort (CPUE) and number of fish $\cdot$ net^{-1} $\cdot$ day^{-1} by age and size group for the three fixed-gear fisheries.

Results

By-catch

By-catch of adult (≥35.5 cm) striped bass in gaspereau traps was generally higher during May-June in the NW Miramichi section than elsewhere in the estuary (Table 2). This observation corroborated previous logbook reports, 1981-1989 (Chaput and Randall 1990). Striped bass < 35.5 cm were evenly distributed throughout the estuary (Table 2). Striped bass ≥35.5 cm accounted for 42% of the total by-catch in gaspereau traps. Catches were 1354 large bass and 1718 small bass. (Catch per visited trap adjusted to total traps fished per day and summed for the entire fishing season.) Striped bass 2+ years of age (20 cm to 33 cm) dominated the legal by-catch of gaspereau fishers (Fig. 2a) and accounted for 57% of the total by-catch (Fig. 2a).

Mature males comprised >90% of the large bass. Males as small as 28 cm were also sampled. This observation differed from the 35 cm minimum length at maturity for striped bass in the Kouchibouguac River estuary (Hogans and Melvin 1984). Under regulations in place in 1993, about 3% of the spawning stock, i.e., striped bass >33 cm, could be removed (Fig. 2a). However, because of imprecise measures of length by fishers about 10% of the spawning stock was removed (Unpublished data), virtually all of which were males.

Seventy-seven of 272 large striped bass tagged within the NW Miramichi River Estuary (15 May – 18 June) were recaptured in the vicinity of the spawning grounds. Potential exploitation rate of gaspereau traps was about 25%. Any mortality of marked fish, tag loss, or non-reporting of tags would make this rate higher.

Striped bass were captured in low numbers in eel traps throughout the Miramichi River estuary. Logbooks for 83 nets, examined every 48h, over a 100 day period, indicated a by-catch of 910 striped bass <15 cm (ages 0^+, 1^+) and 93 striped bass ≥ 15 cm (Table 2), CPUE was always < 1 bass per day. Fyke nets in the NW Miramichi River section had similar low CPUEs: age 0^+, 0.05; age 1^+, 0.50; and age 2^+, 0.02 bass per day (Fig. 2b). Mortality of striped bass was probably low as only 3 of the 89 bass sampled were dead in the net.

Smelt traps caught striped bass mainly during October to December; Table 2). CPUE exceeded 100 bass per day in mid-October and declined to <1 in early December. Age 0^+ striped bass accounted for 84% of the by-catch (Fig. 2c) and few of these fish survived capture because they were not sorted from the catch until taken to shore. Similarly, mortality of age 1^+ (14% of by-catch) was also high. Many of the age 2^+ and older fish (2% of by-catch) were released live. Size composition of the catch ranged between 7 cm and 45 cm FL (Fig. 2c). Estimated total by-catch of striped bass in the autumn fishery (900 net days) was:

Age Group	*Thousands*
0^+	52
1^+	6.5
2^+ or older	0.6

The winter fishery caught few striped bass with a total observed by-catch (55 trap records; 113 net days; 6 January – 4 March) of:

Age Group	*Fish*
0^+	10
1^+	3
2^+ or older	1

All by-catch occured between Chatham and East Point (Fig. 1) even though effort during the winter smelt fishery was distributed throughout Miramichi Inner Bay (Fig. 1). Total fishing effort during winter was not precisely determined but license records indicated more than 300 nets were fished.

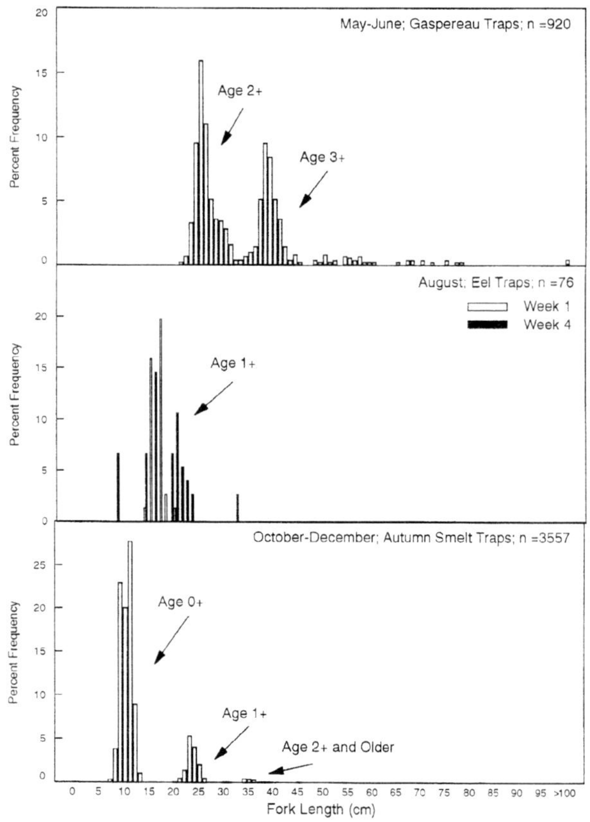

Fig. 2. Length frequency distribuion (FL cm) of striped bass captured in (a) gaspereau traps, (b) eel traps and (c) smelt traps (autumn).

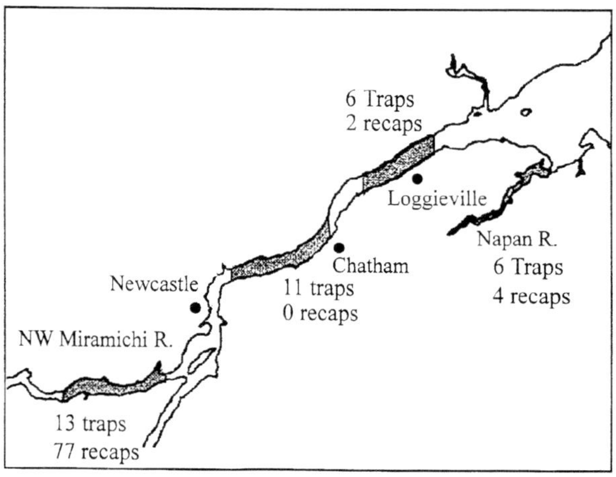

Fig. 3. Summary of recapture data for 272 striped bass marked in the region of the NW Miramichi River Estuary. The distribution and number of traps in the four regions of the estuary where gaspereau are fished are shown. Recaptures occurred between 10 May and 20 June, 1993.

Seasonal and Annual Movements of Striped Bass

Few (7%) tagged striped bass were recaptured away from the spawning grounds during May-June (Fig. 3), with only 6 of 83 recaptures reported from outside the tagging area. Two bass were recovered near Loggieville during the 3rd week of June and four bass were recovered in Napan River after 20 June. The Napan River gaspereau fishery was 2 weeks later than the Miramichi River fishery. One striped bass marked on 15 June in the Northwest Miramichi was recaptured the same week by an angler at the head of tide in the Southwest (SW) Miramichi River (Fig. 4).

Four recaptures occurred during autumn (Fig. 4); two from the Tabusintac River and one from Grande Anse, Chaleur Bay, during September; and one recapture from the Miramichi River Estuary during November.

Adults marked during spawning in the Kouchibouguac River (Hogans and Melvin 1984) have been recaptured in Miramichi River. Some tags were recovered later the same year in the NW Miramichi River section during the spawning season. Other tags were recovered from fish overwintering within the Miramichi River Estuary (Table 3). Juvenile striped bass (25-28 cm) tagged in the Kouchibouguac River estuary have also been recaptured in the Miramichi River Estuary. Tags have been recovered from adults on the spawning grounds and from juveniles and adults throughout the spring, autumn, and winter (Table 3).

Discussion

Striped bass are susceptible to incidental capture during their spawning, summer feeding, and overwintering periods. Gaspereau traps (May-June) caught both juvenile (age 2^{+}) and adult bass.

Table 2. CPUE (fish·net^{-1}·24 h^{-1}) of striped bass by fishery and location with method of sampling, and net days of fishing.

Location	Sample	Net Days Method	Length (cm)	CPUE (fish·net^{-1}·24 h^{-1})
Gaspereau				
NW Miramichi	Direct	73 (13)	≥35.5	2.0±2.3
			<35.5	3.8±5.1
Estuary	Direct	12 (3)	≥35.5	0.3±0.3
			<35.5	3.5±5.3
Eel				
NW Miramichi	Logbook	2 200 (44)	≥15	0.0±0.2
			<15	0.1±0.2
	Direct	176 (44)	≥15	0.0±0.1
			<15	0.5±0.2
Loggieville	Logbook	750 (15)	≥15	0.0±0.1
			<15	0.2±0.1
Napan	Logbook	1 200 (24)	≥15	0.0±0.1
			<15	0.0±0.1
Smelt				
Autumn	Direct	900 (21)	≥35.5	0.5±0.9
			<35.5	28.8±29.0
Winter	Direct	113 (55)	≥35.5	0.0±0.1
			<35.5	0.0±0.1

Age 1^+ fish may escape these traps because of the relatively large mesh size (3.1 cm; Table 1). However, by-catch is greatest for age 0_+ striped bass during the autumn smelt fishery. The eel fishery is not likely to significantly impact on the striped bass population. Total by-catch was estimated to be about 3000 adolescent and adult fish (age 2^+ and older), about 7000 age 1^+ fish, and about 53000 age 0^+ fish.

The potential impact of these by-catches on the Miramichi population remains unknown. Adult striped bass captured in the 1993 gaspereau fishery were predominantly males. If females were present in similar numbers to males but for some reason not caught, the exploitation rate for the spawning population as a whole would be only half that estimated or about 12% of adults. Striped bass exhibit sex-specific schooling except during the actual period of spawning (Schofield 1931; Hogans and Melvin 1984). Furthermore, if striped bass spawn, undisturbed by commercial fisheries, in the SW Miramichi River Estuary the exploitation rate would be lower. However, a similar, high exploitation rate (21%) was estimated for spawning striped bass in the Kouchibouguac River, 1983; but fishing mortality was due almost completely to poaching with gill nets (Hogans and Melvin 1984).

Whatever the true population size and exploitation rate in the Miramichi River, the release of adult striped bass captured incidentally in the 1993 gaspereau fishery increased spawning escapement by at least 1000 fish. Striped bass of commercial size (Age 2 and older) are thought to number only in the tens of thousands within the estuary, even during periods of high abundance (Chaput and Randall 1990; Chaput 1995).

The impact of the high incidental catch and mortality of young-of-the-year striped bass in the smelt fishery is also unknown. Young-of-the-year and juvenile striped bass may be distributed in the portion of the lower Miramichi River Estuary downstream of the autumn smelt fishery and our beach-seining (Fig. 1b). First-year survival is important for recruitment, particularly when there are few age classes in the spawning population and when little opportunity exists for multiple spawning

Table 3. Summary of recaptures in the Miramichi River Estuary of striped bass marked in the Kouchibouguac River Estuary (Type =life-history stage when marked, Season-Yr = season and year of marking, *n* = number of recaptures, Date =month and year of recapture, NWM = Northwest Miramichi River Estuary, MRE =Miramichi River Estuary). Recaptures were from gaspereau traps, smelt traps, anglers or unknown.

Marked	Kouchibouguac		Recaptured-Miramichi		Remarks
Type	Season-Yr	*n*	Date	Location	
Adult	Spring 83	2	June 1983	NWM-Gaspereau	Marked–recaptured: 2 spawning areas
		2	May-June 1983	MRE-Gaspereau	Marked in spawning area
		7	May-June 1983	MRE-Unknown	Marked in spawning area
		2	June 1983	MRE-Angled	Marked in spawning area
		1	September 1983	MRE-Unknown	Marked in spawning area
		1	February 1984	MRE-Unknown	Marked: spawning area; overwinter: MRE
Juvenile-	Autumn 91	1	June 1993	NWM-Gaspereau	Recaptured as adult
		1	November 1991	MRE-Smelt	Recaptured as juvenile
	Autumn 92	1	May 1993	NWM-Gaspereau	Recaptured as juvenile?
		1	October 1992	MRE-Smelt	Within–season migration to MRE
		1	January 1994	MRE-Smelt	Overwintering in MRE

(Rago and Goodyear 1987). However, Huntsman (1945) argued that Miramichi striped bass could be maintained without changes to the smelt fishery, and prefered regulation changes on the adults. He noted that landings of striped bass increased to record levels despite increased effort in smelt fisheries. He attributed the increase to a three-year closure of adult bass fisheries.

Evidence for spawning fidelity or philopatry is equivocal for Gulf of St. Lawrence striped bass. Juveniles after age 2 can migrate among estuaries. Individuals marked as juveniles in one estuary were recovered as spawners in another estuary. Lack of philopatry was also supported by a lack of variability in mitochondrial DNA between juvenile striped bass (ages 0^+ and 1^+) sampled from the estuaries of the Miramichi and adjacent Tabusintac River (Wirgin et al. 1993).

Alternatively, striped bass could exhibit philopatry. The mark-recapture data and lack of genetic difference in mixed age-group samples (i.e., Wirgin et al. 1993) may indicate an ability to migrate beyond the natal estuary at an earlier age than previously thought (i.e., Hogans and Melvin 1984). In this case, recaptures in the Miramichi River Estuary of juvenile striped bass tagged elsewhere (Table 3) may reflect their geographic position at the onset of winter conditions. Northern

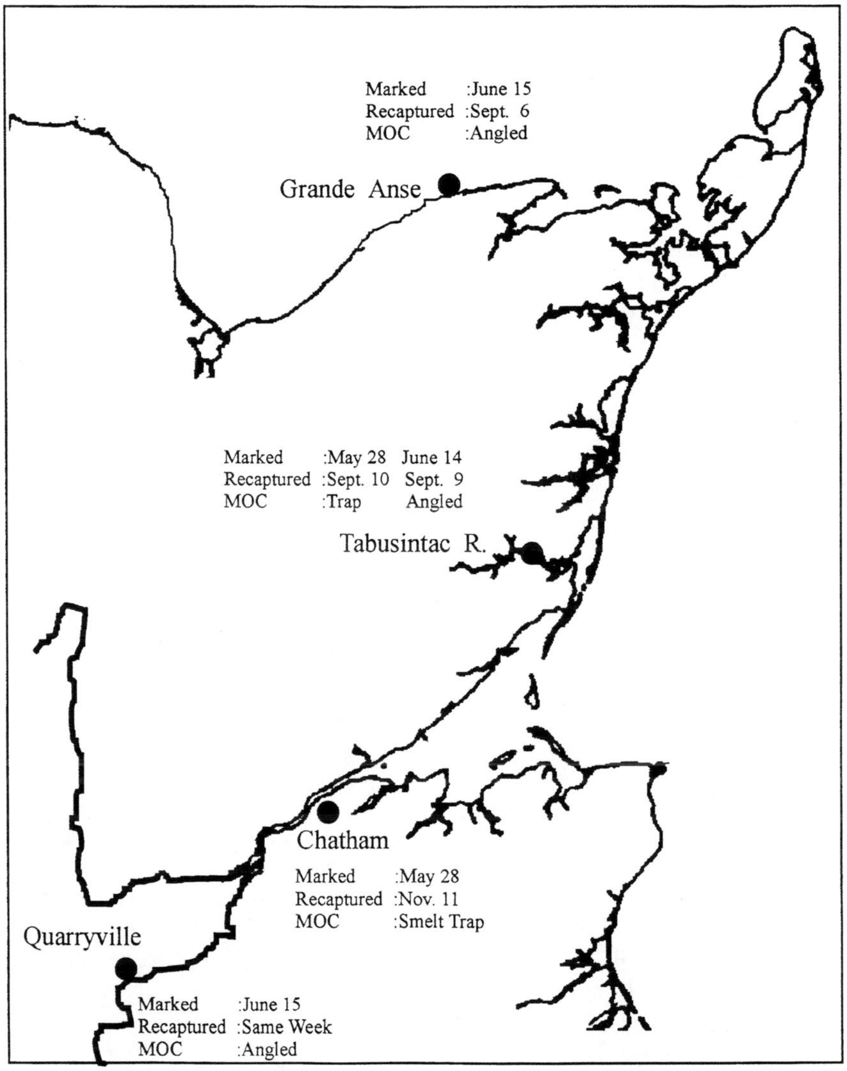

Fig. 4. Summary of recapture data for fish marked in the Northwest Miramichi River Estuary but recaptured by another means other than a gaspereau trap.
The date marked, date and location of recapture and method of recapture (MOC) are given for each return ($n = 5$).

populations of striped bass seek freshwater refugia during winter (Hogans and Melvin 1984; Hanson and Courtenay 1995; Bradford unpublished data), presumably to avoid winter sea temperatures (Drinkwater et al. 1993) which fall below the freezing point of unprotected fish blood (-0.7°C; Fletcher 1977). Some of the adult fish tagged in the Kouchibouguac River estuary, in a spawning area and during the spawning season (April), may have been overwintering Miramichi fish. Opportunistic overwintering could explain recaptures one month later in the spawning area and during the spawning season for the Miramichi population (Table 3).

By-catch provides useful information on the current health of the striped bass resource. Direct sampling of the NW Miramichi gaspereau fishery supports current perceptions that striped bass abundance is low relative to the long term average. The observed catch of approximately 3000 market-sized fish (about 33 cm and larger) was less than the average 1981-1989 catch of 5099 bass and an order of magnitude lower than the maximum catch of 21861 fish in 1981 (Chaput and Randall 1990). Continued sampling of the NW Miramichi gaspereau fishery for estimates of numbers, age and sex ratio could provide useful data on changes in adult population size. Continued sampling of the by-catch of age 0^+ striped bass could prove to be indicative of annual juvenile production.

The impact of fishing mortality on the abundance of Miramichi striped bass remains an open question, for reasons cited above, and because tag recapture information shows that striped bass migrate throughout the southern Gulf of St. Lawrence. Interceptions of striped bass with Miramichi origins in commercial, Aboriginal food and recreational fisheries elsewhere in the southern Gulf of St. Lawrence cannot be easily quantified without considerable expansion of sampling effort. In the absence of knowledge of the structure of striped bass populations in the southern Gulf of St. Lawrence it appears appropriate to manage at the regional level and thereby reduce fishing mortality on adults regardless of the season and sector of the fishery in which they are intercepted.

Acknowledgements

We thank the commercial fishers of the Miramichi River estuary for their co-operation during the course of this study and for sharing in their knowledge of the river and its fisheries. We also thank Mr. Kevin Davidson (DFO, Gulf Region, Moncton, N.B.) and Mr. Mark Hambrook (DFO) and the additional staff of the Miramichi Fish Culture Station, Newcastle, N.B. for logistic support. Mr. Eric Tremblay (Kouchibouguac National Park, Kouchibouguac, N.B.) made available mark-recapture information. We thank Mr. Bob Carroll (deceased), Mr. Fred Wheaton and Mr. Hal Wood of the New Brunswick Wildlife Federation (NBWF) and Mr. Dave Dunn and Mr. Victor Gionet of the DFO (Moncton) for their support and encouragement. Mr. Marc-Andre Bernard and Mr. Andy Bahm provided valuable assistance in the field. Dr. R. Peterson provided useful comments as did an anonymous reviewer. This study was funded through the Canada-New Brunswick Co-operation Agreement on Recreational Fisheries Development and was administered through the NBWF. The ichthyoplankton and juvenile striped bass distribution information summarized in this manuscript are parts of a Master of Science thesis being conducted by Kimberly Robichaud of the University of New Brunswick.

References

Chadwick, E.M.P., and Claytor, R.R. 1989. Run timing of pelagic fishes in Gulf of St. Lawrence: area and species effects. J. Fish. Biol. 35(Suppl. A): 215-223.

Chaput, G.J. 1995. Temporal distribution, spatial distribution, and abundance of diadromous fish in the Miramichi River watershed, p. 121–139. *In* E.M.P. Chadwick [*editor*]. Water, science, and the public: the Miramichi ecosystem. Can. Spec. Publ. Fish. Aquat. Sci. 123.

Chaput, G.J., and Randall, R.G. 1990. Striped bass (*Morone saxatilis*) from the Gulf of St. Lawrence. Can. Atl. Fish. Sci. Adv. Comm. Res. Doc. 90/71: 29p., Department of Fisheries and Oceans, Dartmouth, N.S.

Chaput, G.J., and Robichaud, K.A. 1995. Size and growth of striped bass, (*Morone saxatilis*), from the Miramichi River, Gulf of St. Lawrence, Canada, p. 161–176. *In* E.M.P. Chadwick [*editor*]. Water, science, and the public: the Miramichi ecosystem. Can. Spec. Publ. Fish. Aquat. Sci. 123.

Drinkwater, K.F., Petrie, B., and Narayanan, S. 1993. Overview of environmental conditions in the northwest Atlantic in 1991. NW Atl. Fish. Org. Sci. Counc. Stud. 20: 19-46.

Dunning, D.J., Ross, Q.E., Waldman, J.R., and Mattson, M.T. 1987 Tag retention by, and tagging mortality of, Hudson River striped bass. N. Amer. J. Fish. Manage. 7: 535-538.

Hanson, J.M., and Courtenay, S.C. 1995. Seasonal abundance and distribution of fishes in the Miramichi Estuary, p. 141–160. *In* E.M.P. Chadwick [*editor*]. Water, science, and the public: the Miramichi ecosystem. Can. Spec. Publ. Fish. Aquat. Sci. 123.

Hogans, W., and Melvin, G. 1984. Kouchibouguac National Park Striped Bass (*Morone saxatilis*) Fishery Survey. Aquatic Industries Limited, St. Andrews, N.B.

Hooper, W.C. 1991. Striped bass management in New Brunswick, p. 29-40. *In* R.H. Peterson [*editor*]. Proceedings of a workshop on biology and culture of striped bass (*Morone saxatilis*). Can. Tech. Rep. Fish. Aquat. Sci. 1832: vi + 66p.

Huntsman, A.G. 1945. Miramichi fisheries investigations twenty- five years ago. In: The Commercial and the World, February 8, 1945. Page 6. Chatham, N.B.

LeBlanc, C.H., and Chaput, G.J. 1991. Landings of estuarine fishes in the Gulf of St. Lawrence 1917-1988. Can. Data Rep. Fish. Aquat. Sci. 842: 101p.

Loftus, K.K., Greig, L.A., Pinfold, T.A., Kilfoil, M., and Meisner, J.D. 1993. Comprehensive strategy for the anadromous and inland recreational fisheries of New Brunswick. Can. Manuscr. Rep. Fish. Aquat. Sci. 2216: xxii + 235p.

Melvin, G.D. 1991. A review of striped bass, *Morone saxatilis*, population biology in eastern Canada, p. 1-11. *In* R.H. Peterson [*editor*]. Proceedings of a workshop on biology and culture of striped bass (*Morone saxatilis*). Can. Tech. Rep. Fish. Aquat. Sci. 1832: vi + 66p.

Norton, V., Smith, T., and Strand, I. 1983. The economic value of the Atlantic Coast commercial and recreational striped bass fisheries. Maryland Sea Grant Publication No. VM-SG-TS-83: 12. University of Maryland, College Park, MD. 55p.

Perley, M.H. 1852. Report on the sea and river fisheries of New Brunswick. Queens Printer, Fredericton, N.B. 294p.

Rago, P.J., and Goodyear, C.P. 1987. Recruitment mechanisms of striped bass and Atlantic salmon: comparative liabilities of alternative life histories. Amer. Fish. Soc. Symp. 1: 402-416.

Ricker, W.E. 1975. Computation and interpretation of biological statistics of fish populations. Bull. Fish. Res. Board Can. 191.

Schofield, E.C. 1931. The striped bass of California. Div. Fish Game Calif. Fish Bull. 29: 83p.

Scott, W.B., and Scott, M.G. 1988. Atlantic fishes of Canada. Can. Bull. Fish. Aquat. Sci. 219: 731p.

Venning, W.H. 1885. Annual report of the Department of Fisheries, Dominion of Canada for the year 1884. MacLean, Roger & Company, Ottawa, Canada. 301p.

Waldman, J.R., Dunning, D.J., Ross, Q.E., and Mattson, M.T. 1990. Range dynamics of Hudson River striped bass along the Atlantic coast. Trans. Am. Fish. Soc. 119: 910-919.

Wirgin, I.I., Ong, T.-L., Maceda, L., Waldman, J.R., Moore, D., and Courtenay, S.C. 1993. Mitochondrial DNA variation in striped bass (*Morone saxatilis*) from Canadian rivers. Can. J. Fish. Aquat. Sci. 50: 80-87.

SUMMARY AND SELECTED ABSTRACTS

"The sense of environmental stewardship, community betterment, and readiness to take action on our own initiative will make a difference in the health and welfare of the Miramichi River, and especially in the quality of life for all those who are fortunate to live along its banks."

Premier Frank McKenna
April 13, 1994, Miramichi Environmental Workshop

Gaspereau is a term given to two closely-related species of fish, blueback herring and alewife. Commercial gaspereau fisheries are an important industry on the Miramichi River.

Miramichi Environmental Science Workshop: Round-Table Summary

Harry Collins
Coordinator, MREAC

Purpose

Following three days of scientific papers and public involvement, the workshop concluded with a plenary session designed to chart a course for the future. The number of participants was very encouraging for the organizers. A cross section of interest was represented at the table by research scientists, industry representatives, to various levels of governments, universities, and private citizens. The following summarizes the discussion and results gleaned from this session.

Opening remarks by the chair of MREAC set the stage for the discussion. Mr. Goodfellow highlighted the need to more than recap the three days of science. MREAC had previously completed and circulated an 'Environmental Action Plan' that charted its environmental activities over the short term. Were these actions on target with the pressing issues of the Miramichi watershed? Were there other environmental issues that MREAC should be addressing? Finally as a result of the workshop, had the scientific community identified research gaps?

Water Quality

MREAC has dedicated a great deal of time to problems related to waste water on the Miramichi. It was expressed that this forum might provide insights into the issue of the large quantities of untreated waste water entering the Miramichi River from sewage treatment plants and other point sources (95% of effluent entering the river is from a pulp mill). A question was raised as to the expectations for improving water quality by both MREAC and the community at large. In response, MREAC is hoping to improve the water quality everywhere on the Miramichi for recreational purposes and ultimately to have the shellfish closure line moved further down river or eliminated.

Primary treatment of municipal effluent was not considered to be enough. The ongoing move toward a regional sewage treatment plant for Miramichi municipalities, to be incorporated into the City of Miramichi, was hoped to solve much of the problem of by-pass and infiltration. The value of tertiary treatment was questioned due to climatic conditions. Good secondary treatment may be the better solution. Miramichi waters, it was noted, tend to be nutrient poor. Nitrogen was identified as a potential problem in sewage effluent, in part because of its relationship with the production of phycotoxins. This discussion was balanced by the potential problems connected with the use of chlorine for sterilization. It was noted that in the United Kingdom sterilization is not used. Problems related to rural, waste-water treatment were discussed briefly. One delegate commented that the problem in the rural setting was more likely lack of education than lack of technology. MREAC's experience provided some support to that claim. Tying septic system permits closer to building permits may be a partial solution to provide greater control in rural settings. Pump-out risers were recommended to encourage septic tank maintenance.

Forestry and Cutting Practices

The debate over the Christmas Mountains harvesting issue raised the question of the need for identifying and protecting reasonable amounts of land in the Miramichi watershed for the purpose of benchmark or control studies. No organization, including MREAC, was examining this issue, which may have long-term consequences for science on the Miramichi. Forestry practices in the Miramichi permit annual harvests of 2-5% of public lands. This practice raised the question about long-term wood supply and the danger of compromising management controls currently in place. Noted as an example was the use of buffer strips that are required along rivers and streams. There is some pressure to allow limited harvesting within these strips. Are such compromises, often made at a political level, acceptable to the environment? Do they fly in the face of the principle of sustainability? Who should be answering these and other such questions?

Science

Science in winter was lacking and may be increasingly an important factor in environmental monitoring. Projects such as the Catamaran Brook study are addressing this on a limited scale but there is room for more work under ice and snow.

Despite the large amount of science that was taking place, there was a need for continuous monitoring at several locations in the Miramichi to better understand the full range of environmental impacts. Important parameters such as pH, salinity and temperature were often not being collected at strategic locations. Further to this, much attention has been focused on the Northwest Miramichi, leaving data collection in the Southwest Miramichi wanting. Preventative environmental action was suggested as being equally as important as monitoring. Monitoring will often explain reasons for mortality while prevention can often prevent mortalities.

Communications

Although communications are assisted by workshops, there is considerable opportunity and need for continuous and broader communication. The New Brunswick Community College would welcome an opportunity to be more involved in Miramichi science and in MREAC, especially through the environmental technologies program. School groups have shown a similar high interest level. Among the scientists, there could be greater coordination of how data were collected, particularly to allow comparisons. There was also a need for better communication of ongoing studies. MREAC offered to be a clearing house and to coordinate such a service. The workshop was noted as being exclusively a hard-science forum designed for government and universities. It was suggested that in future MREAC should consider opening workshops to commercial fishers and industry. Future meetings, if broader or different in scope, will need to be well-designed to achieve objectives.

Research Gaps

Despite the large quantity of fisheries research there seems to be more room for work in other areas of estuarine science. The dynamics of the Miramichi Bay and estuary are less well understood. It was recommended that a fisher's logbook program for estuarine species would be helpful. The productivity of the system needs to be better understood, especially at the bottom of the food chain.

Environmental Changes

Environmental improvements in the Miramichi may result in long-term release and redistribution of contaminants stored in the sediments. Improved water quality should result in better aerobic conditions in the estuarine benthos, resulting in increased biological activity. This may result in contaminants, formerly trapped under anaerobic conditions, being released and becoming more widely distributed in the environment. This potential threat may warrant greater monitoring effort, particularly with regard to characterizing deposits of contaminants.

Solid Wastes

Solid waste management has not been an issue to MREAC or the scientific community in general. Closed sites do not have follow-up monitoring connected with them. MREAC did some initial work for its 1992 final report but found no problem and has not followed this as an issue. New standards that require remediation of closed dump sites should result in better leachate control in future. MREAC could map closed sites and conduct periodic follow-up sampling. Hazardous wastes are currently disposed out-of-province. Although this issue needed to be handled at a different level, it did not appear to present a problem to the water quality of the Miramichi.

Regrouping

Future meetings hosted by MREAC should probably be on a 2-3 year basis. MREAC has an opportunity to coordinate scientific research on the Miramichi River but will need the cooperation of the scientific community to keep informed about activities and opportunities. The steering committee may wish to change the nature of future workshops.

Selected Abstracts

Back to Basics: Acid Rain in New Brunswick

D. Albin

New Brunswick Department of the Environment, P.O. Box 6000, Fredericton, N.B., E3B 5H1

The New Brunswick Department of the Environment (NBDOE) has operated its own network of acid deposition sites since 1980. These sites complement acid-rain monitoring efforts of federal departments by providing improved spatial resolution of deposition patterns. Results have indicated that this additional monitoring work is justified, and strong regional deposition gradients have been identified.

From 1980 to 1987, from 3 to 6 deposition sites were operated by NBDOE, in both urban and rural locations. In the years 1988-90, a period of especially detailed monitoring, the aim was to use data from a dense, 18-site network to evaluate and tune a regional acid-deposition model. The model evaluation project was a joint undertaking with N.B. Power. This work has now been completed and a final report is available from N.B. Power. Since 1990 the network has been reduced in size, but a lesser number of sites in the south of the province has been compensated by the addition of three new stations in Chaleur Bay region as well as sites in Fundy National Park and Canterbury in southwest NB. These sites are located with special reference to the operation of a new thermal generating station at Belledune, which began operation in November 1993. Thus the network continues under the joint funding of NBDOE and N.B. Power.

Since 1988, both sulphate concentrations (volume-weighted means) and sulphate deposition have declined. In all cases, sulphate concentrations and deposition refer to values corrected for the influence of sulphates from marine origin. Absolute sulphate concentrations in selected southern sites are higher than those found in the north of the province. Values from northeastern sites typically lie in the range 1.0-1.5 mg/L sulphate, compared with 1.5-2.0 mg/L in the south. However, there is some evidence that values are tending to approach 1.0-1.5 mg/L in the south of the province. Five and a half years is a short time over which to construe trends; however, the overall trend is downward and the pattern seems to be consistent with observations elsewhere in eastern Canada.

Point Source Discharges to the Miramichi River

Neil Brodie

New Brunswick Department of the Environment, P.O. Box 6000, Fredericton, N.B., E3B 5H1

The paper presents information on point source discharges of wastes from human activities in the Miramichi watershed for the benefit of researchers interpreting and planning studies. Point sources are confined by definition to intentional, specific discharge points or very restricted area sources. Diverse landwash sources such as farming, timber harvesting or acid precipitation are not considered here.

Contaminant Types and Influence on the Receiving Water

The types of contaminants which were found during the river study completed in 1992 include biological oxygen demand, suspended solids, heavy metals, coliform bacteria, biodegradable organics, and persistent and bioaccumulative organics. In assessing the impact of a pollutant on the ecosystem, the concentration, the rate of deposition or flow and the type of pollutant must be considered. Some pollutants are more easily degraded or treated by the receiving water, while some are highly-resistant to degradation and may exert toxic effects on living things in either the short term or the long term and may sometimes bio-accumulate in the food chain.Another factor in considering the effects of point-source discharges is the mixing zone. The size of the zone depends upon the loading of the contaminant and the assimilative capacity of the receiving water. Mixing zones are areas where the higher concentrations of the discharge pipe are diluted down to some lower level that is considered acceptable for minimal alteration of the water body.

Assimilative capacity is another subjective concept like mixing zone. Assimilative capacity is roughly defined as the amount of foreign material a receiving water can accept before its value is degraded. With a mixing-zone decision, someone must decide how much of the river can be devoted to diluting a pollutant to acceptable levels. With an assimilative-capacity decision, someone must decide how much the level of suspended solids can rise or how much dissolved oxygen concentrations can drop before the value of a river is degraded.

Some other influences must be considered when assessing the importance of point-source discharges. The effects of point-source discharges are not usually constant. Sources often have varying flow rates and concentrations over the course of a year and even greater variation takes place in the flow rate and water quality of the receiving water at the same time. Sediment deposition areas, chemistry changes in intertidal zones and other important natural factors must be considered. Finally, contaminants must be reviewed in terms of the potential receptors. Will they affect the quality of the water column, sediments or biota or the humans who contact the water or ingest the biota?

Groupings of Point Sources on the Miramichi

The basic groups of point sources are industrial, municipal sewage treatment plants, private sewage treatment plants, and solid waste disposal sites or dumps. Industrial sources and municipal sewage treatment plants have the largest potential impact due to their chemical make-up and bacteria content respectively. They also have the largest discharge volumes of the various sources.

The industrial sources of greatest significance are the Miramichi Pulp & Paper Kraft Mill complex in Newcastle, the Miramichi Pulp & Paper Groundwood Mill in Nelson-Miramichi and the Heath Steele Mine complex. The municipal sewage treatment plants of greatest significance are the towns of Newcastle and Chatham and the Village of Douglastown. The other sources are of much less significance to the Miramichi River, primarily due to their much smaller volumes. However, even low-volume discharges when improperly treated, can have important localized effects on the river.

Monitoring Follow up and Concentrated Study Areas

Areas which deserve continued monitoring are: mercury levels in white suckers and striped bass at the confluence of the NW and SW rivers; the levels of metals and PAH at the mouth of the NW; and dioxin levels in white sucker and lobster tomalley. Areas that could benefit from inter-disciplinary studies are: the upper reaches of the Northwest Miramichi; the inter-tidal area of the Northwest; and, the confined area of Strawberry Marsh.

The Influence of Dredging and Spoils Deposition on Sediment Stability in the Miramichi Inner Bay

M. Brylinsky[1], C.A. Amos[2], H. Christian[2], and G.R. Daborn[1]

[1]Acadia Centre for Estuarine Research, Acadia University and [2]Atlantic Geoscience Centre, Bedford Institute of Oceanography, Dartmouth, N.S., B2Y 4A2

During the fall of 1991 and the summer of 1993, studies were carried out to assess the impact of dredging and spoils deposition on sediment stability in the Miramichi Inner Bay. Sediments within the bay range from loose to extremely loose organic silty clays with exceptionally high organic contents, high water content and low bulk density. The seabed is poorly consolidated and close to liquefaction. These characteristics suggest the seabed is very susceptible to collapse when subjected to shear forces. Measurement of sediment settling rates, however, indicate that the sediments are not generally of a nature that would lead to hindered settling and the development of fluid mud layers. Examination of a spoils deposit one year after deposition indicated it to have similar biological characteristics and to be at least as stable, if not more, as a natural undisturbed control site. The least stable seabed was located within the navigation channel. Studies of recently deposited spoils indicated that they quickly (within days) developed strength equal to that of a natural control site, but that this strength appeared to be limited to the upper surface layer. Continuous monitoring of suspended sediment concentrations indicated little evidence of either naturally-occurring fluid-mud layers or that dredging and disposal activities result in chronically-high suspended-sediment concentration outside of the immediate area of the channel where dredging is taking place. One question that remains to be addressed is that of the stability of recently deposited spoils under conditions of strong storm events.

Double Stress: Impact on Fish

M.D.B. Burt and W. Pilgrim

Biology Department, University of New Brunswick and N.B. Department of Environment, Fredericton, N.B., E3B 6E1

It is well known that animals under stress are not as healthy as those that are not subject to stress. Stress in fish can be the result of: physical factors (such as abnormal temperatures, handling and crowding), chemical factors (such as abnormal acidity or pH or elevated levels of heavy metals or organic pollutants), and biological factors (such as parasites and other pathogenic organisms). Where stress is limited, few animals actually die but if sufficient stress occurs, due to a combination of different factors, this can result in mortalities.

All species of fish are naturally infected with various endo- and ectoparasites. Such parasitic infections seldom cause the death of their natural host, even if the parasites are present in relatively large numbers. However, if some other stress factor is also present, then those fish with higher numbers of parasites become more vulnerable and can die. Such a situation seems to have occurred in the past in the Miramichi when large numbers of fishes of different species died. All of these dead and dying fishes were reported to have heavy infections of parasites, though not sufficiently heavy to cause mortality of their hosts under normal circumstances. However, the incursion of highly acidic tailings from a metal mine into the river system at that time would not only have caused additional stress due to the lowered pH but this would also have increased the concentrations of copper and zinc in the water causing yet more additional stress resulting in the death of those fish under greatest stress from parasites.

A similar situation may occur in Scandinavia where juvenile Atlantic salmon have been killed in Norwegian rivers by a small ectoparasite, *Gyrodactylus salaris*. Fish in all rivers did not appear to be affected to the same degree, however, and it was assumed initially that certain "strains" of fish showed more, or less, innate genetic resistance. Many of the fish, from certain Swedish rivers for example, did not die whereas all those from Norwegian rivers did. More recently, however, a positive correlation has been established between mortalities of juvenile Atlantic salmon and river waters with low pH. This suggests that susceptibility or resistance to infection with G. salaris is more a result of higher acidity (low pH) than any genetic factors. Experimental work in our laboratory has shown clearly that if freshwater amphipods (*Hyalella azteca*) are infected with the larval stage (cysticercoid) of a tapeworm (*Hymenolepis hopkinsi*) that is commonly found as adults in ducks, those amphipods that are infected die sooner than the non-parasitised controls if the pH of the water is lowered. These and other double stresses that have an impact on fish and other aquatic organisms are under study.

Health Hazard Assessment of the Findings of Chemical Contaminants in Fish and Shellfish Collected in Miramichi River

C.F. Dalpé

Chemical Evaluation Division, Bureau of Chemical Safety, Food Directorate, Health Protection Branch, Health Canada, Ottawa, Canada

The determination of the human health significance of the presence of chemical contaminants in foods is a procedure identified by Health Canada as Health Hazard Assessment (HHA). This procedure is a scientific process based on the careful consideration of two basic parameters, the toxicity of the substances and the potential exposure to these chemicals, and includes the following steps: determination of the Tolerable Daily Intake (TDI); determination of the Probable Daily Intake (PDI); and, comparison of the TDI and PDI.

If a hazard is identified, some form of intervention is warranted in order to lower human exposure to these contaminants. In such a case, a risk management analysis helps in determining the most suitable measures to reduce exposure to the identified hazard. This analysis is conducted on a case-by-case basis, taking into consideration the severity of the hazard to human health, the importance of the food in the diet of the concerned populations, the availability of substitutes, technological and socio-economic factors.

More recently, the Health Protection Branch has assessed the human-health hazard related to measured levels of dioxins and furans in fish and shellfish collected in the Miramichi River in 1991 under the National Dioxin Program and to measured levels of various trace metals in species collected in 1991 as part of a monitoring program conducted by the Miramichi River Environmental Assessment Committee (MREAC).

Although levels of dioxins and furans were measured in the digestive glands of lobsters collected near Chatham and levels of cadmium, copper, lead, and zinc were measured in oysters, mussels, clams, lobster, flounder, shad, striped bass, gaspereau, eel, and mudsuckers collected in the Miramichi, consumption of these fish and shellfish was not considered to pose a health hazard.

The Environmental Implications of the New Brunswick Spruce Budworm Spray Program

W.R. Ernst

Environment Canada, Atlantic Region, Dartmouth, N.S., B2Y 2N6

The province of New Brunswick has the distinction of mounting the largest and longest-running aerial spray program in North America. The spray program, directed at the spruce budworm (*Choristoneura fumiferana*), was initiated in 1952 and has continued to this day, with only one year of interruption (1959). The insecticides used operationally have included DDT, Trichlorfon, Phosphamidon, Aminocarb, Fenitrothion and *Bacillus thuringiensis* var. *kurstaki* (B.t.k.). The latter three have been the pesticides of choice in most recent years. A comparison of the environmental implications of the use of those three pesticides is made. Justification is made for ranking the environmental acceptability (from least to most) as fenitrothion, aminocarb and B.t.k.

One of those pesticides, fenitrothion, has recently been subject to an intensive special review under the authority of the Pest Control Products Act and the outcome of that review is discussed.

Distribution and Seasonal Patterns of Aquatic Insects in Catamaran Brook, a Tributary of the Little Southwest Miramichi River

Donna J. Giberson and Heather Garnett

Department of Biology, University of Prince Edward Island, Charlottetown, P.E.I.

Aquatic invertebrates are a major component of the diets of many freshwater fish, including juvenile Atlantic salmon and brook trout. The majority of the invertebrates collected from Catamaran Brook have been aquatic insects, and include representatives from all aquatic insects groups. Fish may consume aquatic insects from two sources: directly from stream bottom substrates to which they cling or are attached, and from the water column when insects are drifting or emerging. Aquatic insects are particularly vulnerable to predation by fish during emergence.

Aquatic invertebrates have been monitored in Catamaran Brook since 1990 as part of a larger study investigating potential logging impacts on the biology, chemistry and hydrology of a small Miramichi catchment. Excluding the Chironomidae, 93 genera in 43 aquatic insect families have been found in the brook. The greatest diversity (excluding Chironomidae) has been found in the Ephemeroptera (25 genera in 8 families) and the Trichoptera (19 genera in 12 families). Seasonal patterns of emergence varied with insect taxa, stream reach, and stream habitat type. For example, the aquatic Diptera (primarily Simuliidae and Chironomidae) showed major emergence peaks in early July and the end of August, whereas most Plecoptera and Ephemeroptera emerged during mid-summer, and Trichoptera emergence occurred fairly consistently from mid-June to the end of August. Diptera were by far the most common insects emerging from the stream, reaching a peak emergence of nearly 3000/m2 at one lower reach site in early July (primarily consisting of Simuliidae). In this paper, the distributions and phenologies of the major groups of aquatic insects in Catamaran Brook will be presented, and related to physical factors such as water temperatures, stream flow, and habitat type.

Physical Features of the Miramichi Estuary

Diane Lavoie

Department of Oceanography, Université du Québec à Rimouski, 300 Allée des Ursulines, Rimouski, Québec G5L 3A1

Description of the Miramichi Estuary

The Miramichi Estuary is located in the eastern part of the province of New Brunswick and communicates with the Gulf of Saint Lawrence. The estuary consists of two distinct parts: the Miramichi River, which in geomorphological terms is a coastal-plain estuary, and the bay, which is a bar-built estuary (St-Hilaire 1993). The river varies in width up to 1 km, and its estuary is 40 km in length. The bay is triangular and covers an area of nearly 300 km2; it is 20 km long and about 20 km wide at its mouth. The bay is shallow (3-5 km deep) except for a navigable channel which exceeds 10 m in depth at some points. The flow of water between the bay and the Gulf is impeded by a series of small barrier islands.

The river estuary is highly stratified between Chatham and Bartibog Island; stratification decreases as the river widens after Bartibog Island. The bay is relatively well mixed.

These conditions do not however prevail year-round but change frequently in response to environmental conditions. The tides, freshwater inflow, wind and ice cover strongly influence the estuary's characteristics (salinity, stratification, circulation). As the two parts of the estuary have different characteristics and are affected in different ways by the above environmental factors, we will deal with them separately where appropriate in the remainder of this text.

Tides

The tides in the Miramichi are of the mixed type, primarily semi-diurnal. Mean tidal range varies from 0.2 to 1.2 m, with a mean period of 12.4 h. There is a 70-min phase lag between the mouth of the estuary and Chatham; tidal range is slightly greater upstream (St-Hilaire 1993). The tides migrate over 10 km up the Northwest Miramichi and over 20 km up the Southwest Miramichi.

The salt-water wedge shifts with the tide (migrating downstream at ebb tide and upstream at flood tide) over a distance equal to the tidal excursion (approximately 7-8 km), which itself varies with the spring-tide/neap-tide cycle.

Tidal currents in the central part of the bay are circular and tend towards cyclonic rotation. Approaching the river estuary and channels, the tidal ellipse lengthens and the currents tend to reverse rather than to rotate (Vilks and Krauel 1982). Maximum current speeds are ebb-directed at the river estuary's surface and flood-directed at the bottom.

Temperature and Salinity

The river's waters are generally 1-3°C warmer than the waters of the Gulf, except around the months of October and April when the Gulf is warmer.

Tide-related temperature variance is in the order of 1-2°C. Diurnal variance is also 1-2°C but can be considerably higher in shallow areas, rising as high as 15°C during the summer (Vilks and Krauel 1982). Temperature plays a minor role however in determining the estuary's dynamics. As is typical of estuaries, water density is primarily a function of salinity, which ranges from 0 at the head of the estuary to a mean of 24.5 (at the surface) and 26 (at the bottom) at the barrier islands.

Ice

Freeze-up influences the estuary's dynamics in two ways. First, it creates a boundary similar to the bottom, creating additional friction for surface currents. Secondly, it creates a kind of protective screen against the wind, reducing mixing of the water column. This has a strong influence on water column stratification. Under equal freshwater inflow conditions, stronger stratification has been observed in the river estuary when there is ice cover (Vilks and Krauel 1982). In the bay, salinity varies little with depth (1 to 3) due to mixing by the wind and tide. A halocline moving with the tide has however been observed during the period of ice cover. This stronger stratification is probably due to reduced mixing, for while friction between the current and the ice does create turbulence, it is not as strong as the turbulence created by the wind.

Ice cover has also been observed to influence speed gradients. In the river estuary, maximum currents are ebb-directed at the surface and flood-directed at the bottom; however, during the period of ice cover, maximum currents occur at midwater depth (although circulation remains estuarine), illustrating the influence of the ice as a second boundary, in addition to the bottom. It has also been suggested that the salt-water wedge does not rise as high under ice cover conditions.

Freshwater Inflow

The Miramichi River drainage basin covers an area of 13 779 km^2 (St-Hilaire 1993). The two main tributaries (accounting for 80% of freshwater inflow) are the Northwest Miramichi and Southwest Miramichi, which join at Beaubears Island to form the Miramichi River. Average annual flow at the river's mouth is approximately 250 m^3/s but daily peaks of 6 300 m^3/s can be reached during the spring freshet.

Freshwater inflow plays a predominant role in determining estuary conditions. In the river estuary, increased flow results in increased stratification. The surface layer is diluted and there is a slight reduction in salinity at the bottom. It has also been observed that, during periods of strong inflow, ebb tide currents and their periods increase, while flood tide currents remain fairly constant and their period is reduced. In the bay, which is relatively well mixed, there is a general drop in salinity.

In stratified estuaries, the depth of the surface layer generally increases with freshwater inflow (Kjerfve 1988). In the Miramichi River Estuary, these conditions are most intense in the spring, when inflow is very strong and salt water is forced entirely out of the river estuary. The salt-water wedge is then near Bartibog Island. This leads to an increase in horizontal salinity gradients in the bay.

Wind

According to Vilks and Krauel's (1982), southwesterly winds tend to prevail in the summer and early fall, with a more westerly component in late fall and winter. Mean wind speed observed at Chatham is 16 km/h but winds of 20-25 km/h are very frequent in the fall.

The winds are strongest in the bay, due to its shallowness and greater fetch, as is evident from the fact that the water column in the centre of the bay is well mixed and there is no well-defined circulation pattern. One indication of the extent of wind mixing appears in the spring during break-up: despite increased freshwater inflow, which generally increases stratification, de-stratification of the water column is observed due to increased wind mixing, which was inhibited by the ice cover. Vilks and Krauel (1982) have also reported that significant influence of the wind on circulation in the bay has been observed on a number of occasions; currents acquire a strong component moving in the direction of the wind. Given the shallowness of the bay, the wind may be expected to act on the entire water column, even at low velocities.

Currents

Currents in the Miramichi Estuary vary primarily as a function of the tides and freshwater inflow, except in the bay where there is no clear relationship between the speed of the current and freshwater inflow. In addition to these two main factors, there is the wind, which primarily affects the bay (where there is no precise circulation pattern), and ice cover. Only the bay, however, is well mixed vertically.

Variance in freshwater inflow has a bearing on circulation in the bay, although it does not seem to influence current speed. Vilks and Krauel (1982) have suggested that the Coriolis effect deflects the flow of freshwater at the mouth of the river southward, which in conjunction with a stronger flood-tide stream along the northern shore produces cyclonic circulation in the bay. Indeed, current profiles suggest predominant flood circulation through Portage Gully and along the northern shore, and predominant ebb current through Portage Channel. It has been observed, however, that freshwater inflow does not always deviate rightward at the river mouth but sometimes continues straight along the northern shore. This could be due to the wind counterbalancing the Coriolis effect.

Mean tidal currents decrease from the river estuary through the bay and increase in the channels between the barrier islands. Speed measurements taken at Loggieville show currents of 0 to 70 cm/s at the surface and 0 to 30 cm/s at the bottom. In the bay, currents are weaker, rarely exceeding 30 cm/s, and are relatively uniform with depth. Measured speeds at spring tide are approximately twice the speeds at neap tide.

Freshwater Residence Time

In a stratified estuary, the depth of the halocline — i.e., the thickness of the surface layer — is relatively constant from the head to the mouth, at any given rate of flow. On the basis of horizontal and vertical salinity profiles from Vilks and Krauel (1982), we may estimate the thickness of the top layer in the river estuary and of the bottom layer at approximately 3 m each during the month of October, for a total mean depth of 6 m. It has been determined that the salt-water wedge can be located 40 km upstream of the river's mouth under low-flow conditions. Considering the month of October, when the river's flow is weak (X = 120 m^3/s), we can estimate the area of the river estuary at 28 km^2 (with a mean width of 0.7 km) and the volume of the surface layer (V) at 0.084 km^3. The salinity of the surface layer at the mouth of the estuary (S_0) can be estimated at 18 and of the bottom layer (S_1) at 21. It is then possible to determine the residence time of fresh water in the estuary as well as the mean speed in the two layers (U_1 at the surface and U_2 at the bottom). As mean water volume passing through a surface per unit of time equals surface area multiplied by mean current speed, the volume of outflow V_0 is given by

(1) $$V_0 = 0.8 \text{ km} \times 3 \text{ m} \times U_1$$
$$= 2400 \text{ m}^2 \times U_1$$

The volume of outflow must be equal to the volume of inflow at the head of the river and in the bottom layer (V_1). Leaving aside precipitation, evaporation and density differences, we get

(2) $$V_0 = X + V_1$$

Moreover, to maintain salinity, the seaward saline flow in the surface layer must be replaced by up-estuary saline flow in the bottom layer. Leaving aside the density differential between the two layers, we get

(3) $$V_0 S_0 = V_1 S_1$$

Combining equations (2) and (3), we get

(4) $$V_0 = \frac{XS_1}{S_1 - S_0}$$

$$2400\text{m}^2 \times U_1 = \frac{120\text{m}^3/\text{s} \times 21}{21 - 18}$$

$$U_1 = 35 \text{ cm/s}$$
$$V_0 = 840 \text{ m}^3/\text{s}$$

Equation (3) gives us $V_1 = 720$ m^3/s, which with equation (1) gives us a mean speed U_2 in the lower layer of 30 cm/s.

The volume of the surface layer is 0.084 km^3. As salinity is 0 at the head of the estuary and 18 at the mouth, we can estimate mean salinity at 9 in the surface layer. Given that the salinity of inflow at the bottom is 21, we may conclude that surface volume is 43% fresh water. Assuming that freshwater flow is restricted to the surface layer, residence time for fresh water (in October) is given by:

(5) $$t = \frac{V}{X}$$

$$t = \frac{0.084 \text{ km}^3 \times 43\%}{120 \text{ m}^3/\text{s}}$$

$$t = 301\ 000 \text{ s} = 3.48 \text{ days}$$

Vertical Mixing

The main factors responsible for the flow or displacement of suspended particles are tide-driven and wind-driven currents. Given the back-and-forth nature of tidal currents, their net displacement capacity can be assumed to be nil (in the absence of interaction with the topography), for a particle will be returned to its original position at the completion of a tidal cycle (Kjerfve 1988). It is rather the combination of tide-induced vertical mixing and speed variances at different depths which cause net displacement. Vertical mixing is related to turbulence. Winds create a well mixed turbulent layer at the surface while tidal currents create a well mixed bottom layer.

Available turbulent energy for wind mixing (*Pw*) and tidal mixing (*Pt*) in the bay may be compared using the following equations (Schroeder et al. 1990):

(6) $$Pw = \rho_e[\rho_\Delta/\rho_e)C_D]^{3/2}W^3$$

where ρ_e is water density (1020 kg/m^3), ρ_Δ is air density (1.25 kg/m^3), C_D is the friction coefficient at the surface (0.00125) (Bowden 1983) and W is mean wind speed (7 m/s). We thus get the result Pw = 0.66×10^3 Wm^{-2}.

(7) $$Pt = \rho_e\, C_D^{3/2}\, U_m^{\ 3}$$

where U_m is the mean speed of the tidal current at the bottom (0.10 m/s) and C_D is the friction coefficient at the bottom (0.0025). We then get the result $Pt = 0.13 \times 10^3\ Wm^{-2}$.

We see therefore that wind mixing is more significant in the bay than tidal mixing, as has been previously observed.

References

Bowden, K.F. 1983. Physical oceanography of coastal waters. Ellis Horwood Ltd., England. 302pp.

Kjerfve, B. 1988. Hydrodynamics of estuaries. Vol. 1, Estuarine Physics. CRC Press, Florida. 163pp.

Shroeder, W.W., Dinnel, S.P, and Wiseman, W.J., Jr. 1990. Salinity stratification in a river-dominated estuary, Estuaries 13(2): 145-154.

St-Hilaire, A. 1993. Étude hydrodynamique de l'estuaire de la rivière Miramichi durant la saison sans glace — 1991. Master's thesis, Université de Moncton, N.B.

Vilks, G., and Krauel, D.P. 1982. Environmental geology of the Miramichi Estuary: physical oceanography. Geol. Surv. Can. Pap. 81-24: 53pp.

Sediment Contamination in the Miramichi

Scott McKnight

President, Land & Sea Environmental Consultants Ltd., 33 Aldernay Dr., Dartmouth, N.S., B2Y 2N4

The Miramichi River drains a major watershed within eastern and central New Brunswick and empties through an extensive estuary into the Gulf of St. Lawrence. The Estuary serves two important functions in addition to acting as the interface between the riverine and ocean environments: (1) it acts as an important fish habitat; and, (2) it acts as a navigational access to ports and industrial facilities 40-60 km inland.

Improvements proposed for the navigational channel in the early 1970's initiated a series of studies which showed that the Estuary component plays a key role in regulation of contaminant transport from the watershed to the Gulf. While much of the contaminant loading is transported as dissolved compounds, the interaction of fresh water and sea water in the Estuary promotes the deposition of much of the suspended solids and associated contaminants in the Estuary sediments, particularly near Loggieville.

Dredging to provide and maintain the navigational channel alters the estuarine processes, both through re-distribution of the contaminated materials and alteration of the interaction zone. The question to be evaluated is how important are these changes to the overall contaminant budget for the Miramichi and what impact will there be on the resident biota?

An Assessment of the Environmental Quality of Sediments in the Miramichi Estuary

W.R. Parker, G. Lindsay, and K. Doe

Environment Canada, Environmental Protection Branch, 45 Aldernay Dr., Dartmouth, N.S., B2Y 2N6

Previous assessments conducted on Miramichi River sediments: (MREAC 1992) indicated some areas with elevated levels of heavy metals and an isolated sample with elevated PAH concentrations. Benthic community assessments indicated that the stretch of estuary between Chatham and Newcastle was "impoverished" biologically. As a result of these findings, a study was initiated in 1993 to further evaluate the quality of the sediments in the Miramichi estuary between Chatham and Newcastle.

Sediment samples were collected from eight locations in the estuary and analyzed for toxicity, chemical/physical properties and benthic invertebrates. Toxicity testing utilized two species of invertebrate and marine bacteria and examined both pore water and solid phase fractions of the samples. Chemical-physical analyses included grain size, PAH's, PCB's, extractable organohalogens, total chlorophenols and metals. The benthic community was enumerated in terms of numbers of organisms and number of species present in the samples. The data was evaluated to identify any relationships between contaminant concentrations and toxicity or impact on the benthic community.

Reference

MREAC. 1992. Summary final report, 1989-1992, Miramichi River Environmental Assessment Committee, 157p. (mimeo).

An In-situ Evaluation of the Aquatic Toxicity of the Tomogonops River System to Juvenile Atlantic Salmon

W.R. Parker

Environment Canada. Environmental Protection Branch, 45 Aldernay Dr., Dartmouth, N.S., B2Y 2N6

The Tomogonops River, a tributary to the Northwest Miramichi River, has been adversely impacted by heavy-metal contamination as a result of 30 years of operations at the Heath Steele base-metal mine. Previous studies have identified elevated levels of metals, the absence of resident fish and acute toxicity to caged salmon parr in some sections of the receiving waters. As a result of ongoing remedial activities at the Heath Steele Mine and a noticeable reduction in heavy metal contamination in receiving water quality as a result of these improvements, an in-site toxicity study using caged Atlantic salmon parr was conducted in June, 1993. This paper will provide the results of that study.

DFO Phycotoxins Research in the Miramichi

J.C. Smith

Department of Fisheries and Oceans, Gulf Region, Moncton, N.B., E1C 9B6

Since 1988, DFO has been operating a national phycotoxins monitoring program, including several stations in the region of the Miramichi Estuary. It appears that the area is at risk from both Paralytic Shellfish Poisoning (PSP) and Diarrhetic Shellfish Poisoning (DSP). To protect the fishing industry and consumers of seafood from these threats, DFO recently initiated a research program in the Miramichi estuary and near offshore areas; this work is designed to determine the factors governing the populations, distribution and toxicity of phytoplankton which produce PSP and DSP toxins and to show how these toxins are accumulated and depurated by shellfish and other organisms. There are also concerns about possible toxic effects on important larval fish populations which inhabit the area. The background to this work and preliminary results of the new program will be presented.

The Greater Kouchibouguac Ecosystem

E. Tremblay, K.A. Leonard, and M. Savoie

Kouchibouguac National Park, N.B. E0A 2A0

Kouchibouguac National Park (KNP) is located on the East Coast of New Brunswick at the head of the Northumberland Strait. Situated about 100 km north of Moncton and 50 km south of Chatham, the area covered by the park is 238.8 km2. The principal feature of the physiography of the park is that the land surface slopes gently from west to east at a rate of less than 5 m/km (> 0.5 %) (Beach 1988). Eight principal ecosystems compose the park; forest (52%), fields (4%), bogs (21%), salt marshes (3%), dunes (2%), estuaries (18%) and freshwater (1%) (Desloges 1980). The majority of watersheds are located only partly within park boundaries. Only the smallest streams have their drainage basins within the park. Eight major drainage basins affect the park (Beach 1988).

The park contains elements important to two regional ecosystems, one terrestrial and the other marine. It was established to provide representation of the Appalachian - Maritime Plain Natural Region in the Canadian National Park System (Environment Canada 1993). It is possible to view this natural region as an ecosystem with Kouchibouguac National Park as an island of wilderness, selected to provide protection to as wide a range of the natural features characteristic of the ecosystem as possible.

Kouchibouguac's coastal location places it clearly on the boundary of terrestrial and marine ecosystems. As such, it is composed of a blend of marine and terrestrial ecosystem elements. The shoreline is characterized by sandy-mud and eelgrass lagoons, soft sandstone bedrock and beautiful sandy beaches (Dunbar et al. 1977). This marine ecosystem directly accounts for about 24% of the surface area of the park and most of Kouchibouguac's outstanding natural features. It is composed of several important sub systems: shallow coastal lagoons and estuaries, salt marshes, and barrier island sand dunes.

There is considerable overlap between the terrestrial and marine regional ecosystems. This is not surprising given the coastal location of the park. Sand dunes and spits, salt marshes and shallow water coastal lagoons, and the organisms living in them, are all important components of both the marine and terrestrial systems. Since Kouchibouguac was established as a National Park representative of the Appalachian - Maritime Plain Natural Region, the ecosystem objectives for the park should focus on the ecosystems components of this regional ecosystem. Because the National Parks System Plan places high significance on the coastal features of the Appalachian - Maritime Plain, the characteristics of importance in the park from the regional marine ecosystem are also all given high conservation priority (Beach 1988).

References

Beach, H. [Editor]. 1988. The Resources of Kouchibouguac National Park: Resource Description and Analysis, KNP, Natural Resource Conservation.

Desloges, C. 1980. Les ressources naturelles du parc national Kouchibouguac. Parcs Canada, région de l'Atlantique, Éd: Le Groupe Dryade, 136p.

Dunbar, M.J., MacLellan, D.C., Filion, A., and Moore, D. 1979. The bio-geographic structure of the Gulf of St. Lawrence. Unpublished report to Parks Canada from the Marine Sciences Centre, McGill University, Montreal, P.Q.

Environment Canada. 1993. Kouchibouguac National Park Management Plan, Parks Service, Atlantic Region, 71 p.

An Innovative Use of Industrial Waste from the Paper Making Process

Jiri Zhor[1], T.W. Bremner[1], Jario Lora[2], and Kendall Pye[2]

[1]Department of Civil Engineering, University of New Brunswick, Fredericton, NB; [2]REPAP Technologies, 2650 Eisenhower Ave., Valley Forge, PA 19482, USA

Historically river valleys by their very nature became the sites of industrial development. The production of a given industry normally was one primary product with several by-products which were treated as industrial waste.

These tended to be abandoned, stock piled, or dumped in the river, with negative effects on the ecosystem. However in the second half of the Twentieth Century, engineering technology has focused on constructive uses for the by-products from industry.

The pulp and paper industry is a case in point. In the production of paper an almost equal volume of lignin is produced, which with conventional technology is heavily contaminated with other materials that limit its usefulness.

At the REPAP Mill at Newcastle using the ALCELL process a high purity lignin is produced for which a multitude of high value uses are being developed. The uses range from brake pads and adhesives for wood to an admixture for concrete. The story of how suitably sulphonated high-purity lignin was developed as a superplasticizer for concrete is the subject of this paper. It can serve as a model for other technologies that will use other by-products from the forest industry, all of which have implications for the protection and preservation of the river valley and the tidal estuary.

Subject Index

Species Index